Karl-Heinz Ohrbach

Dictionary of Ecology

Wörterbuch Ökologie

parat

© VCH Verlagsgesellschaft mbH, D-6940 Weinheim (Federal Republic of Germany), 1991

Distribution:
VCH, P.O. Box 101161, D-6940 Weinheim (Federal Republic of Germany)
Switzerland: VCH, P.O. Box, CH-4020 Basel (Switzerland)
United Kingdom and Ireland: VCH (UK) Ltd., 8 Wellington Court, Cambridge CB1 1HZ (England)
USA and Canada: VCH, Suite 909, 220 East 23rd Street, New York, NY 10010-4606 (USA)

ISBN 3-527-28175-4 (VCH, Weinheim)   ISSN 0930-6862
ISBN 0-89573-989-5 (VCH, New York)

Karl-Heinz Ohrbach

Dictionary of Ecology
English/German
German/English

Wörterbuch Ökologie
Englisch/Deutsch
Deutsch/Englisch

Weinheim · New York · Basel · Cambridge

Dr. Karl-Heinz Ohrbach
Hachmannstraße 22
4291 Altenbeken-Buke
Germany

This book was carefully produced. Nevertheless, author and publisher do not warrant the information contained therein to be free of errors. Readers are advised to keep in mind that statements, data, illustrations, procedural details or other items may inadvertently be inaccurate.

Published jointly by
VCH Verlagsgesellschaft mbH, Weinheim (Federal Republic of Germany)
VCH Publishers, Inc., New York, NY (USA)

Editorial Director: Dr. Hans-Dieter Junge
Production Manager: Dipl.-Wirt.-Ing. (FH) Bernd Riedel

Library of Congress Card No. 91-30095

British Library Cataloguing-in-Publication Data
Dictionary of ecology / Woerterbuch oekologie :
English/German / German/English / Englisch/Deutsch
/ Deutsch/Englisch. – (Parat dictionaries)
I. Ohrbach, Karl-Heinz II. Series
574.503
ISBN 3-527-28175-4

Die Deutsche Bibliothek – CIP-Einheitsaufnahme
Ohrbach, Karl-Heinz
Dictionary of ecology : English/German; German English =
Wörterbuch Ökologie / Karl-Heinz Ohrbach. – Weinheim ;
New York ; Basel ; Cambridge ; VCH 1991
(Parat)
ISBN 3-527-28175-4 (Weinheim ...) Gb.
ISBN 0-89573-989-5 (New York) Gb.
NE: HST

© VCH Verlagsgesellschaft mbH, D-6940 Weinheim (Federal Republic of Germany), 1991
Printed on acid-free paper/Gedruckt auf säurefreiem Papier
Alle Rechte, insbesondere die der Übersetzung in andere Sprachen, vorbehalten. Kein Teil dieses Buches darf ohne schriftliche Genehmigung des Verlages in irgendeiner Form – durch Photokopie, Mikroverfilmung oder irgendein anderes Verfahren – reproduziert oder in eine von Maschinen, insbesondere von Datenverarbeitungsmaschinen, verwendbare Sprache übertragen oder übersetzt werden. Die Wiedergabe von Warenbezeichnungen, Handelsnamen oder sonstigen Kennzeichen in diesem Buch berechtigt nicht zu der Annahme, daß diese von jedermann frei benutzt werden dürfen. Vielmehr kann es sich auch dann um eingetragene Warenzeichen oder sonstige gesetzlich geschützte Kennzeichen handeln, wenn sie nicht eigens als solche markiert sind.
All rights reserved (including those of translation into other languages). No part of this book may be reproduced in any form – by photoprinting, microfilm, or any other means – nor transmitted or translated into a machine language without written permission from the publishers. Registered names, trademarks, etc. used in this book, even when not specifically marked as such, are not to be considered unprotected by law.
Composition: U. Hellinger, D-6901 Heiligkreuzsteinach
Printing: betz-druck gmbh, D-6100 Darmstadt 12
Bookbinding: Verlagsbuchbinderei Kränkl, D-6148 Heppenheim
Printed in the Federal Republic of Germany

Für Ruth und Tina

Geleitwort

Eintrag und Verhalten von Chemikalien in Ökosystemen sind Forschungsgegenstände der Ökologischen Chemie, einer Fachrichtung der Chemie, die vor ca. 20 Jahren begründet wurde.

Da chemische Verbindungen durch Adsorption/Desorption bzw. Verdampfung und Deposition zwischen den Kompartimenten (Luft, Wasser, Boden) eines Ökosystems verteilt werden, darüber hinaus mikrobiell abgebaut oder abiotisch zersetzt werden können, wird deutlich, daß ein Fach allein zur Erfoschung dieser komplexen Vorgänge nicht ausreicht, sondern daße eine interdisziplinäre Kooperation erforderlich ist.

Weiterhin sind in diesem Zusammenhang Wirkungen von Chemikalien auf Individuen, Populationen und Lebensgemeinschaften zu untersuchen, um eine ökotoxikologische Bewertung von Chemikalien vornehmen zu können.

Das vorliegende Fachwörterbuch enthält die Grundbegriffe sowie die spezifische Terminologie der angeführten Gebiete sowie darüber hinaus Begriffe des Umweltschutzes, der Umweltanalytik sowie der Umweltverfahrenstechnik (z. B. Luftreinhaltung, Abwasserbehandlung, Altlastensanierung).

Zur Äquivalenzfindung der Zielbegriffe wurde neueste Standardliteratur verwendet. Das Wörterbuch ist vor allem für praktische Ziele der Sprachmitteilung entwickelt. Es wird somit sowohl dem Studierenden der umweltbezogenen Fachdisziplinen, dem in Forschung und Anwendung tätigen Wissenschaftler als auch dem interessierten Laien gerecht.

Prof. Dr. A. A. F. Kettrup
GSF-Forschungszentrum für Umwelt und Gesundheit GmbH,
Institut für Ökologische Chemie

Vorwort

Die Bestandsaufnahmen in Sachen Ökologie zeigen immer stärker die in den Vordergrund tretenden Auswirkungen ökologischer Ungleichgewichte und damit die Belastung und die Probleme, die sich daraus ergeben. Nahezu alle wissenschaftlichen Disziplinen, die an das Fachgebiet der Ökologie, zum Teil auch nur im weitesten Sinne angrenzen, sind derzeit mit ökologischen Fragestellungen befaßt. Der ständig ansteigende, immense Zuwachs an Wissen auf diesem Gebiet, läßt trotz Einsatz moderner Versorgungs- und Entsorgungstechnologien sowie leistungsfähiger Datenverarbeitungs- und Kommunikationssysteme nur Teillösungen ökologischer Probleme zu. Das vorliegende Wörterbuch Ökologie nimmt daher unter den Wörterbüchern eine gewisse Sonderstellung ein. Der in den Gesamtkomplex – Mensch, Umwelt, Versorgung und Entsorgung – einzuordnende aktuelle Stand von Fachausdrücken ist in diesem Wörterbuch zu einem großen Teil zusammenfassend dargestellt.

Danksagung

Mein besonderer Dank gilt Marianne Näther für die ständige Bereitschaft an der Fertigstellung des vorliegenden Buches mitzuarbeiten, sowie allen Kollegen(innen), die mich durch ihre indirekte oder direkte Hilfe unterstützt haben.

Altenbeken-Buke Karl-Heinz Ohrbach
September 1991

Instructions

This dictionary consists of two parts, i.e. German to English and English to German, each arranged alphabetically. Combinations of words, uniting more basic expressions, are also included, so that the user is able to identify the linguistic rules of word combinations. Boldface text is used for the source language notes. Parenthetical remarks relate the entries to special uses or fields. American spelling is given preferences.

m masculinum; f femininum; n neutrum

Benutzungshinweise

Das parat-Taschenwörterbuch besteht aus zwei Teilen entsprechend den beiden Sprachrichtungen.
Innerhalb der beiden Teile sind die Eintragungen in alphabetisch geordneten Wortnestern zusammengestellt, wobei darauf Wert gelegt ist, möglichst unterschiedliche Kombinationen des Nestwortes aufzunehmen, um die Bildungsgesetze der Wortkombinationen sichtbar werden zu lassen.
Innerhalb der Wortstellen sind die Eintragungen in der Quellsprache hervorgehoben. Eintragungen in Klammern helfen bei der Zuordnung des Begriffes zu einem speziellen Anwendungsfall oder zu einem bestimmten technischen Bereich. Es sei noch darauf hingewiesen, daß es im amerikanischen Englisch zahlreiche Abweichungen gegenüber dem „klassischen" Englisch gibt. Wie in der technisch-wissenschaftlichen Literatur üblich, wird im Zweifelsfall der amerikanischen Schreibweise der Vorzug gegeben.

Englisch/Deutsch
English/German

A

abiogenesis Abiogenese *f*
abiogenic abiogen
abiosis Abiose *f*, Lebensunfähigkeit *f*
abiotic abiotisch
abiotic degradation abiotischer Abbau *m*
abortion Abort *m*, Fehlgeburt *f*
above ground oberirdisch, an der Erdoberfläche
abrasion Abrieb *m*
abrasive sand Schleifsand *m*
abrasive wear Abnutzungsabrieb *m*
abrasives industry Schleifmittelindustrie *f*
absolute value Absolutgröße *f*
absorb/to absorbieren, aufsaugen, einsaugen; resorbieren
absorbate Absorbat *n*, Absorptiv *n*, absorbierter Stoff *m*
absorbency Absorptionsvermögen *n*, Absorptionsfähigkeit *f*
absorbent Absorbens *n*, Absorptionsmittel *n*, absorbierender Stoff *m*
absorbing layer absorbierende Schicht *f*
absorption Absorption *f*; Resorption *f*
absorption bottle Absorptionsflasche *f*
absorption capacity Absorptionskapazität *f*
absorption capacity for water Wasseraufnahmevermögen *n*
absorption cell Absorptionsküvette *f*
absorption chamber Absorptionskammer *f*
absorption system Absorptionssystem *n*
absorptive absorptiv, absorptionsfähig, aufnehmend, saugfähig
absorptive capacity Absorptionskapazität *f*
absorptive power Absorptionskraft *f*
absorptivity Absorptionsfähigkeit *f*, Aufnahmefähigkeit *f*
abundance abundance Abundanz *f*, Häufigkeit *f*, Fülle *f*, Überfluß *m*
abundance curve Häufigkeitsverteilungskurve *f*
abundance ratio Häufigkeitsverhältnis *n*
abundant reich, reichlich, reich an
abuse Mißbrauch *m*
acaricide Akarizid *n*, Milbenbekämpfungsmittel *n*
accelerated test Kurzversuch *m*, Kurztest *m*
acceptable limit annehmbare Grenze *f*, akzeptabler Grenzwert *m*
acceptance certificate of the waste disposal plant Annahmeerklärung *f* der Abfallbeseitigungsanlage
acceptance condition Annahmebedingung *f*
acceptance of life Lebensbejahung *f*
acceptance of technology Technikakzeptanz *f*
accident Panne *f*, Zufall *m*, Unfall *m*, Zwischenfall *m*
accident instruction sheet Unfallmerkblatt *n*
accident prevention programme Unfallverhütungsprogramm *n*
accident report Unfallbericht *m*
accident risk Unfallgefahr *f*
accidental effect Unfallfolge *f*
accidental value Zufallswert *m*

acclimatization Akklimatisation f, Akklimatisierung f
acclimatize/to akklimatisieren, gewöhnen, anpassen
accomodate/to anpassen an, einstellen auf, akkomodieren
accompanying substance Begleitsubstanz f, Begleitstoff m
accountability for the disposal Nachweispflicht f für die Entsorgung, Entsorgungsnachweispflicht f
accumulate/to akkumulieren, anhäufen, ansammeln, speichern, anlagern
accumulation Akkumulation f, Anhäufung f, Ansammlung f, Akkumulierung f, Anreicherung f, Speicherung f
accumulation of energy Energieaufspeicherung f
accumulation of harmful substances in human tissues Schadstoffanreicherung f in menschlichem Gewebe
accumulation potential Akkumulationspotential n
accumulation process Anreicherungsvorgang m
accumulation property Akkumulationseigenschaft f
accumulation rate Akkumulationsrate f
accumulator Akkumulator m, Akku m, Speicher m, Stromspeicher m, Batterie f
accumulator acid Akkusäure f
accumulator plate Akkumulatorplatte f
accumulator tester Akkuprüfer m
acid Säure f
acid anhydride Säureanhydrid m

acid-base equilibrium Säure-Basen-Gleichgewicht n
acid-base metabolism Säure-Basen-Stoffwechsel m
acid bath Säurebad n
acid-proof säurebeständig
acid-proof hose säurebeständiger Schlauch m
acid-resistant säurebeständig, säureresistent
acid strength Säurestärke f
acidification Ansäuerung f, Ansäuern n, Säurebildung f
acidify/to ansäuern, säuern, sauer einstellen
acidimetry Azidimetrie f, Säuremessung f
acidity Azidität f, Säuregrad m, Säuregehalt m, Säurekonzentration f
acidogeneous microorganism acidogener Mikroorganismus m
acoustic measurement Schallmessung f
acoustic monitoring system Schallüberwachungssystem n
acoustic warning and report system akustisches Warnmeldesystem n
acquired immunity erworbene Immunität f
actinium Aktinium n
actinochemistry Aktinochemie f, Strahlenchemie f
activate/to aktivieren, anregen, beleben
activated aktiviert, belebt, angeregt
activated carbon Aktivkohle f
activated carbon powder Aktivkohlepulver n
activated carbon technology Aktivkohletechnik f

activated coke Aktivkoks *m*
activated sludge Belebtschlamm *m*, belebter Abwasserschlamm *m*, Bioschlamm *m*
activated-sludge basin Belebtschlammbecken *n*, Belebungsbekken *n*
activated sludge flocs Belebtschlammflocken *fpl*
activated sludge microbiology Belebtschlammikrobiologie *f*
activated sludge plant Belebungsanlage *f*
activated sludge process Belebtschlammverfahren *n*
activation Aktivierung *f*
activation analysis Aktivierungsanalyse *f*
activation energy Aktivierungsenergie *f*
active aktiv, wirksam
active carbon Aktivkohle *f*
active component Wirkstoff *m*
activity Aktivität *f*, Wirksamkeit *f*
actual situation of burden (load, impact) aktuelle Belastungssituation *f*
actual value Istwert *m*
acute toxicity akute Toxizität *f*
acyclic interrupt processing azyklische Interruptverarbeitung *f*
adapt/to anpassen, sich anpassen, adaptieren, einstellen
adaption Anpassung *f*, Angleichung *f*, Adaption *f*
adaption expenditure Anpassungsaufwand *m*
adaptive controller adaptiver (selbstanpassender) Regler *m*
adaptive controller adjustment (setting) adaptive Reglereinstellung *f*

adaptive difficulty Anpassungsschwierigkeit *f*
addition reaction Additionsreaktion *f*
additive Beimengung *f*, Zusatz *m*, Zusatzmittel *n*, Zusatzstoff *m*, Additiv *n*
additive flue gas filter technology additive Rauchgasfiltertechnik *f*
addressed running time system adressiertes Laufzeitsystem *n*
addressing into the [receive] mailbox Adressierung *f* in das Empfangsfach
adhesion Anhaftung *f*, Ankleben *n*, Adhäsion *f*, Haftung *f*, Haftfähigkeit *f*
adhesion friction Haftreibung *f*
adhesive drying Klebstofftrocknung *f*
adiabatic adiabatisch
adipose tissue Fettgewebe *n*
adrenal gland Nebenniere *f*
adsorb/to adsorbieren, aufnehmen
adsorbability Adsorbierbarkeit *f*
adsorbable adsorbierbar
adsorbate Adsorbat *n*, adsorbierter Stoff *m*
adsorbent Adsorbens *n*, Adsorptionsmittel *n*
adsorbent bed Adsorptionsschicht *f*
adsorbent layer Adsorptionsschicht *f*
adsorber Adsorber *m*
adsorbing capacity Adsorptionskapazität *f*, Beladefähigkeit *f*
adsorption Adsorption *f*
adsorption analysis Adsorptionsanalyse *f*

adsorption column Adsorptionssäule *f*
adsorption compound Adsorptionsverbindung *f*
adsorption cycle Adsorptionskreislauf *m*
adsorption displacement Adsorptionsverdrängung *f*
adsorption layer Adsorptionsschicht *f*
adsorption process Adsorptionsvorgang *m*
adsorption resin Adsorberharz *n*
adsorption system Adsorptionssystem *n*
adsorption technology Absorptionstechnik *f*
adsorptive Adsorptiv *n*, Adsorbat *n*, adsorbierter Stoff *m*
adsorptive capacity Adsorptionskapazität *f*
adsorptive purification adsorptive Reinigung *f*
adult erwachsen; Erwachsene *f*, Erwachsener *m*
aerate/to belüften, durchlüften, mit Luft versetzen
aerated heap belüftete Miete *f*
aeration Lüftung *f*, Belüftung *f*
aeration aggregate Belüfteraggregat *n*
aeration device Belüftungseinrichtung *f*
aeration disc Belüftungsplatte *f*
aeration impeller Belüftungskreiselrührer *m*
aeration measure Belüftungsmaßnahme *f*
aeration period Belüftungsphase *f*
aeration stone Belüftungsstein *m*
aeration tank Belüftungstank *m*

aeration time Belüftungszeit *f*
aerator Belüfter *m*
aerator group Belüftergruppe *f*
aerator performance Belüfterleistung *f*
aerial luftig, zur Luft gehörend, in der Luft befindlich
aerial nitrogen Luftstickstoff *m*
aerial oxygen Luftsauerstoff *m*
aerial photo Luftaufnahme *f*
aeroallergene forecast Pollenvorhersage *f*
aerobe Aerobe *f*, Aerobier *m*
aerobic aerob
aerobic bacteria aerobe Bakterien *fpl*
aerobic cell aerobe Zelle *f*
aerobic degradation aerober Abbau *m*
aerobic stabilization of sludge aerobe Schlammstabilisierung *f*
aerobiosis Aerobiose *f*
aerodynamic aerodynamisch
aeromedicine Aeromedizin *f*, Luftfahrtmedizin *f*
aerometer Aerometer *n*, Dichtemesser *m*
aerosol Aerosol *n*
aerosol can Sprühdose *f*
aerosol propelling gas Treibgas *n*
aerosol separator Aerosolabscheider *m*
aerosol-smog chamber Aerosol-Smogkammer *f*
aerospace Weltraum *m*
affinity Affinität *f*, Verwandschaft *f*
afforestation Aufforstung *f*
aflatoxin Aflatoxin *n*
after-burning chamber Nachbrennkammer *f*

after-care Nachbehandlung *f*, Nachsorge *f*
after-clarification Nachklärung *f*
after-clarification plant Nachkläranlage *f*
after-clarification with precipitation Nachklärung *f* mit Fällung
after-composting Nachkompostierung *f*
after-purification plant Nachkläranlage *f*
after-run Nachlauf *m*
after-smell Nachgeruch *m*
after-treatment Nachbehandlung *f*
agar Agarnährboden *m*
agar plate count Agarkeimzahl *f*
agent Agens *n*, Mittel *n*
agglomerate Agglomerat *n*, Zusammenballung *f*, Zusammenhäufung *f*; Sinterstoff *m*
agglomerate/to agglomerieren, zusammenballen; sintern
agglomeration Agglomerierung *f*, Agglomeration *f*
aggregate Aggregat *n*
aggregate dimensioning Aggregatauslegung *f*
aggregate for oxygen entry Sauerstoffeintragsaggregat *n*
aggregate group Aggregatgruppe *f*
aggregate type Aggregatetyp *m*
aggregation Aggregation *f*
aging Altern *n*, Alterung *f*, Ermüdung *f*, Vergüten *n*
aging test Alterungstest *m*, Alterungsprüfung *f*
agricultural landwirtschaftlich
agricultural chemical Agrochemikalie *f*
agricultural chemistry Agrikulturchemie *f*

agricultural-ecological model agrarökologisches Modell *n*
agricultural ecology Agrarökologie *f*, Landwirtschaftsökologie *f*
agricultural-economical demand agrarökonomische Forderung *f*
agricultural ecosystem Agrarökosystem *n*
agricultural hydrology Landwirtschaftshydrologie *f*
agricultural law Landwirtschaftsgesetz *n*
agricultural stock-breeding landwirtschaftliche Tierhaltung *f*
agriculture Agrikultur *f*, Ackerbau *m*, Landwirtschaft *f*
agrobiology Agrobiologie *f*
agronomic agronomisch
agronomy Agronomie *f*, Ackerbaukunde *f*
agrotechnology Agrartechnik *f*
aid Hilfsstoff *m*, Hilfsmittel *n*, Hilfe *f*
aim of protection Schutzziel *n*
aim of redevelopment Sanierungsziel *n*
air Luft *f*
air admission tube Belüftungsstutzen *m*
airborne freischwebend, in der Luft enthalten
airborne dust Flugstaub *m*
airborne contamination Luftverseuchung *f*
airborne pollutant Luftschadstoff *m*, in der Luft enthaltener Schadstoff *m*
airborne radioactivity Radioaktivität *f* der Luft
air chamber Luftkammer *f*
air circulation Luftzirkulation *f*

air circulation system Umluftverfahren *n*
air-conditioning Klimatisierung *f*
air investigation Luftuntersuchung *f*
air monitoring Luftüberwachung *f*
air ozonization Luftozonisierung *f*
air plant Luftreinigungsanlage *f*
air poisoning Luftvergiftung *f*
air pollutant Luftschadstoff *m*
air pollution control Luftverschmutzungskontrolle *f*
air pollution forecast Luftverschmutzungsvorhersage *f*
air pollution Luftverunreinigung *f*
air purification system Luftreinhaltungssystem *n*
air quality Luftqualität *f*
air quality data Daten *pl* zur Luftqualität, Luftqualitätsdaten *pl*
air sampling Luftprobenahme *f*, Probenahme *f* aus der Luft
air-tight seal luftdichter Verschluß *m*
air traffic Luftverkehr *m*, Flugverkehr *m*
airborne in der Luft enthalten, durch die Luft übertragen
aircraft disinfection Flugzeugdesinfektion *f*
aircraft noise Fluglärm *m*
aircraft spray equipment Flugzeug-Sprühausrüstung *f*
airplane fertilizing Flugzeugdüngung *f*
alarm device Alarmeinrichtung *f*
alarm level Auslöseschwelle *f* für einen Alarm, Alarmauslöseschwelle *f*
alarm preparation Meldungsvorbereitung *f*
alarm switch Alarmschalter *m*
alarm system Alarm[ierungs]system *n*
alcaline chloride electrolysis Alkalichloridelektrolyse *f*
alcohol test Alkoholtest *m*
alcoholysis Alkoholyse *f*
aldehyde Aldehyd *m*
alga Alge *f*
algae control (removal) Algenbekämpfung *f*
algae growth Algenwachstum *n*
algicide Algenbekämpfungsmittel *n*, Algizid *n*
alimentary canal Verdauungskanal *m*
alimentary deficiency Nahrungsmangel *m*
alimentary research Ernährungsforschung *f*
alimentary substance Nährstoff *m*
alimentary tract Verdauungstrakt *m*
aliphatic aliphatisch
aliphatic hydrocarbon aliphatischer Kohlenwasserstoff *m*
alkali Alkali *n*, Lauge *f*
alkali fastness Alkalibeständigkeit *f*
alkali-free laugenfrei, alkalifrei
alkali lye Alkalilauge *f*
alkalify/to alkalisieren
alkaline alkalisch, alkalihaltig, basisch
alkaline earth Erdkali *n*
alkalinity Alkalität *f*, Basizität *f*
alkaloid Alkaloid *n*
alkyl Alkyl *n*
alkyl halide Alkylhalogenid *n*
alkylation Alkylierung *f*
all-clear Entwarnung *f*

allelopathy Allelopathie *f*
allergen Allergen *n*, Allergiestoff *m*
allergic allergisch, überempfindlich
allergic reaction allergische Reaktion *f*
allergic rhinitis Heuschnupfen *m*
allergology Allergologie *f*
allergy Allergie *f*, Überempfindlichkeit *f*
alleviation Erleichterung *f*, Linderung *f*
allocation Allokation *f*
allopathy Allopathie *f*
allotropic allotrop
allowance Toleranz *f*, Zugabe *f*
alloy Legierung *f*, Mischmetall *n*
alloying component Legierungsbestandteil *m*
alluvial alluvial
alluvial soil Auenboden *m*
alpha decay Alphazerfall *m*
alpha-active alpha-radioaktiv
alpha emitter Alphastrahler *m*
alpha emitting alphastrahlend
alpha particle Alphateilchen *n*
alpha radioactivity Alpha-Radioaktivität *f*
alpha ray Alphastrahl *m*
alphanumeric keyboard alphanumerische Tastatur *f*
alternative stress Wechselbeanspruchung *f*
altimetry Höhenmessung *f*
alumina Tonerde *f*, Aluminiumoxid *n*
alumina ceramic Aluminiumoxidkeramik *f*
aluminium Aluminium *n*
aluminium alloy Aluminiumlegierung *f*

aluminium can producer Aluminiumdosenhersteller *m*
amalgam Amalgam *n*, Quecksilberlegierung *f*
amalgam of lead Bleiamalgam *n*
amalgam process Amalgamverfahren *n*
amalgamation Amalgamierung *f*
amalgamation process Amalgamverfahren *n*
ambience Umgebung *f*, Atmosphäre *f*
ambient umgebend
ambient air Umgebungsluft *f*
ambient atmosphere Umgebungsluft *f*, umgebende Atmosphäre *f*
ambient noise Umweltlärm *f*
ambient pressure Umgebungsdruck *m*
amendment of the chemical law Novelle *f* des Chemikaliengesetzes
americium Americium *n*
amide Amid *n*
amine Amin *n*
amino acid Aminosäure *f*
amino sequence Aminosäuresequenz *f*
aminobenzene Aminobenzol *n*, Anilin *n*
aminoplast Aminoplast *n*
aminoplast resin Harnstoffharz *n*, Aminoplastharz *n*
amitosis Amitose *f*
ammonia Ammoniak *n*
ammonia excretion Ammoniakausscheidung *f*
ammonium analyzer Ammoniumanalysator *m*
ammonium-controlled nitrification ammoniumgesteuerte Nitrifikation *f*

ammonium inlet Ammoniumeinleitung *f*
ammunition Munition *f*
amoeba Amöbe *f*
amount of heat Wärmebetrag *m*
amount of rainfall Niederschlagshöhe *f*
ampholyte Ampholyt *m*
amphoteric amphoter
amplification Verstärkung *f*
amplifier Verstärker *m*
amplitude Amplitude *f*
amylase Amylase *f*
anabiosis Anabiose *f*
anabiotic anabiotisch
anabolic anabolisch
anaerobe Anaerobier *m*
anaerobic anaerob
anaerobic bacteria anaerobe Bakterien *fpl*
anaerobic cell anaerobe Zelle *f*
anaerobic fixed-bed reactor anaerober Festbettreaktor *m*
anaerobic pond Faulteich *m*
anaerobic soil anaerober Boden *m*
anaerobic-thermophilic / mesophilic stabilization anaerob-thermophile/mesophile Stabilisation *f*
anaerobiosis Anaerobiose *f*
analogue input analoger Eingang *m*
analogue output analoger Ausgang *m*; Analogwertausgabe *f*
analogue output module Analogausgabebaugruppe *f*
analogue recorder connection Analogschreiberanschluß *m*
analogue value Analogwert *m*
analogue-value processing Analogwertverarbeitung *f*
analyse/to analysieren, bestimmen

analysis Analyse *f*, Gehaltsbestimmung *f*
analysis by weight Gewichtsanalyse *f*
analysis calibration function Analyseneichfunktion *f*, Kalibrierfunktion *f*
analysis method Analysenmethode *f*
analysis of domestic waste Hausmüllanalyse *f*
analysis of variance Varianzanalyse *f*
analysis result Analysenergebnis *n*
analyst Analytiker *m*
analyte Analyt *m*
analytical analytisch
analytical balance Analysenwaage *f*
analytical method analytisches Verfahren *n*, analytische Methode *f*
analytical result analytisches Ergebnis *n*, Analysenergebnis *n*
analytical test Analysenprobe *f*
analytically pure analysenrein
analytics Analytik *f*
analyzer Analysator *m*, Meßzelle *f*, Analysengerät *n*
anatomy Anatomie *f*
anemometer Anemometer *n*, Windgeschwindigkeitsmesser *m*
angiography Angiographie *f*
anhydrous nicht wäßrig, wasserfrei, entwässert
anhydrous salt wasserfreies Salz *n*
anilide Anilid *n*
animal Tier *n*
animal experiment Tierversuch *m*, Tierexperiment *n*
animal fat Tierfett *n*
animal feed Tierfutter *n*

animal protection Tierschutz *m*
anion Anion *n*
anion-active anionaktiv
anion exchange Anionenaustausch *m*
anisotropic anisotrop
anisotropy Anisotropie *f*
annual load curve Jahresganglinie *f*
annual output curve Jahresganglinie *f*
annual [plant] Einjahrespflanze *f*, Annuelle *f*
annular furnace Ringofen *m*
anode Anode *f*
anode discharge Anodenentladung *f*
anode slime Anodenschlamm *m*
anode sludge Anodenschlamm *m*
anodic anodisch
anodic treatment Eloxieren *n*
anoxic anoxisch
antagonist Antagonist *m*
antagonistic antagonistisch, entgegenwirkend
anthracene Anthracen *n*
anthracite Anthrazit *m*
anthracosis Anthrakose *f*, Kohlenstaublunge *f*
anthropobiology Anthropobiologie *f*
anthropogenic anthropogen, durch menschliche Einwirkung entstanden
anthropogenic chemical anthropogene Chemikalie *f*
anthropogenic impact anthropogene Einwirkung *f*
anthropogenic substance anthropogene Substanz *f*, anthropogener Stoff *m*
anti-ager Alterungsschutzmittel *n*
anti-humanitarian antihumanitär

anti-inflammatory entzündungshemmend, entzündungswidrig
anti-static antistatisch
antiacid säurewidrig
antiallergic agent Antiallergikum *n*
antiasthmatic antiasthmatisch
antibacterial antibakteriell
antibiotic antibiotisch; Antibiotikum *n*
antibody Abwehrstoff *m*, Antikörper *m*
antibody technology Antikörpertechnik *f*
anticancer drug Antikrebsmittel *n*, Antikrebswirkstoff *m*, Antikrebsarzneimittel *n*
anticorrosion additive Korrosionsschutzmittel *n*
anticorrosive korrosionsverhindernd
antidotal giftabtreibend
antidote Gegengift *n*, Gegenmittel *n*
anti-fouling bewuchsverhindernd
antifoam Antischaummittel *n*
antifogging beschlagsverhindernd
antifreeze Frostschutzmittel *n*
antifungin Antifungin *n*
antigen Antigen *n*
antihormone Gegenhormon *n*, Antihormon *n*
antiknock Antiklopfmittel *n*
antimony Antimon *n*
antioxidizing agent Antioxidationsmittel *n*
antiputrefactive fäulnisverhindernd
antirabic vaccination Tollwutschutzimpfung *f*
antiradiation therapy Strahlenschutztherapie *f*
antiseptic antiseptisch, fäulnisverhindernd, fäulniswidrig

antiserum Antiserum *n*, Immunserum *n*
antitoxin Gegengift *n*, Antitoxin *n*
antivirus agent Antivirusmittel *n*
antizymotic gärungshemmend
apparent density Schüttgewicht *n*
applicability of waste gas Anwendbarkeit *f* von Deponiegas
applied research angewandte Forschung *f*
applied science angewandte Wissenschaft *f*
approval right Genehmigungsrecht *n*
approximate calculation Näherungsrechnung *f*
approximate method Näherungsverfahren *n*, Näherungsmethode *f*
approximate value angenäherter Wert *m*, Näherungswert *m*
approximation Annäherung *f*, Näherung *f*, Näherungslösung *f*, Approximation *f*
aquatic aquatisch, im Wasser lebend
aquatic field system aquatisches Feldsystem *n*
aquatic flora aquatische Flora *f*, Wasserpflanzenwelt *f*
aquatic system Meeressystem *n*, aquatisches System *n*
aquatic toxicology Wassertoxikologie *f*
aquatic weed Wasserunkraut *n*
aqueous wäßrig
aqueous-organic wäßrig-organisch
aqueous solubility Wasserlöslichkeit *f*
aqueous solution wäßrige Lösung *f*
arable area Anbaufläche *f*
arable land Ackerland *n*
arable soil Ackerboden *m*

arboreous baumreich, bewaldet
arboriculture Baumzucht *f*
arboriform baumförmig
arc furnace Lichtbogenofen *m*
arc welding Bogenschweißen *n*, Lichtbogenschweißen *n*
architectural accomodation and protection of the countryside architektonische Anpassung *f* und Landschaftsschutz
area-covering registration of the estate waste flächendeckende Erfassung *f* der Siedlungsabfälle
area display Bereichsanzeige *f*
area for ocean dumping Verklappungsplatz *m*
area of application Anwendungsbereich *m*
area of danger Gefahrgebiet *n*, Gefahrenzone *f*
area of development Entwicklungsgebiet *n*
area of pure air Reinluftgebiet *n*
area of remedial Sanierungsgebiet *n*
area of remediation Sanierungsgebiet *n*
area of suspicion Verdachtsfläche *f*
area overview display Bereichsübersichtsanzeige *f*
areation intensity Belüftungsintensität *f*
argillaceous material Tonmaterial *n*
argon Argon *n*
arid dürr, trocken, arid
arithmetic processor Arithmetikprozessor *m*
aromatic aromatisch, wohlriechend; aromatische Substanz *f*
aromatic hydrocarbon aromatischer Kohlenwasserstoff *m*

arsenic Arsen *n*
arsenic dust Arsenstaub *m*
arsenic sludge Arsenschlamm *m*
artefact Artefakt *n*
article of daily use Gebrauchsgegenstand *m*
artificial künstlich, synthetisch
artificial manure Kunstdünger *m*
asbestos Asbest *m*
asbestos board Asbestpappe *f*
asbestos cement Asbestzement *m*
asbestos concentration Asbestkonzentration *f*
asbestos-containing asbesthaltig
asbestos dust exposed employee asbeststaubgefährdeter Arbeitnehmer *m*
asbestos fibre Asbestfaser *f*
asbestos fibre dust Asbestfaserstaub *m*
asbestos filler Asbestfüllstoff *m*
asbestos joint Asbestdichtung *f*
asbestos sampling Asbestprobenahme *f*
asbestos suit Asbestanzug *m*
asbestosis Asbestose *f*
ash Asche *f*
ash content Aschegehalt *m*
ash determination Aschebestimmung *f*
ash heap Aschenhalde *f*
ash pit Aschengrube *f*
ash removal Entaschung *f*
ash utilization Ascheverwertung *f*
ash/to veraschen, einäschern
asphalt Asphalt *m*, Bitumen *n*
asphalt concrete mix Asphaltbetonmischgut *n*
asphalt supporting layer mix Asphalttragschichtmischgut *n*

asphalted cardboard Dachpappe *f*
asphalted paper Asphaltpappe *f*, Bitumenpapier *n*
aspirate/to ansaugen
aspiration Ansaugen *n*, Atmen *n*, Aufsaugen *n*
aspirator Luftsauger *m*, Luftabsauger *m*, Sauger *m*, Aspirator *m*
assay Analyse *f*, Gehaltsbestimmung *f*, Probe *f*, Erprobung *f*, Test *m*
assay sensitivity Nachweisempfindlichkeit *f*
assay/to untersuchen, erproben, probieren, prüfen
assembly line work Fließbandarbeit *f*
assessment Abschätzung *f*, Bemessung *f*, Beurteilung *f*, Einschätzung *f*
assessment of damage Schadensfeststellung *f*, Schadenserfassung *f*
assessment of exposure Expositionsbewertung *f*
assimilate/to assimilieren, angleichen, anpassen, umsetzen
assimilation Angleichung *f*, Assimilation *f*, Aufnahme *f*
assimilation process Assimilationsvorgang *m*
atmosphere Atmosphäre *f*, Außenluft *f*, Luft *f*
atmospheric atmosphärisch
atmospheric conditions Witterungsverhältnisse *npl*
atmospheric layer atmosphärische Schicht *f*
atmospheric moisture Luftfeuchtigkeit *f*
atmospheric nitrogen Luftstickstoff *m*
atmospheric oxygen Luftsauerstoff *m*

atmospheric pollution Luftverschmutzung f
atmospheric precipitation atmosphärischer Niederschlag m
atmospheric pressure Umgebungsdruck m
atom Atom n
atom fission Atomspaltung f
atomic atomar, atomisch
atomic absorption analysis Atomabsorptionsanalyse f
atomic age Atomzeitalter n
atomic clock Atomuhr f
atomic energy Atomenergie f
atomic explosion Atomexplosion f
atomic mass Atommasse f
atomize/to atomisieren, zerstäuben
atomizer Zerstäuber m, Atomizer m, Sprühapparat m
atrophy Atrophy f, Schwund m, Verkümmerung f
atropine Atropin n
attack Attacke f, Anfall m, Befall m, Angriff m
attack/to anfallen, angreifen, befallen
attackable angreifbar
attenuate/to abschwächen, dämpfen
attenuation Dämpfung f, Schwächung f, Abschwächung f
attenuation constant Dämpfungskonstante f
attenuation region Dämpfungsbereich m
attraction of gravity Gravitationskraft f
attrition Attrition f
attrition washing drum Attritionswaschtrommel f
audible alarm hörbarer Alarm m
autarkic system autarkes System n
authority of approval Genehmigungsbehörde f
authorized person for hazardous goods Gefahrgutbeauftragter m, Beauftragter m für Gefahrgut
authorizing procedure Genehmigungsverfahren n
auto-regulation Selbstregulation f
autocatalysis Autokatalyse f
autogenesis Selbstentstehung f
automatic test routine automatische Prüfroutine f
automation component Automatisierungskomponente f
automation function Automatisierungsfunktion f
automation task Automatisierungsaufgabe f
automobile catalyst Autoabgaskatalysator m
automobile exhaust gas Autoabgas n
automobile finish Autolack m
autoradiogram Autoradiogramm n
autosampler Autosampler m, automatisch arbeitende Probeaufgabeeinrichtung f
autotrophy Autotrophie f
autoxidation Autoxidation f
auxiliary agent Hilfsstoff m, Hilfsreagens n
auxiliary product Hilfsmittel n
avalanche Lawine f
average Mittelwert m, Durchschnitt m; Havarie f
average concentration Mittelwertkonzentration f
average deviation mittlere Abweichung f
average of an oil tanker Öltankerhavarie f, Öltankerunfall m

average value Mittelwert *m*
avoidance of emissions Emissionsvermeidung *f*
axial-flow pump Axialpumpe *f*
azeotrope Azeotrop *n*, azeotropes Gemisch *n*
azeotropic distillation azeotrope Destillation *f*

B

baby food Säuglingsnahrung *f*
bacillary stäbchenförmig, bazillär
bacillary spore Bazillenspore *f*
bacill focus Bazillenherd *m*
bacillus Bazillus *m*
bacillus-free bazillenfrei
back diffusion test Rückdiffusionstest *m*
back flow Rückfluß *m*
backfiring safeguard Flammenrückschlagsicherung *f*
backflow ratio Rücklaufverhältnis *n*
backflushable filter rückspülbares Filter *n*
bacteria resistance Bakterienresistenz *f*
bacterial bakteriell
bacterial contamination bakterielle Verschmutzung *f*, Bakterienverseuchung *f*
bacterial count Keimzahl *f*
bacterial counting Keimzählung *f*
bacterial flora Bakterienflora *f*
bacterial growth Bakterienwachstum *n*
bacterial nitrification bakterielle Nitrifikation *f*
bacterial pollution bakterielle Verunreinigung *f*

bacterial toxin Bakteriengift *n*
bactericidal bakterizid, antibakteriell, bakterienabtötend
bactericide Bakterizid *n*, Bakteriengift *n*
bacteriological bakteriologisch
bacteriologist Bakteriologe *m*
bacteriology Bakteriologie *f*
bacterium Bakterie *f*, Bakterium *n*
badlands Ödland *n*, Brachland *n*, vegetationsloses Land
baffle Prallblech *n*, Trennblech *n*, Leitblech *n*, Schallschluckhaube *f*, Umlenkblech *n*
baffle plate Stauklappe *f*
bag filter Sackfilter *n*, Beutelfilter *n*
bag sealing plant Beutelschweißanlage *f*
baking finish Einbrennfarbe *f*, Einbrennlack *m*
bale cutter Ballenschneider *m*
bale press Ballenpresse *f*
baling press Ballenpresse *f*
ball mill Kugelmühle *f*
ball stop-cock Kugelhahn *m*
ball valve Kugelventil *n*
ballast Ballast *m*
ballast concrete Schotterbeton *m*
ballast material Ballaststoff *m*
ballast tank Ballastbehälter *m*
bank filtration Uferfiltration *f*
bank formation Uferausbildung *f*
bank stabilization Uferbefestigung *f*
banquette Bankett *n*, steile Böschung *f*
bar-curve graph Balken-Kurven-Diagramm *n*
bar graph Balkendiagramm *n*
bar probe Stabsonde *f*

bark Rinde *f*
bark compost Rindenkompost *m*
bark processing Rindenaufbereitung *f*
barometer Barometer *n*, Luftdruckmesser *m*
barometric variation Luftdruckschwankung *f*
barrage Talsperre *f*, Damm *m*, Staudamm *m*
barrage aeration Talsperrenbelüftung *f*
barrage water Talsperrenwasser *n*
barrel mill Trommelmühle *f*
barrenness Unfruchtbarkeit *f*
barrier layer Sperrschicht *f*
basal metabolism Basalmetabolismus *m*, Basalstoffwechsel *m*
base Base *f*, Lauge *f*; Basis *f*, Grundlage *f*, Fundament *n*
base catalysis Basenkatalyse *f*
base coating Grundbeschichtung *f*
base component Basiskomponente *f*
base exchanger Basenaustauscher *m*
basic basisch
basic capacity Basizität *f*
basic chemical Grundchemikalie *f*
basic research Grundlagenforschung *f*
basicity Basizität *f*
basify/to basisch machen
basin Bassin *n*, Becken *n*
basin of the clarification plant Klärwerksbecken *n*
bast fibre Bastfaser *f*
batch Beschickungsansatz *m*, Charge *f*, Satz *m*, Ansatz *m*
batch mixer Chargenmischer *m*
batch process Chargenprozeß *m*

batch production Chargenbetrieb *m*
batch recipe Chargenrezept *n*
bath Bad *n*
bathing prohibition zone Badeverbotzone *f*
battery Akkumulator *m*, Batterie *f*
battery acid Batteriesäure *f*, Akkusäure *f*, Akkumulatorsäure *f*
battery capacity Batterieleistung *f*, Batteriekapazität *f*
battery scrap Batterieschrott *m*
bayou sumpfiger Flußarm *m*, Ausfluß *m* aus einem See
beach Badestrand *m*
beach plant Dünenpflanze *f*
beam scale Balkenwaage *f*
bedrock Felsuntergrund *m*
belt conveyor Bandförderer *m*, Förderband *n*, Transportband *n*
belt filter Bandfilter *n*
belt filter press Bandfilterpresse *f*
bench test Laborversuch *m*
benign tumour gutartiger Tumor *m*
bentonite Bentonit *m*
benzene Benzol *n*
benzopyrene Benzpyren *n*
berkelium Berkelium *n*
beryllium Beryllium *n*
beta-active beta-aktiv, beta-radioaktiv
beta-active emitter Betastrahler *m*
beta decay Betazerfall *m*
beta emission Betastrahlung *f*
beta-radioactive beta-radioaktiv
beta x-ray Betastrahl *m*
beverage packing Getränkeverpackung *f*
big industry Großindustrie *f*
bilge water Bilgewasser *n*
bilharziosis Bilharziose *f*

bimetal switch Bimetallschalter *m*
binary mixture binäres Gemisch *n*
binder Bindemittel *n*
binding agent Bindemittel *n*
binding energy Bindungsenergie *f*
binding force Bindungskraft *f*
binding material Bindemittel *n*
bio-assay Bioassay *n*, Bioanalyse *f*, biologische Analyse *f*
bio-bin Biotonne *f*
bio-overgrown area biobewucherte Fläche *f*
bio-purification mechanism Bio-Reinigungsmechanismus *m*
bio-sludge Bio-Schlamm *m*
bioaccumulation Bioakkumulation *f*
bioaccumulative bioakkumulativ
bioactive bioaktiv
bioavailability Bioverfügbarkeit *f*
biocatalyst Biokatalysator *m*
biocenosis Biozönose *f*, Lebensgemeinschaft *f*
biochemical biochemisch; Biochemikalie *f*
biochemical degradation biochemischer Abbau *m*
biochemically active biochemisch aktiv
biochemist Biochemiker *m*
biochemistry Biochemie *f*
biocide Biozid *n*, Schädlingsbekämpfungsmittel *n*
bioclimatology Bioklimatologie *f*
biodegradability biologische Abbaubarkeit *f*, Bioabbaubarkeit *f*
biodegradable biologisch abbaubar
biodegradation Bioabbau *m*, biologischer Abbau *m*
biodegradation rate biologische Abbaurate *f*
biodynamic biodynamisch

bioecology Bioökologie *f*
bioenergetic bioenergetisch
bioenergy Bioenergie *f*
bioengineering Biotechnik *f*
bioequivalence study Bioäquivalenzstudie *f*
biofilter Biofilter *n*
biofiltration Biofiltration *f*
biogas Biogas *n*, Faulgas *n*
biogas control Biogassteuerung *f*
biogas pipe Faulgasleitung *f*
biogas power plant Biogaskraftwerk *n*
biogas processing Biogasaufbereitung *f*
biogas production Biogasproduktion *f*
biogas reactor Biogasreaktor *m*
biogas reservoir Biogasspeicher *m*
biogas utilization Biogasverwertung *f*
biogenic chemical biogene Chemikalie *f*
biogeography Biogeographie *f*
biogeology Biogeologie *f*
biogeosphere Biogeosphäre *f*
bioindicator Bioindikator *m*
bioleaching Bioauslaugung *f*
biological biologisch
biological activity biologische Aktivität *f*
biological clarification plant biologische Kläranlage *f*
biological cycle biologischer Kreislauf *m*
biological degradation biologischer Abbau *m*
biological filter biologisches Filter *n*
biological filtration biologische Filterung *f*

biological in-situ redevelopment biologische in-situ-Sanierung *f*
biological monitoring biologische Überwachung *f*
biological nutrient elimination biologische Nährstoffelimination *f*
biological purification biologische Reinigung *f*
biological purification stage biologische Reinigungsstufe *f*
biological sewage sludge stabilization biologische Klärschlammstabilisation *f*
biological sewage treatment biologische Abwasserbehandlung *f*
biological slime biologischer Rasen *m*
biological waste Bioabfall *m*, Biomüll *m*
biological waste collection Biomüllsammlung *f*
biological waste compost plant Biomüllkompostwerk *n*
biological waste composting Biomüllkompostierung *f*
biologically degradable biologisch abbaubar
biology Biologie *f*
biomanipulation Biomanipulation f
biomass Biomasse *f*
   biomass concentration Biomassenkonzentration *f*
   biomass synthesis Biomassesynthese *f*
biomedicine Biomedizin *f*
biometric evaluation biometrische Auswertung *f*
biometry Biometrie *f*
biopharmaceutics Biopharmazie *f*
biopolymer Biopolymer *n*
bioprocess Bioprozeß *m*

bioreactor Bioreaktor *m*
biorhythm Biorhythmus *m*
biosludge Bioschlamm *m*
biosphere Biosphäre *f*
biostimulation Biostimulation *f*
biosynthesis Biosynthese *f*
biosynthetic biosynthetisch
biosystem Biosystem *n*
biotechnological biotechnologisch
biotechnology Biotechnologie *f*
biotest Biotest *m*
biotic biotisch
   biotic degradation biotischer Abbau *m*
   biotic potential biotisches Potential *n*
biotope Biotop n(m), Lebensraum *m*
biotransformation Biotransformation *f*, Bioumwandlung *f*
biowasher Biowäscher *m*
biowasher system Biowäschersystem *n*
biowaste Biomüll *m*
biowaste compost Biomüllkompost *m*
bird migration Vogelzug *m*
bird protection Vogelschutz *m*
birth defect Geburtsfehler *m*
bismuth Wismut *n*, Bismut *n*
bituminous bituminös, asphalthaltig, pechhaltig
bituminous coal Fettkohle *f*, backende Kohle *f*; Steinkohle *f*
bituminous coal tar Steinkohlenteer *m*
blade dryer Schaufeltrockner *m*
blank Blindwert *m*
blank test Blindwertmessung *f*
blast air Gebläseluft *f*
blast flare-unit Gebläse-Abfackelanlage *f*

blast furnace Hochofen *m*
blast furnace slag Hochofenschlacke *f*
blaze loderndes Feuer *n*
bleach/to bleichen, weißen, entfärben
bleaching agent Bleichmittel *n*
bleaching effect Bleichwirkung *f*
bleed valve Auslaßventil *n*
bleeder screw Ablaßschraube *f*
block valve Absperrventil *n*
blocking layer Sperrschicht *f*
blood Blut *n*
blood analysis Blutanalyse *f*
blood bank Blutbank *f*
blow-off-pipe Abblaserohr *n*
blower Gebläse *n*, Gebläsemaschine *f*
blower aggregate Belüfteraggregat *n*
blower group Belüftergruppe *f*
blower performance Belüfterleistung *f*
blowing Blasen *n*, Verblasen *n*
blowing agent Blähmittel *n*, Treibmittel *n*
blue asbestos Blauasbest *m*
blue disease Blaufäule *f*
blue-green alga Blaualge *f*
body fluid Körperflüssigkeit *f*
body heat Körperwärme *f*
body liquid Körperflüssigkeit *f*
body temperature Körpertemperatur *f*
body weight Körpergewicht *n*
boiler equipment Kesselanlage *f*
boiler explosion Kesselexplosion *f*
boiler injection method Kesselinjektionsverfahren *n*
boiler insulation Kesselisolierung *f*
boiler output Kesselleistung *f*
boiler pump Kesselpumpe *f*
boiler scale Kesselstein *m*
boiler steam Kesseldampf *m*
boiling heat Siedehitze *f*
boiling temperature Siedetemperatur *f*
boiling water reactor Siedewasserreaktor *m*, Verdampfungsreaktor *m*
bomb calorimeter Bombenkalorimeter *n*
bond Bindung *f*
bond breaking Bindungsbruch *m*
bond energy Bindungsenergie *f*
bone dust Knochenmehl *n*
bone formation Knochenbildung *f*
bone glue factory Knochenleimfabrik *f*
bone injury Knochenschaden *m*
bone marrow Knochenmark *n*
bone marrow cell Knochenmarkzelle *f*
bone meal Knochendünger *m*, Knochenmehl *n*
border case Grenzfall *m*
border region Grenzgebiet *n*
borehole Bohrloch *n*
boring tower Bohrturm *m*
boron Bor *n*
botanical botanisch
botany Botanik *f*
bottle cap Flaschenverschluß *m*
bottle cleaner Flaschenreinigungsmaschine *f*
bottom flue Sohlenkanal *m*
bottom heating Bodenheizung *f*
bottom hole Bodenöffnung *f*
bottom of a disposal site Deponiesohle *f*
bottom of a refuse dump Deponiesohle *f*

bottom sealing Basisabdichtung f, Sohlenabdichtung f
bottom sediment Bodensediment n, Bodenablagerung f
boulder Felsblock m
boundary condition Randbedingung f
boundary diffusion Grenzflächendiffusion f
boundary layer determination Grenzflächenerfassung f
boundary problem Randproblem n
bracing climate Reizklima n
brackish water Brackwasser n
branch cock Verteilungshahn m
branch pipe Abzweigleitung f
branch point Verzweigungsstelle f
break of the flow Abreißen n der Strömung
breakdown Zusammenbruch m, Zusammenfall m, Ausfall m
breakdown strength Durchschlagsfestigkeit f
breaker Brecher m
breakthrough Durchbruch m
breakthrough curve Durchbruchskurve f
breast cancer Brustkrebs m
breathing capacity Atemgrenzwert m
breathing protection Atemschutz m
breathing protection system Atemschutzsystem n
breathing resistance Atmungswiderstand m
brewery Brauerei f
brief-case size Aktentaschenformat n
brightening agent Aufheller m
briquet Brikett n, Preßkohle f
briquetting plant Brikettieranlage f

bromine Brom n
bronchial asthma Bronchialasthma n
bronchitic bronchitisch
bronchitis Bronchitis f, Bronchialkatarrh m
bronchoscopy Bronchoskopie f
bronze Bronze f
brown alga Braunalge f
brown coal Braunkohle f
brown coal mining Braunkohleförderung f
browncoal coke Braunkohlekoks m
bubble meter Blasenzähler m
bubble tray Glockenboden m
bucket Becher m, Kübel m, Eimer m
bucket conveyor Becherwerk n
bucket wheel Becherrad n, Schöpfrad n
bud Knospe f
buffer Puffer f, Pufferlösung f
buffer action Pufferwirkung f
buffer effect Puffereffekt m
buffer system Puffersystem n
building materials testing Baustoffprüfung f
building-physical bauphysikalisch
building preservative agent Bautenschutzmittel n
building project Bauvorhaben n
building regulation Bauvorschrift f
building rubble Bauschutt m
building technical situation bautechnischer Zustand m
bulk density Schüttdichte f
bulk goods Schüttgut n
bulk manufacture Massenfertigung f
bulk material Grobstoff m, Schüttgut n

bulkage Ballaststoff *m*
bulkage capacity Füllstoffkapazität *f*
bulking of sludge Schlammblähung *f*
bulky material Grobmaterial *n*, Schüttmaterial *n*
bulky refuse Grobmüll *m*, Sperrmüll *m*
bulldozer Planierraupe *f*
bunker Bunker *m*
burn-off device Abfackelungsanlage *f*
burn-off equipment Abfackelanlage *f*
burner Brenner *m*, Ofen *m*, Verbrennungsofen *m*
burner nozzle Brennerdüse *f*
burning behaviour of waste Abfallbrennverhalten *n*
burnt lime gebrannter Kalk *m*
burnup Abbrand *m*
butterfly valve Drosselventil *n*
by-effect Nebenwirkung *f*
by-product Nebenprodukt *n*, Abfallprodukt *n*
bypass Nebenleitung *f*, Ableitung *f*, Nebenrohr *n*
bypass valve Bypass-Ventil *n*

C

cabinet dryer Trockenschrank *m*
cable probe Seilsonde *f*
cable waste Kabelabfall *m*
cadaver Kadaver *m*, Tierleiche *f*
cadmium Cadmium *n*
cadmium reduction column Cadmium-Reduktionssäule *f*
caesium Cäsium *n*
calander Kalander *m*
calciferous kalkhaltig
calcify/to verkalken
calcination plant Kalzinieranlage *f*
calcining furnace Brennofen *m*, Kalzinierofen *m*
calcite solubility Calcitlösevermögen *n*
calcium Calcium *n*
calibration constant Kalibrierfaktor *m*, Eichfaktor *m*
calibration curve Kalibrierkurve *f*, Eichkurve *f*
calibration device Kalibriereinrichtung *f*
calibration factor Kalibrierfaktor *m*, Eichfaktor *m*
calibration mixture Kalibriermischung *f*
californium Californium *n*
calming agent Beruhigungsmittel *n*
caloric kalorisch
caloric efficiency Wärmewirkungsgrad *m*
caloric receptivity Wärmeaufnahmefähigkeit *f*
calory Kalorie *f*
calorific power Heizwert *m*, Heizkraft *f*
calorific value Heizwert *m*, Brennwert *m*
calorific wärmeerzeugend
calorimeter Kalorimeter *n*
calorimetric kalorimetrisch
calory Kalorie *f*
can Blechdose *f*, Dose *f*, Kanister *m*
canal Kanal *m*, Graben *m*
canal system Grabensystem *n*, Kanalsystem *n*
canalization f Kanalisierung *f*, Kanalisation *f*

canalization system Kanalisationssystem *n*, Kanalisation *f*
cancer Krebs *m*, Karzinom *n*
cancer cell Krebszelle *f*
cancer research centre Krebsforschungszentrum *n*
cancer control Krebsbekämpfung *f*
cancer diagnosis Krebserkennung *f*
canceration Krebsbildung *f*
cancerogenic Kanzerogen *n*, kanzerogen
cancerogenity Kanzerogenität *f*
cancerologist Krebsforscher *m*
cancerous ulcer Krebsgeschwür *n*
canceroid krebsähnlich
canned food Dosenkonserve *f*
canned fruit Obstkonserve *f*
canned milk Dosenmilch *f*
cap with bayonet lock Kappe *f* mit Bajonettverschluß
capability of disposal Entsorgungsfähigkeit *f*
capacity of sludge incineration Schlammverbrennungskapazität *f*
capillarity Kapillarität *f*, Kapillarwirkung *f*
capillary Kapillare *f*, Saugröhrchen *f*
capillary action Kapillarwirkung *f*
capillary active kapillaraktiv
capillary column Kapillarsäule *f*
capillary combustion tube Verbrennungskapillare *f*
capillary depression Kapillardepression *f*
capital contribution to connection cost Anschlußkostenbeitrag *m*
capture cross section Einfangquerschnitt *m*
capture probability Einfangswahrscheinlichkeit *f*
carbide Karbid *n*

carbide lamp Karbidlampe *f*
carbohydrate Kohlenhydrat *n*
carbohydrate metabolism Kohlenhydratstoffwechsel *m*
carbon Kohlenstoff *m*
carbon-adsorbed precoat layer kohlenstoffangereicherte Precoatschicht *f*
carbon black Lampenruß *m*, Ruß *m*
carbon compound Kohlenstoffverbindung *f*
carbon dioxide Kohlendioxid *n*
carbon metabolism Kohlenstoffkreislauf *m*
carbon monoxide Kohlenmonoxid *n*
carbonization Verkokung *f*, Verkohlung *f*, Inkohlung *f*
carbonization process Verschwelung *f*
carbonize/to karbonisieren, verkohlen, auskohlen, verkoken, inkohlen
carbonizing plant Kokerei *f*
carbonyl Carbonyl *n*
carbonyl chloride Phosgen *n*, Carbonylchlorid *n*
carcass of used tyre Altreifenkarkasse *f*
carcinoembryonic antigen karzinoembryonales Antigen *n*
carcinogen Krebserreger *m*, Krebserzeuger *m*, Karzinogen *n*
carcinogenesis Krebsentstehung *f*, Karzinogenese *f*
carcinogenic krebserzeugend, karzinogen
carcinogenic potential krebserzeugendes Potential *n*
carcinolysis Karzinolyse *f*
carcinoma Karzinom *n*, Krebs *m*, Krebsgeschwür *n*, Krebsgeschwulst *n*

carcinoma formation Krebsbildung *f*
carcinosarcom Karzinosarkom *n*
carcinostasis Karzinostase *f*, Krebshemmung *f*
carcinostatic krebshemmend; krebshemmende Substanz *f*
cardboard packing Kartonverpakkung *f*
cardiac asthma Herzasthma *n*
cardiac disease Herzkrankheit *f*
cardiac infarct Herzinfarkt *m*
cardiac infarction Herzinfarkt *m*
cardiac poison Herzgift *n*
cardiovascular disease Herzgefäßkrankheit *f*
carditis Karditis *f*, Herzentzündung *f*
carelessness Fahrlässigkeit *f*, Achtlosigkeit *f*, Unachtsamkeit *f*, Unvorsichtigkeit *f*, Sorglosigkeit *f*
caries Karies *f*, Zahnfäule *f*
cariogenic karieserzeugend
carious kariös
carpeting of algae Algenteppich *m*
carrier Träger *m*
 carrier gas Trägergas *n*
 carrier gas line Trägergasleitung *f*
 carrier substance Trägersubstanz *f*, Trägerstoff *m*
car scrap Autoschrott *m*
car wreck Autowrack *n*
cartridge Patrone *f*, Kartusche *f*, Kassette *f*
cascade Kaskade *f*
 cascade configuration Kaskadenkonfiguration *f*
 cascade dryer Kaskadentrockner *m*
 cascade effect Kaskadenwirkung *f*
 cascade mill Kaskadenmühle *f*
 cascade of reactors Kaskade *f* von Reaktoren
 cascade preheater Kaskadenvorwärmer *f*
 cascade test Kaskadentest *m*
case of accident Störfall *m*
case of application Anwendungsfall *m*
case of disturbance Störfall *m*
case of trouble Störfall *m*
case study Fallstudie *f*
cast aluminium Aluminiumguß *m*
cast iron Gußeisen *n*
casting scrap Gußschrott *m*
casuality station Unfallstation *f*
catabolism Katabolismus *m*
catabolite Katabolit *m*
catalogue of demands Forderungenkatalog *m*
catalogue of special waste types Sonderabfallartenkatalog *m*
catalysis Katalyse *f*, Reaktionsbeschleunigung *f*
catalyst Katalysator *m*, Reaktionsbeschleuniger *m*
 catalyst lifetime Katalysatorlebensdauer *f*
 catalyst mass Katalysatormasse *f*
 catalyst poison Katalysatorgift *n*
 catalyst poisoning Katalysatorvergiftung *f*
 catalyst surface Katalysatoroberfläche *f*
 catalyst technology Katalysatorentechnologie *f*
 catalyst type Katalysatortyp *m*
 catalyst volume Katalysatorvolumen *n*
catalytic katalytisch
 catalytic acceleration katalytische Beschleunigung *f*
 catalytic decomposition katalytische Zersetzung *f*

catalytic degradation katalytischer Abbau *m*
catalytic effect katalytische Wirkung *f*
catalytic reaction katalytische Reaktion *f*
catalyze/to katalysieren
catastrophe Katastrophe *f*, Verhängnis *n*
caterpillar Raupe *f*, Raupenfahrzeug *n*
cation Kation *n*
cation exchange Kationenaustausch *m*
cation exchange resin Kationenaustauschharz *n*
cattle breeding Rinderzucht *f*
cattle manure Rinderdung *m*
cattle plague Rinderpest *f*
causative agent verursachendes Agens (Mittel) *n*
cause of a disease Krankheitsursache *f*
cause of cancer Krebsursache *f*
cause of (disturbance) malfunction Störungsursache *f*
cause of emission Emissionsursache *f*
caustic kaustisch, ätzend; Ätzmittel *n*, Ätzstoff *m*, Beizmittel *n*
caustic alkali Ätzkali *n*
caustic liquid Ätzflüssigkeit *f*
caustic liquor Beizflüssigkeit *f*
caustic lye Beizlauge *f*
caustic-resisting ätzbeständig
caution Vorsicht *f*, Behutsamkeit *f*
caution signal Warnsignal *n*
cave[rn] inhabitant Höhlenbewohner *m*
cavern Kaverne *f*

cavity Hohlraum *m*, Aussparung *f*, Loch *n*
cell Zelle *f*, Hohlraum *m*
cell activity Zellaktivität *f*
cell-biological testing model zellbiologisches Testmodell *n*
cell culture Zellkultur *f*
cell division Zellteilung *f*
cell formation Zellbildung *f*
cell substrate Zellsubstrat *n*
cellular zellulär, zellenartig, zellig, zellenförmig
cellular body Zellkörper *m*
cellular concrete Porenbeton *m*
cellular constituent Zellbestandteil *m*
cellular membrane Zellmembran *f*
cellular necrosis Zellnekrose *f*
cellulitis Zellgewebeentzündung *f*
cellulose Cellulose *f*, Zellstoff *m*
cellulose acetate Celluloseacetat *n*
cellulose bleaching Zellstoffbleichen *n*
cellulose decomposition Celluloseabbau *m*
cellulose disintegration Celluloseaufschluß *m*
cellulose hydrolysis hydrolytischer Celluloseaufschluß *m*
cellulose ion exchanger Celluloseionenaustauscher *m*
cellulose powder Cellulosepulver *n*
cement Zement *m*, Bindemittel *n*, Steinmörtel *m*, Kitt *m*, Klebekitt *m*
cement slab Betonplatte *f*
cement stone Zementstein *m*
cement/to zementieren, kitten, verkitten
cementation Zementierung *f*, Verkittung *f*, Zementeinspritzung *f*
cementation furnace Härteofen *m*

central heating network Fernheiznetz *n*
central heating Zentralheizung *f*
central nervous system Zentralnervensystem (ZNV) *n*
central pretreatment plant zentrale Vorbehandlungsanlage *f*
central processing plant zentrale Aufbereitungsanlage *f*
centre for raw material recovery Rohstoffrückgewinnungszentrum *n*
centre of infection Ansteckungsherd *m*
centrifugal aerator Kreiselbelüfter *m*, Zentrifugallüfter *m*
centrifugal extractor Zentrifugalextraktor *m*
centrifugal filter Siebschleuder *f*
centrifugal hydroextractor Schleudertrockner *m*
centrifugal pump Kreiselpumpe *f*
centrifugal separator Zentrifugalabscheider *m*
centrifugation Zentrifugierung *f*
centrifuge Zentrifuge *f*
ceramic regenerator keramischer Regenerator *m*
cereal pest Getreideschädling *m*
certifiable anzeigepflichtig, zu bescheinigen
certificate Attest *n*, Zeugnis *n*, Gutachten *n*
certification Beurkundung *f*
certified gas mixture Gasmischung *f* mit definierter (gesicherter) Zusammensetzung
certify/to beglaubigen, bescheinigen
cessation of exposure Einstellung *f* der Exposition
chain conveyor Kettenförderer *m*
chain reaction Kettenreaktion *f*
chain wheel Kettenrad *n*
chalk Kreide *f*, Kalk *n*
chalk bed Kreideschicht *f*
chamber acid Kammersäure *f*
chamber filter press Kammerfilterpresse *f*
chamber furnace Kammerofen *m*
chance of living Lebenschance *f*
change of climate Klimawechsel *m*
characteristic curve Kennlinie *f*
charcoal Holzkohle *f*
charcoal burning Holzverkohlung *f*
charcoal filter Kohlefilter *n*
charcoal pile Meiler *m*, Kohlemeiler *m*
charcoal tablet Kohletablette *f*
charcoal wood Meilerholz *n*
charge Charge *f*, Beschickung *f*, Einsatz *m*, Einsatzmaterial *n*; Ladung *f*; Füllgut *n*, Beladung *f*; Preis *m*, Gebühr *f*; Last *f*
charge capacity Ladekapazität *f*, Ladevermögen *n*
charge door Ladetür *f*, Beschickungstür *f*
charge quantity Chargengröße *f*, Chargenmenge *f*
charge/to beladen, bepacken, chargieren; belasten
charging device Beschickungseinrichtung *f*
charging the furnace Ofenbeschickung *f*
chart of measuring (locations) points Meßstellenblatt *n*
chemical Chemikalie *f*
chemical compound chemische Verbindung *f*
chemical consumption Chemikalienverbrauch *m*

chemical degradation chemischer Abbau *m*
chemical emergency protection chemischer Katastrophenschutz *m*
chemical engineering chemische Verfahrenstechnik *f*
chemical industry chemische Industrie *f*
chemical recovery Chemikalienrückgewinnung *f*
chemical tanker Chemikalientanker *m*
chemisorption Chemisorption *f*
chemoresistance Chemoresistenz *f*
chemosensitive chemosensitiv
chemosensor Chemosensor *m*
chimney flue Rauchkanal *m*, Abzugskanal *m*
chimney gas Abzugsgas *n*
chimney hole Rauchabzugsöffnung *f*
chimney hood Rauchfang *m*
chimney soot Kaminruß *m*
chip wood Spanholz *n*
chipper Hackmaschine *f*
chloracne Chlorakne *f*
chlorate Chlorat *n*
chloride Chlorid *n*
  chloride of lime bleaching Chlorkalkbleiche *f*
  chloride of lime plant Chlorkalkanlage *f*
chlorinated hydrocarbon Chlorkohlenwasserstoff *m*, chlorierter Kohlenwasserstoff *m*
chlorinating Chlorierung *f*
  chlorinating agent Chlorierungsmittel *n*
  chlorinating plant Chlorierungsanlage *f*
chlorination Chlorierung *f*

chlorination degree Chlorierungsgrad *m*
chlorination of waste water Abwasserchlorierung *f*, Chlorierung *f* von Abwasser
chlorine Chlor *n*
chlorite Chlorit *n*
chlorophyll Chlorophyll *n*, Blattgrün *n*
chloroplast Chloroplast *n*
chromatogram Chromatogramm *n*
chromatograph Chromatrograph *m*
chromatographic chromatographisch
  chromatographic column Chromatographiesäule *f*
  chromatographic packing Füllstoff *m*, Adsorptionsmittel *n*
chromatography Chromatographie *f*
chromium Chrom *n*
  chromium plating Verchromung *f*
  chromium dust Chromstaub *m*
  chromium sewage Chromabwasser *n*
chromophore Chromophor *m*
chromosome Chromosom *n*
chronic concentration chronische Konzentration *f*
chronic disease chronische Krankheit *f*
chronic liver damage chronischer Leberschaden *m*
chronically affecting chronisch schädigend
cinder Abbrand *m*, Kohlenschlacke *f*, Schlacke *f*, ausgeglühte Kohle *f*
circuit control Kreislaufsteuerung *f*
circular tank Kreisbecken *n*
circulating air Umluft *f*
  circulating air conveyor Umluftförderanlage *f*

circulating cooling water Umlaufkühlwasser *n*
circulating dryer Umlufttrockner *m*
circulating evaporator Umlaufverdampfer *m*
circulating performance Umwälzleistung *f*
circulating pump Umlaufpumpe *f*
circulating reactor Umlaufreaktor *m*
circulating system Umlaufsystem *n*, Zirkulationssystem *n*
circulation Kreislauf *m*, Umlauf *m*
circulation heat exchanger Umlaufwärmetauscher *m*
cistern Zisterne *f*
citizens' action committee Bürgerinitiative *f*
civil defence Zivilschutz *m*
civil law company for waste especially to be supervised zivilrechtliche Sonderabfallgesellschaft *f*
claim for compensation Ersatzanspruch *m*
claim of refund Erstattungsanspruch *m*
clarification Klären *n*, Klärung *f*, Abklärung *f*, Läuterung *f*
clarification basin Klärbassin *n*
clarification effect Klärwirkung *f*
clarification plant Kläranlage *f*, Klärwerk *n*
clarifier Klärmittel *n*; Kläranlage *f*
clarifier inlet Kläranlageneinlauf *m*
clarifier tank Klärgefäß *n*, Läutergefäß *n*, Klärtank *m*
clarify/to klären, abklären, abschlämmen; reinigen
class of risk Gefahrenklasse *f*
classification plant Sortieranlage *f*
classifier Sichter *m*
classifying Sortieren *n*; Einteilen *n*, Aufgliedern *n*, Auslesen *n*, Klassieren *n*
classifying by weight Gewichtsauslese *f*
classifying technique Sortiertechnik *f*, Klassiertechnik *f*
clay Tonerde *f*, Lehm *m*
clay-immobilized pesticide tonimmobilisiertes Pestizid *n*
clay layer Tonschicht *f*
clay marl Tonmergel *m*
clay powder Tonmehl *n*, Tonpulver *n*
clay soil Tonboden *m*
clean-up Reinigungsaufbereitung *f*
clean-up of contaminated soil Reinigungsaufbereitung *f* von kontaminiertem Erdreich
cleaning Reinigung *f*; Spülung *f*; Waschung *f*, Waschen *n*
cleaning automatization Reinigungsautomatisierung *f*
cleaning ball Reinigungskugel *f*
cleaning basin Reinigungsbecken *n*
cleaning effect Reinigungswirkung *f*; Wascheffekt *m*
cleaning kit Reinigungsausrüstung *f*
cleaning of the exhaust air Abluftreinigung *f*
cleaning residue Reinigungsrückstand *m*
cleaning trench Reinigungsgraben *m*
cleaning vehicle for refuse bins Müllbehälterreinigungsfahrzeug *n*
cleanse/to reinigen; putzen; abspülen, abwaschen

cleansing Reinigung *f*; Abspülen *n*
clear-cutting Kahlschlag *m*
clear-felling Kahlschlag *m*
clear/to klären, läutern, abklären
cleared woodland Rodung *f*
clearer Klärbecken *n*
clearing Klärung *f*, Abklärung *f*, Klären *n*, Läuterung *f*; Rodung *f*, Leerung *f*
clearing authority Clearingstelle *f*
clearing bath Klärbad *n*
clearing tank Klärtank *m*
cleavage product Spaltprodukt *n*
cliff Klippe *f*
climate Klima *n*
climate endangering Klimagefährdung *f*
climate system Klimasystem *n*
climatic klimatisch
climatic active klimawirksam
climatic catastrophe Klimakatastrophe *f*
climatic chamber Klimaprüfschrank *m*, Klimakammer *f*
climatic change klimatische Veränderung *f*, Klimawechsel *m*
climatic control Klimakontrolle *f*
climatic data Klimadaten pl
climatic disaster Klimakatastrophe *m*
climatic factor Klimafaktor *m*
climatic rule Klimaregel *f*
climatic zone Klimazone *f*
climatologist Klimatologe *m*
climatology Klimatologie *f*, Klimakunde *f*, Klimalehre *f*
climax Höhepunkt *m*, Gipfel *m*
clinic-specific waste krankenhausspezifischer Abfall *m*
clinic[al] waste Klinikabfall *m*, Klinikmüll *m*, Krankenhausmüll *m*

clip-on-cap Kapselverschluß *m*
closed circuit geschlossener Kreislauf *m*
closed conveyor Kreiselförderer *m*
closed loop Regelkreis *m*
closed-loop adaptive control block adaptiver Reglerbaustein *m*
closed production system geschlossenes Produktionssystem *n*
closed system geschlosssenes System *n*
closure Deckel *m*; Bedeckung *f*, Abdeckung *f*; Stillegung *f*
cloud Wolke *f*
cloud formation Wolkenbildung *f*
cloud point Trübungspunkt *m*
cloudburst Wolkenbruch *m*
cloudy wolkig, trübe
coagulability Gerinnbarkeit *f*, Gerinnungsfähigkeit *f*
coagulable koagulierbar, gerinnbar
coagulant Gerinnungsmittel *n*, Koagulierungsmittel *n*
coagulation Koagulation *f*, Gerinnung *f*, Ausflockung *f*, Flockenbildung *f*
coagulation factor Gerinnungsfaktor *m*
coal Kohle *f*
coal analysis Kohleanalyse *f*
coal battery Kohlebatterie *f*
coal carbonization Kohleentgasung *f*
coal carbonizing plant Kokerei *f*
coal consumption Kohleverbrauch *m*
coal power station Kohlekraftwerk *n*
coal upgrading Kohleveredelung *f*

coalescence separation Koaleszenzabscheidung f
coalescence separator Koaleszenzabscheider m
coalescence separator system Koaleszenzabscheidersystem n
coalminer's pneumoconiosis Bergwerksarbeiterpneumokoniose f
coarse filtration Grobfiltration f
coarse rake Grobrechen m
coarse screening Grobsiebung f
coarse separation Grobabtrennung f, Grobabscheidung f
coast Küste f
coast pollution Küstenverschmutzung f
coastal area Küstengebiet n
coastal erosion Küstenerosion f
coat of tar Teeranstrich m
cobalt Cobalt n
code number Kennummer f
coefficient of thermal expansion Wärmeausdehnungskoeffizient m
coefficient of variation Variationskoeffizient m
coffee roaster Kaffeerösterei f
coil condenser Kühlschlange f, Schlangenkühler m
coke Koks m
  coke dust Koksstaub m
  coke gas Koksgas n
  coke oven Koksofen m
  coke-oven plant Kokerei f
  coke washer Kokswäscher m
  coke waste Koksabfall m
coke/to verkoken, backen, carbonisieren
coking Verkokung f, Backen, Kohlevergasung f
  coking plant Kokerei f, Verkokungsanlage f, Koksanlage f

cold air flow Kaltluftstrom m
cold milling Kaltvermahlung f
cold sterilization Kaltentkeimung f, Kaltsertilisation f
coli-bacillus Kolibazillus m
coli-titer Kolititer m, Kolitest m
collapse Kollaps m, Zusammenbruch m
collapse/to kollabieren, zusammenbrechen, einstürzen
collecting-chamber Sammelraum m
collecting flask Auffangkolben m
collecting vessel Sammelbehälter m
collection container Sammelbehälter m
collection goods Sammelware f
collection of domestic refuse Hausmüllsammlung f
collection place Sammelstelle f
collection tank Auffangwanne f, Sammeltank m
collection vessel Sammelbehälter m
collector Sammler m, Abscheider m, Kollektor m
colloid Kolloid n
colloid chemistry Kolloidchemie f
colloidal kolloidal
colloidal particle kolloidales Teilchen n
colloidal state Kolloidzustand m
colloidal substance Kolloid n
colloidchemical kolloidchemisch
colorimeter Kolorimeter n
colonization Besiedlung f
colour and dye industry Farbenindustrie f
colour monitor Farbsichtstation f
colour monitoring station Farbsichtstation
colour-separated collecting of used

glass farbgetrennte Sammlung f von Altglas
column loading Säulenbeladung f
combat against cancer Krebsbekämpfung f
combatting of traffic noise Verkehrslärmbekämpfung f, Bekämpfung f von Verkehrslärm
combination power plant Kombi-Kraftwerk n
combining nozzle Mischdüse f
combustibility Brennbarkeit f
combustible brennbar; Brennmaterial n, Brennstoff m
combustion Verbrennung f
combustion analysis Verbrennungsanalyse f
combustion chamber Verbrennungskammer f
combustion energy Verbrennungsenergie f
combustion enthalpy Verbrennungsenthalpie f
combustion plant Verbrennungsanlage f
combustion product Verbrennungsprodukt n
combustion residue Verbrennungsrückstand m
commercial packaging handelsübliche Verpackung f
commercial product Handelsprodukt n, Handelsware f
commercial quality Handelsqualität f
communal authority kommunale Aufsichtsbehörde f
communal budget Kommunalhaushalt m
communal clarification plant kommunale Kläranlage f

communal power-heat combination kommunale Kraft-Wärme-Kopplung f
communal sewage sludge kommunaler Klärschlamm m
communal vehicle Kommunalfahrzeug n
communal waste disposal öffentliche Müllentsorgung f
communicable disease übertragbare Krankheit f
compact clarifier Kompaktkläranlage f
compact dewatering facility Kompaktentwässerer m
compact shaft Kompaktschacht m
compact/to kompaktieren, verdichten, zusammendrücken
compacter Verdichter m
compacting capacity Verdichtungsleistung f
comparative value Vergleichswert m
compartment concentration Kompartimentkonzentration f
compatible to disposal entsorgungsgerecht
compensation claim Schadenersatzanspruch m
competition oxidation value Oxidations-Konkurrenzwert m
complete analysis Vollanalyse f
complete disposal Totalentsorgung f
complete fertilizer Volldünger m
complete monitoring lückenlose Überwachung f
complex formation Komplexbildung f
complex fertilizer Komplexdünger m

complexing agent Komplexbildner *m*
component Bestandteil *m*
compost Kompost *m*, Düngeerde *f*
compost aeration Kompostbelüftung *f*
compost application Kompostanwendung *f*
compost area Kompostplatz *m*
compost energy plant Kompostenergieanlage *f*
compost heap Komposthaufen *m*, Komposthalde *f*
compost material Kompostmaterial *n*
compost of baby's nappies Kompost *m* aus Baby-Windeln
compost of plant waste Pflanzenabfallkompost *m*
compost of waste paper Altpapierkompost *m*, Kompost *m* aus Altpapier
compost pile Komposthaufen *m*
compost product Kompostprodukt *n*
compost quality Kompostqualität *f*
compost raw material Kompostrohmaterial *n*
compost/to kompostieren
compostable kompostierbar
composting Kompostierung *f*
composting method Kompostierungsverfahren *n*
composting of biological waste Biomüllkompostierung *f*
composting of green waste Grünabfallkompostierung *f*
composting plant Kompostaufbereitungsanlage *f*, Kompostierungsanlage *f*
composting project Kompostierungsprojekt *n*
composting system Kompostierungssystem *n*
compressed air Druckluft *f*, Pressluft *f*
compressed air bottle Druckluftflasche *f*
compressed-air filter Druckluftfilter *n*
compressed-air plant Druckluftanlage *f*
compressed fittings Druckluftarmaturen *fpl*
compressed gas Druckgas *n*
compressed tubing Druckluftschlauch *m*
compression Kompression *f*
computer-controlled emission calculator computergesteuerter Emissionsrechner *m*
computer scrap Computerschrott *m*
computerized network control system rechnergeführtes Netzleitsystem *n*
concentrate/to konzentrieren, anreichern; eindicken; einengen; verdichten
concentrate by evaporation/to eindampfen
concentration Konzentration *f*, Konzentrierung *f*, Anreicherung *f*; Eindickung *f*
concentration change Konzentrationsänderung *f*
concentration excess Konzentrationsüberschuß *m*
concentration gradient Konzentrationsgefälle *n*, Konzentrationsgradient *m*
concentration profile Konzentrationsprofil *n*, Konzentrationsverlauf *m*

concentration range Konzentrationsbereich *m*
concentrative effect Packungseffekt *m*
concrete Beton *m*
concrete construction Betonkonstruktion *f*
concrete encasement Betonumhüllung *f*
condensate Kondensat *n*, Dampfwasser *n*, Kondenswasser *n*
condensate accumulator vessel Kondensatsammelgefäß *n*
condensate collector Kondensatsammler *m*
condensate separator Kondensatabscheider *m*
condensation Kondensation *f*, Kondensierung *f*, Niederschlag *m*
condensation core Kondensationskern *m*
condensation point Kondensationspunkt *m*
condensation product Kondensationsprodukt *n*
condensation water Kondenswasser *n*
condense/to kondensieren, niederschlagen; verdichten
condenser Kondensationsapparat *m*, Kondenser *m*, Kondensator *m*
condenser coil Kühlschlange *f*
condensing flue gas purification kondensierende Rauchgasreinigung *f*
condition/to konditionieren
conditioning Konditionierung *f*
conditioning agent Konditionierungszuschlag *m*
conditioning plant Konditionierungsanlage *f*

conditions of approval Genehmigungsvoraussetzungen *fpl*
confiscate/to konfiszieren, beschlagnahmen
congealer Gefrierer *m*, Gefriervorrichtung *f*
congenital immunity angeborene Immunität *f*
congenital malformation angeborene Mißbildung *f*
congestion Ansammlung *f*, Anhäufung *f*, Andrang *m*, Stauung *f*
conglomerate/to konglomerieren, sich zusammenballen
connecting cable Anschlußleitung *f*
connecting lead Anschlußleitung *f*
connecting technique Anschlußtechnik *f*
connection charge Anschlußgebühr *f*
connection conditions Anschlußbedingungen *fpl*
conservation of food Lebensmittelkonservierung *f*
consolidation of arable land Flurbereinigung *f*
constituent Bestandteil *m*, Inhaltsstoff *m*
construction proofing Bautenschutz *m*
constructural noise control Baulärmschutz *m*
consulting for environmental protection Umweltschutzberatung *f*
consumption forecast Verbrauchsprognose *f*
consumption of fuel Brennstoffverbrauch *m*
consumption of raw material Rohstoffverbrauch *m*

consumption of water Wasserverbrauch *m*
contact filtration Kontaktfiltration *f*
contact insectizide Kontaktinsektizid *n*
contact poison Kontaktgift *n*
contact substance Kontaktmittel *n*
contact surface Kontaktoberfläche *f*
contact surface catalysis Oberflächenkatalyse *f*
contact time Kontaktzeit *f*
contactor relay Hilfsschütz *n*
container cleaning Behälterreinigung *f*
container collection Containersammlung *f*
container for materials of value Wertstofftonne *f*
container for recycling materials Wertstofftonne *f*
container system Behältersystem *n*, Containersytem *n*
contaminant Kontaminant *n*, verunreinigende Substanz *f*, Verseuchungsstoff *m*
contaminated area Altlast *f*
contaminated site Örtlichkeit *f* mit Altlast
contaminate/to kontaminieren, verunreinigen, verschmutzen, verderben
contamination Kontaminierung *f*, Verunreinigung *f*, Verseuchung *f*; Unreinheit *f*, Schmutz *m*
contamination control Verseuchungsbekämpfung *f*
contamination problem Kontaminationsproblem *n*
continuous operation Dauerbetrieb *m*

continuous process Kreisprozeß *m*, Dauerprozeß *m*
contraindication Kontraindikation *f*, Gegenanzeige *f*
contrast examination Kontrastuntersuchung *f*
control Regelung *f*, Steuerung *f*
control accuracy Regelgenauigkeit *f*
control loop Regelkreis *m*
control measure Kontrollmaßnahme *f*, Bekämpfungsmaßnahme *f*
control parameter Regelparameter *m*
control relay Steuerrelais *n*, Hilfsschütz *n*
control system flow chart Funktionsplan *m*
control unit Steuereinheit *f*
control/to kontrollieren; regeln; steuern; bekämpfen
controlled system Regelstrecke *f*
convection heating Konvektionsheizung *f*
convection of heat Wärmeübertragung *f*
conversion Konvertierung *f*, Umwandlung *f*, Umformung *f*; Umrechnung *f*; Umsetzung *f*, Umsatz *m*, Stoffumsetzung *f*, Konversion *f*
conversion of the calorific value Brennwertkonvertierung *f*
conversion process Umsetzungsprozeß *m*, Umwandlungsprozeß *m*
conversion product Konvertierungsprodukt *n*
converter Reaktor *m*, Reaktionsofen *m*, Konverter *m*
converter fume Konverterrauch *m*
conveyance platform Förderplattform *f*

conveyor belt Förderband *n*
coolant Kühlmittel *n*
coolant lubricant Kühlschmierstoff *m*
coolant water Kühlwasser *n*
cooler Kühler *m*
cooling Kühlung *f*
cooling cycle Kühlkreislauf *m*
cooling fluid Kühlflüssigkeit *f*
cooling jacket Kühlmantel *m*
cooling lubricant Kühlschmierstoff *m*
cooling water Kühlwasser *n*
copper Kupfer *n*
copper-containing sludge kupferhaltiger Schlamm *m*
coprecipitation Mitfällung *f*
coral island Koralleninsel *f*
coral reef Korallenriff *n*, Korallenbank *f*
cork wood Korkholz *n*
correction of disturbances Störungsbeseitigung *f*
corrode/to korrodieren; rosten
corrosion Korrosion *f*; Rostbildung *f*, Rostfraß *m*
corrosion phenomenon Korrosionserscheinung *f*
corrosion problem Korrosionsproblem *n*
corrosion protection Korrosionsschutz *m*
corrosion resistance Korrosionsbeständigkeit *f*
corrosion-resistant korrosionsbeständig
corrosive korrosiv
corrugated board Wellpappe *f*
cosmic chemistry kosmische Chemie *f*

cosmic radiation Höhenstrahlung *f*, kosmische Strahlung *f*
cosmic rays Höhenstrahlen *mpl*
cosmos Kosmos *m*, Weltall *m*, All *n*
cost-benefit-analysis Kosten-Nutzen-Analyse *f*
cost factor Kostenfaktor *m*
cost-saving kostensparend
countercurrent principle Gegenstromprinzip *n*
countercurrent washing Gegenstromwaschung *f*
counterpoison Gegengift *n*, Antidoton *n*
course of a disease Krankheitsverlauf *m*
course of operation Bedienablauf *m*
course of reaction Reaktionsverlauf *m*
course of rot Rotteverlauf *m*
covering of algae Algenteppich *m*
covering of cables Abdeckung *f* von Kabeln
cracking plant Krackanlage *f*
cracking process Krackverfahren *n*
cream of lime Kalkmilch *f*
cremation Kremierung *f*, Verbrennung *f*, Einäscherung *f*
criterion of assessment Beurteilungskriterium *n*
criterion of compost quality Kompost-Qualitätskriterium *n*
critical region kritischer Bereich *m*
critical situation kritische Situation *f*, kritische Lage *f*
crocidolite fibre Krokydolithfaser *f*
crop protection Pflanzenschutz *m*
crop protection agent Pflanzenschutzmittel *n*

crop protection law Pflanzenschutzgesetz *n*
cross-link/to vernetzen
cross-linked macromolecule vernetztes Makromolekül *n*
cross-linker Vernetzer *m*
cross-linking site Vernetzungsstelle *f*
cross sensitivity Querempfindlichkeit *f*
crude gas Rohgas *n*
crude material Rohstoff *m*
crude material processing Rohstoffaufbereitung *f*
crude metal Rohmetall *n*
crude oil Rohöl *n*
crude oil residue Rohölrückstand *m*
crude rubber Rohgummi *n*
crude soda Rohsoda *f*
crude waste water Rohabwasser *n*
crush section Knautschzone *f*
crusher Zerkleinerungsmaschine *f*, Brecher *m*, Brechwerk *n*
crushing plant Brecheranlage *f*, Zerkleinerungsanlage *f*
crushing process Zerkleinerungsprozeß *m*
crushing technology Zerkleinerungstechnik *f*
crustaceans Krebstiere *mpl*
cryotechnology Kryotechnik *f*
cryotechnology plant Kryotechnikanlage *f*
crystallization evaporator Kristallisationsverdampfer *m*
culture medium Kulturboden *m*
culture tube Kulturröhrchen *n*
curium Curium *n*
current network Stromnetz *n*
current supply Stromversorgung *f*

cutaneous reaction Hautreaktion *f*
cutaneous test Hauttest *m*
cutaneous tolerance Hautverträglichkeit *f*
cutireaction Hautreaktion *f*
cutting oil Schneidöl *n*, Bohröl *n*, Kühlöl *n*
cutt-off current Reststrom *m*
cyanide Cyanid *n*
cyclic hydrocarbon zyklischer Kohlenwasserstoff *m*
cyclone Zyklon[abscheider] *m*
cyclone dust collector Fliehkraftstaubabscheider *m*
cyclone dust separator Zyklonentstauber *m*
cyclone filter Zyklonfilter *n*
cyclone scrubber Zyklonwäscher *m*
cytobiology Zytobiologie *f*
cytogenesis Zytogenese *f*
cytogenetics Zytogenetik *f*
cytogenous zytogen, zellbildend
cytology Zytologie *f*, Zellenlehre *f*
cytolysis Zytolyse *f*
cytomorphology Zellmorphologie *f*
cytomorphosis Zellveränderung *f*
cytophysiologic zellphysiologisch
cytophysiology Zellphysiologie *f*
cytoplasm Zytoplasma *n*, Zellplasma *n*
cytozoon Zellparasit *m*

D

daily exchange täglicher Austausch *m*
daily fluctuations tägliche Schwankungen *fpl*
daily load curve Tagesganglinie *f*

daily output curve Tagesganglinie *f*
dam Damm *m*, Staumauer *f*
dam aeration Talsperrenbelüftung *f*
dam level Aufstauhöhe *f*
damage of forest Waldschaden *m*
damage of trees Baumschaden *m*
damage to the crops Feldschaden *m*
damaged area Schadensbereich *m*
damaged by radiation strahlengeschädigt
damming-up Aufstauen *n*
danger assessment Gefährlichkeitsbewertung *f*, Gefahrenabschätzung *f*
danger of infection Ansteckungsgefahr *f*
danger pay Gefahrenzulage *f*
danger zone Gefahrenzone *f*, Gefahrenbereich *m*
dangerous gefährlich, gefahrvoll
dangerous to the environment umweltgefährlich
dangerousness Gefährlichkeit *f*
daphnia Daphnia *f*, Wasserfloh *m*
data bank Datenbank *f*
data channel Datenkanal *m*
data drain Datensenke *f*
data flow Datenfluß *m*
data integrity Datenintegrität *f*
data memory Datenspeicher *m*
data network Datennetz *n*
data processing Datenverarbeitung *f*
data security Datensicherheit *f*
data storage Datenspeicher *m*
data transfer Datenübertragung *f*
de-aerate/to entlüften, entgasen
de-aerator Entlüfter *m*, Entlüftungsanlage *f*

de-ferrization Enteisenung *f*
de-ironing Enteisenung *f*
deacidify/to entsäuern
deactivate/to desaktivieren, entaktivieren
deactivated column desaktivierte Säule *f*
deactivation Desaktivierung *f*, Entaktivierung *f*
dead water Totwasser *n*
death of fish Fischsterben *n*
death of seals Robbensterben *n*
debarking drum Entrindungstrommel *f*
debark/to entrinden, abschälen
debris box Abfallbehälter *m*, Abfallkasten *m*
decalcification Entkalkung *f*
decalcify/to entkalken
decant/to dekantieren, abdekantieren, abgießen, abschlämmen
decantation Dekantierung *f*, Abklärung *f*, Schlämmung *f*
decantation centrifuge Dekantierungszentrifuge *f*
decanter Dekantiergefäß *n*
decay Zerfall *m*; Fäulnis *f*, Verwesung *f*, Moder, Verderb *m*
decay path Zerfallsweg *m*
decay product Zerfallsprodukt *n*
decay resistance Fäulnisbeständigkeit *f*
decay series Zerfallsreihe *f*
decay/to verfallen; verfaulen, verwesen; vermodern
dechlorination Entchlorung *f*
deciduous forest Laubwald *m*
decolorize/to entfärben, bleichen
decolorizing agent Entfärbungsmittel *n*, Entfärber *m*

decompose/to [sich] zersetzen, abbauen, aufschließen, aufspalten
decomposing agent Zersetzer *m*
decomposition Zersetzung *f*, Abbau *m*, Aufschluß *m*
decomposition product Zersetzungsprodukt *n*, Abbauprodukt *n*, Zerfallsprodukt *n*
decontaminant Entseuchungsmittel *n*, Dekontaminationsmittel *n*
decontaminant technology Entseuchungstechnologie *f*
decontaminate/to dekontaminieren, entgiften, entgasen, entseuchen, entstrahlen
decontamination Dekontamination *f*, Dekontaminierung *f*, Entseuchung *f*, Entgiftung *f*, Entaktivierung *f*
decontamination agent Dekontaminationsmittel *n*, Entseuchungsmittel *n*, Entgiftungsmittel *n*
decrease in value Wertminderung *f*
decrease in weight Gewichtsabnahme *f*
decrease of emission Emissionsminderung *f*
decrease of leaching water Sickerwasserverringerung *f*
deemulsification reaction Demulgierungsreaktion *f*
deep-bed filtration Tiefenfiltration *f*
deep cooling Tiefkühlung *f*
deep-freeze installation Tiefkühleinrichtung *f*
deep sea Tiefsee *f*
deep-ventilation system Tiefenbelüftungssystem *n*
deep well Tiefbrunnen *m*
defoamer Antischaummittel *n*
defoliant Entlaubungsmittel *n*
deforestation Abholzung *f*
degas/to entgasen
degasification Entgasung *f*
degasification measure Entgasungsmaßnahme *f*
degasification of a refuse dump Deponieentgasung *f*
degassing Entgasen *n*
degerminate entkeimen
degermination Entkeimung *f*
degradable abbaubar
degradable packaging material abbaubares Verpackungsmaterial *n*
degradation Abbau *m*, Zersetzung *f*, Degeneration *f*
degradation catalyst Abbaubeschleuniger *m*
degradation constant Abbaukonstante *f*
degradation process Abbauvorgang *m*
degradation property Abbaueigenschaft *f*
degradation rate Abbaurate *f*
degradation test Abbautest *m*
degrade/to abbauen, zerlegen, degradieren
degrease/to entfetten, abfetten
degreasing bath Entfettungsbad *n*
degree of carbonization Inkohlungsgrad *m*
degree of contamination Kontaminierungsgrad *m*
degree of dust removal Entstaubungsgrad *m*
degree of hardness Härtegrad *m*
degree of heat exchange efficiency Wärmetauscherwirkungsgrad *m*
degree of inactivation Inaktivierungsgrad *m*

degree of nitrogen removal Entstickungsgrad *m*
degree of pollution Verunreinigungssgrad *m*, Verschmutzungsgrad *m*
degree of purity Reinheitsgrad *m*
degree of upgrading of the system Ausbaugrad *m* des Systems
dehalogenation Dehalogenierung *f*
dehydration Dehydratation *f*, Dehydratisierung *f*
delivery condition Lieferungsbedingung *f*, Anlieferungsbedingung *f*
delocalization Delokalisierung *f*
delocalize/to delokalisieren
demand of redevelopment Sanierungsbedarf *m*
demanganizing Entmanganung *f*
demineralization Entmineralisierung *f*, Entsalzung *f*
demineralization of water Wasserentsalzung *f*
demineralize/to entmineralisieren, entsalzen
demographic demographisch
demolition Abriß *m*, Abbruch *m*
demolition work Abbrucharbeiten *fpl*
demonstration plant Demonstrationsanlage *f*
demulsification Entmischung *f*
demulsifier Demulgator *m*
denaturation Vergällung *f*, Denaturierung *f*
denitrification Denitrifikation *f*, Entstickung *f*
denitrification basin Denitrifikationsbecken *n*
denitrification zone Denitrifikationszone *f*
denitrify/to denitrifizieren

density determination Dichtebestimmung *f*
density of population Bevölkerungsdichte *f*
deodorant Desodorationsmittel *n*, Deodorant *n*; geruchsbindend
deodorization Desodorierung *f*, Geruchsbekämpfung *f*
deodorization filter Biofilter *n* zur Geruchsbekämpfung
deodorize/to desodorieren
deoxidant Desoxidationsmittel *n*
deoxidize/to desoxidieren
depolarization current Depolarisationsstrom *m*
deposit Deposition *f*, Ablagerung *f*; Absatz *m*, Bodensatz *m*, Niederschlag *m*; Aufschüttung *f*, Lagerstätte *f*; Flaschenpfand *n*
deposit of mud Schlammablagerung *f*
deposit of rust Rostansatz *m*
deposit/to aufbringen, ablagern, absetzen, absitzen, ansetzen
depositing of mud Anschlämmen *n*
depositing tank Absetztank *m*, Reinigungstank *m*
deposition Ablagerung *f*, Absetzung *f*, Niederschlag *m*
deposition of fat Fettablagerung *f*
deposition parameter Depositionsparameter *m*
depot Depot *n*, Ablagerung *f*, Lagerplatz *m*
depot of hazardous substances Gefahrstoffdepot *n*
dermatitis Hautentzündung *f*, Dermatitis *f*
dermatologist Dermatologe *m*, Hautarzt *m*
dermatology Dermatologie *f*

dermatosis Dermatose *f*, Hautkrankheit *f*
derrick Bohrturm *m*
desensitization Desensibilisierung *f*
desert Wüste *f*
desert animal Wüstentier *n*
desert plant Wüstenpflanze *f*
desert soil Wüstenboden *m*
desiccate/to trocknen, austrocknen, ausdörren
desiccation Austrocknung *f*, Eintrocknung *f*
design Auslegung *f*
design engineer Anlagenplaner *m*
design technique Aufbautechnik *f*
desilicate/to entkieseln
desk version Auftischversion *f*
deslag/to entschlacken
desludger Schlammseparator *m*
desludging Schlammabscheidung *f*
desolate öd, verwüstet, einsam, verlassen, desolat
desorption Desorption *f*
destabilization Destabilisierung *f*
destabilize destabilisieren
desulphurization Entschwefelung *f*
desulphurization plant Entschwefelungsanlage *f*
desulphurization process Entschwefelungsprozeß *m*
desulphurize/to entschwefeln
detect/to nachweisen, auffinden, herausfinden, ermitteln, feststellen
detectability Nachweisbarkeit *f*
detectable nachweisbar, feststellbar
detection Nachweis *m*, Beobachtung *f*, Erkennung *f*
detection limit Nachweisgrenze *f*
detection method Nachweismethode *f*

detection of pollutants Schadstoffnachweis *m*, Nachweis *m* von Schadstoffen
detection of troubles Fehlersuche *f*
detector Detektor *m*, Anzeiger *m*
detector fouling Detektorverschmutzung *f*
detergent Detergens *n*, Reinigungsmittel *n*, Waschmittel *n*, Spülmittel *n*
determination error Bestimmungsfehler *m*
determination limit Bestimmungsgrenze *f*
detoxicate/to entgiften
detoxification Entgiftung *f*
detoxification function Entgiftungsfunktion *f*
detoxify/to entgiften
detrimental effect schädlicher Effekt *m*, Schadwirkung *f*
deuterium Deuterium *n*, schwerer Wasserstoff *m*
developer Entwickler *m*, Entwicklerflüssigkeit *f*
development Entwicklung *f*
development of virgin land Neulanderschließung *f*
dewater/to entwässern
dewatering Entwässerung *f*
dewatering sludge Industrieschlamm *m*
dewatering system Entwässerungssystem *n*
diagnosis system Diagnosesystem *n*
diagram of firing capacity Feuerungsleistungsdiagramm *n*
diaphragm Diaphragma *n*, Membran *f*, Membranfilter *n*
diaphragm filter press Diaphragmafilterpresse *f*

diaphragm process Diaphragmaverfahren *n*
diaphragm pump Diaphragmapumpe *f*
diaphragm valve Membranventil *n*
diatomaceous earth Kieselgur *f*
diatomite Diatomit *m*, Berggur *f*
diesel Dieselkraftstoff *m*
diffusibility Diffusionsfähigkeit *f*
diffusion Diffusion *f*, Ausbreitung *f*
diffusion coefficient Diffusionskoeffizient *m*
diffusion heat Diffusionswärme *f*
diffusion path Diffusionsweg *m*
diffusion potential Diffusionspotential *n*
diffusion process Diffusionsprozeß *m*
diffusion pump Diffusionspumpe *f*
diffusive sampler Passivprobenehmer *m*, Diffusionssammler *m*
digest/to verdauen, digerieren
digested sludge Faulschlamm *m*, ausgefaulter Schlamm *m*
digester Faulbehälter *m*, Rotteturm *m*, Rottebehälter *m*
digester gas Faulgas *n*
digester gas pipe Faulgasleitung *f*
digestibility Verdaulichkeit *f*
digestion Verdauung *f*, Aufschluß *m*, Digestion *f*
digestion tank Faulbehälter *m*
digestion tower Faulturm *m*
digestion vessel Faulbehälter *m*
diluent verdünnend; Verdünnungsmittel *n*
dilute/to verdünnen, strecken, wässern
diluting agent Verdünnungsmittel *n*
diluting heat Verdünnungswärme *f*
dilution Verdünnung *f*

dilution analysis Verdünnungsanalyse *f*
dilution effect Verdünnungseffekt *m*
dimensioning Auslegung *f*
dioxin Dioxin *n*
dioxin disaster Dioxinkatastrophe *f*
dioxin emission Dioxinemission *f*
dioxin poisoning Dioxinvergiftung *f*
dioxin sample Dioxinprobe *f*
dioxin scandal Dioxinskandal *m*
dioxin toxicity Dioxintoxizität *f*
dioxin uptake Dioxinaufnahme *f*
direct determination Direktbestimmung *f*
direct introducer Direkteinleiter *m*
direct precipitation Direkt[aus]fällung *f*
direct release Direkteinleitung *f*
direct vapour feeding Direktdampfeinleitung *f*
direction for testing Prüfvorschriften *fpl*
directions for transportation of hazardous materials (harmful goods) Gefahrguttransportvorschriften *fpl*
dirt Schmutz *m*
dirt-dissolving schmutzlösend
dirt trap Schmutzfänger *m*
dirty water Schmutzwasser *n*
disassembling plant for used cars Demontagewerk *n* für Altautos
disaster Katastrophe *f*
disaster relief Katastrophenhilfe *f*
discard Abfall *m*
discharge Ausfluß *m*, Abfluß *m*, Abführung *f*, Ablaß *m*, Ausladung *f*, Auslaß *m*, Entladung *f*
discharge gutter Ablaufrinne *f*
discharge pipe Ausflußrohr *n*

discharge/to abfließen, ablassen, ausfließen, auslassen, entladen, entlassen
discontinuity position Unstetigkeitsstelle *f*
disease carrier Krankheits[über]träger *m*
disequilibrium Ungleichgewicht *n*, gestörtes Gleichgewicht *n*
disinfect/to desinfizieren, entkeimen
disinfectant Desinfektionsmittel *n*, Entgiftungsmittel *n*, Entkeimungsmittel, desinfizierend;
disinfection Desinfektion *f*, Entseuchung *f*
disinfection basin Desinfektionsbecken *n*
disintegration Zersetzung *f*, Abbau *m*; Auflösung *f*; Verfall *m*, Verwitterung *f*
disinterment Exhumierung *f*, Ausgrabung *f*
dispenser Abfüllvorrichtung *f*, Verteiler *m*
dispersant Dispersionsmittel *n*, Dispergens *n*, Verteilungsmittel *n*
dispersion Dispersion *f*, Streuung *f*, Zerstäubung *f*
dispersion mechanism Dispersionsmechanismus *m*, Zerstäubungsmechanismus *m*
dispersion tendency of chemicals Verteilungstendenz *f* von Chemikalien
display unit Sichtbildgerät *n*
disposability Verfügbarkeit *f*
disposable verfügbar; wegwerfbar
disposable container Einwegcontainer *m*, Einwegbehälter *m*
disposable dust mask Einwegstaubmaske *f*
disposable filter Einwegfilter *n*
disposable method Einwegverfahren *n*
disposable package Wegwerfpackung *f*
disposable sludge deponierbarer Schlamm *m*
disposable product Einwegartikel *m*
disposable syringe Einwegspritze *f*
disposable vat Einwegfaß *n*
disposal Beseitigung *f*, Erledigung *f*, Entsorgung *f*
disposal and deposit for used accumulators Entsorgung *f* und Pfand für Altbatterien
disposal behaviour Deponieverhalten *n*
disposal capacity Entsorgungskapazität *f*
disposal concept Entsorgungskonzept *n*
disposal management Ensorgungswirtschaft *f*
disposal method Beseitigungsmethode *f*
disposal network automation Entsorgungsnetzautomatisierung *f*
disposal of bulky refuse Sperrmüllbeseitigung *f*
disposal of residues Rückstandentsorgung *f*, Entsorgung *f* von Rückständen
disposal of special waste Sonderabfallentsorgung *f*
disposal pathway Entsorgungsweg *m*
disposal plant Beseitigungsanlage *f*, Entsorgungsanlage *f*

disposal project Entsorgungsprojekt *n*
disposal security Entsorgungssicherheit *f*
disposal site Mülldeponie *f*, Deponie *f*
disposal site capacity Deponiekapazität *f*
disposal site register Deponiekataster *m*
disposal site structure Deponieaufbau *m*
disposal structure Entsorgungsstruktur *f*
disposal system Entsorgungssystem *n*
disposal technology Deponietechnik *f*
disposal unit Müllschlucker *m*
disposal vessel Entsorgungsgefäß *n*
dispose [of]/to entsorgen, beseitigen
dispose of biologically/to biologisch entsorgen
dispose of chemically/to chemisch entsorgen
disposed films Folienabfall *m*
dissociate/to dissoziieren, abtrennen, zerfallen
dissociation Dissoziation *f*, Zerfall *m*
dissociation constant Dissoziationskonstante *f*
dissolubility Löslichkeit *f*, Auflösbarkeit *f*, Lösbarkeit *f*
dissolution Auflösung *f*, Lösung *f*, Trennung *f*
dissolvable auflösbar, löslich
dissolved oxygen Gelöst-Sauerstoff *m*
dissolve/to lösen, auflösen, in Lösung bringen, in Lösung gehen

dissolvent Lösemittel *n*, Auflösemittel *n*
distance-dependent utilization tariff entfernungsabhängige Nutzungspauschale *f*
distill/to destillieren
distillate Destillat *n*
distillation Destillation *f*
distillation plant Destillationsanlage *f*
distillation product of naphta Erdöldestillationsprodukt *n*
distillation residue Destillationsrückstand *m*
distortion phase Störphase *f*
distribution behaviour Verteilungsverhalten *n*
distribution chromatography Verteilungschromatographie *f*
distribution function Verteilungsfunktion *f*
distribution model Verteilungsmodell *n*
district heating Fernwärme *f*
district heating pipe Fernwärmeleitung *f*
disturbance Störung *f*, Unruhe *f*
disturbance detection Störerkennung *f*
disturbance recorder Störprotokolldrucker *m*
disturbance report Störprotokoll *n*
disturbance signal alarm Störungssignalisierung *f*
disturbance signal processing Störsignalverarbeitung *f*
ditch system Grabensystem *n*
domestic clarification plant Hauskläranlage *f*
domestic coal Hausbrandkohle *f*
domestic fire Hausbrand *m*

domestic waste Hausmüll *m*
domestic waste compost Hausmüllkompost *m*
domestic waste composting Hausmüllkompostierung *f*
domestic waste dump Hausmülldeponie *f*
dosage and screening area Dosier- und Siebbereich *m*
dosage pump Dosierpumpe *f*
dosage station Dosierstation *f*
dosage system Dosiersystem *n*
dosage technique Dosiertechnik *f*
dose Dosis *f*
dose effect relation Dosis-Wirkung-Beziehung *f*
dose-response model Dosis-Wirkungs-Modell *n*
dose/to dosieren, zumessen
dosing Dosierung *f*, Dosieren *n*
dosing interval Dosierungsintervall *n*
double-walled design doppelwandige Bauweise *f*
downpour Platzregen *m*
drain Abflußkanal *m*, Abflußrinne *f*; Abwasserkanal *m*, Abwasserleitung *f*, Entwässerung *f*
drain and sewer system Kanalisationssystem *n*, Kanalsystem *n*
drain layer Dränschicht *f*
drain of the clarification plant Klärwerkskanal *m*
drain pipe Abflußrohr *n*
drain purification Kanalreinigung *f*
drain screw Ablaßschraube *f*
drain shaft Ablaufschacht *m*
drain system Kanalsystem *n*
drain/to ablaufen lassen, drainieren, entleeren; entwässern

drainable entwässerbar
drainage Dränage *f*, Entwässerung *f*, Trockenlegung *f*; Abfließen *n*, Ablaufen *n*
drainage adaption Abflußanpassung *f*
drainage gutter Ablaufrinne *f*
drainage zone Entwässerungszone *f*
draining conditions Abflußbedingungen *fpl*
draining regulation Entwässerungssatzung *f*
dredged material ausgebaggertes Material *n*
dredging Ausbaggern *n*
dressing plant Aufbereitungsanlage *f*
drilling fluid Bohrflüssigkeit *f*
drilling mud Bohrschlamm *m*
drilling rig Bohranlage *f*
drilling test Bohrprobe *f*
drinkable trinkbar
drinking water Trinkwasser *n*
drinking water biology Trinkwasserbiologie *f*
drinking water contamination Trinkwasserverseuchung *f*
drinking water decree Trinkwasserverordnung *f*
drinking water fluoridizing Trinkwasserfluoridierung *f*
drinking water plant Trinkwassergewinnungsanlage *f*
drinking water preparation Trinkwasseraufbereitung *f*
drinking water processing Trinkwasseraufbereitung *f*
drinking water production Trinkwasserproduktion *f*

drinking water protection area Trinkwasserschutzgebiet *n*
drinking water regulation Trinkwasserverordnung *f*
drinking water supply Trinkwasserversorgung *f*
drinking water transport system Trinkwassertransportsystem *n*
drinking waterworks Trinkwasserwerk *n*
drive engineering Antriebstechnik *f*
drive technique Antriebstechnik *f*
driving technique Antriebstechnik *f*
drug Droge *f*; Arzneimittel *n*, Arznei *f*, Medikament *n*; Rauschgift *n*
drug abuse Arzneimittelmißbrauch *m*; Drogenmißbrauch *m*
drug addiction Arzneimittelsucht *f*; Drogensucht *f*
drug consumption Medikamentenverbrauch *m*; Drogenkonsum *m*
drug monitoring Pharmaspiegelkontrolle *f*
drug safety Arzneimittelsicherheit *f*
dry area Trockengebiet *n*
dry battery Trockenbattrie *f*
dry deposition Trockendeposition *f*, Trockenablagerung *f*, trockene Ablagerung *f*
dry distillation Trockendestillation *f*, Trockenentgasung *f*
dry operating rotary pump trokkenlaufende Rotationspumpe *f*
dry purification plant Trockenreinigungsanlage *f*
dry season Trockenzeit *f*
dry sludge disposal site Trockenschlammdeponie *f*
dryer Trockner *m*
drying agent Trocknungsmittel *n*
drying bed Trockenbeet *n*
drying centrifuge Trockenschleuder *f*
drying circuit Trocknungskreislauf *m*
drying energy Trocknungsenergie *f*
drying equipment Trocknungsanlage *f*
dump/to abladen, abwerfen
dump out/to auskippen, ausschütten
dumper Kipper *m*, Muldenkipper *m*
dumping ability Deponierfähigkeit *f*
dumping area Abladegebiet *n*, Müllabladeplatz *m*
dumping ground Abladegebiet *n*, Müllabladeplatz *m*
dune Düne *f*
dune protection Dünenschutz *m*
duration of effect Wirkungsdauer *f*
duration of exposure Dauer *f* der Exposition
dust Staub *m*
dust bin Mülleimer *m*, Abfallbehälter *m*; Staubbeutel *m*, Staubsack *m*
dust collection funnel Staubsammeltrichter *m*
dust conveyor Staubförderer *m*, Staubförderanlage *f*
dust emission Staubemission *f*, Staubauswurf *m*
dust explosion Staubexplosion *f*
dust extraction plant Entstaubungsanlage *f*
dust filter Staubfilter *n*
dust load Staubfracht *f*, Staublast *f*
dust recovery Staubrückgewinnung *f*
dust removal Entstaubung *f*
dust removal technology Entstaubungstechnik *f*
dusting Bestäubung *f*
dusty staubig

duty of notification Meldepflicht *f*
dwarfism Zwergwuchs *m*
dye house Färberei *f*
dye liquor Farbbrühe *f*, Färbeflotte *f*
dye machine Färbereimaschine *f*
dye stuff Farbstoff *m*
dye works Färberei *f*
dysfunction Dysfunktion *f*, Funktionsstörung *f*
dysplasia Fehlentwicklung *f*, Unterentwicklung *f*, Dysplasie *f*

E

early-alarming system for ice Glatteisfrühwarnsystem *n*
early detection Früherkennung *f*, Frühentdeckung *f*
early stage Frühstadium *n*
earth atmosphere Erdatmosphäre *f*
earth crust Erdkruste *f*
earth pressure Erddruck *m*
earth surface Erdoberfläche *f*
earth heaping Erdaufschüttung *f*
earth tube Masserohr *n*
earth works Erdarbeiten *fpl*
earthquake Erdbeben *n*
earthquake protection Erdbebenschutz *m*
easily soluble leichtlöslich
easy-care finishing Pflegeleichtausrüstung *f*
eccentric press Exzenterpresse *f*
eccentric single-rotor screw pump Exzenterschneckenpumpe *m*
eccentric worm pump Exzenterschneckenpumpe *f*
eco-freak Umweltfanatiker *m*, Öko-Freak *m*
ecocatastrophe Umweltkatastrophe *f*
ecochemistry Ökochemie *f*

ecocidal umweltzerstörend
ecodestruction Umweltzerstörung *f*
ecodiagnosis Ökodiagnose *f*
ecofactor Ökofaktor *m*
ecological ökologisch
ecological action plan ökologischer Handlungsplan *m*
ecological advantage ökologischer Vorteil *m*
ecological agriculture ökologische Landwirtschaft *f*
ecological aim ökologisches Ziel *n*
ecological balance ökologisches Gleichgewicht *f*
ecological chemistry ökologische Chemie *f*, Ökochemie *f*
ecological crisis Umweltkrise *f*
ecological efficiency ökologische Wirksamkeit *f*
ecological limit ökologische Grenze *f*
ecological model ökologisches Modell *n*
ecological niche ökologische Nische *f*
ecological structure ökologische Struktur *f*
ecological valence ökologische Wertigkeit *f*
ecologically acceptable ökologisch vertretbar
ecologically benefical umweltfreundlich
ecologism Ökologismus *m*
ecologist Ökologe *m*, Umweltexperte *m*
ecology Ökologie *f*
ecomedical ökomedizinisch
economical energy utilization rationelle Energieverwendung *f*

economizer Abgasvorwärmer *m*, Heizgasvorwärmer *m*
ecosalvatology Ökosalvatologie *f*
ecosize/to sich der neuen Umgebung anpassen
ecosocial ökosozial
ecosphere Ökosphäre *f*
ecosystem Ökosystem *n*
ecotax Ökosteuer *f*
ecotoxicity Ökotoxizität *f*
ecotoxicological ökotoxikologisch
ecotoxicology Ökotoxikologie *f*
ecotype Ökotyp *m*
eczema Ekzem *n*
effect diagnosis Folgendiagnose *f*
effective heat Nutzwärme *f*
effective range Wirkungsbereich *m*
efficiency of adsorption Adsorptionsleistung *f*
effluent ausströmend, ausfließend; Abfluß *m*, Abwasser *n*
effluent control Abwasserkontrolle *f*
effluent disposal Abwasserbeseitigung *f*, Abwasserentsorgung *f*
effluent water testing Abwasseruntersuchung *f*
effluent water treatment Abwasserbehandlung *f*
efflux Ausfluß *m*, Ausströmung *f*
einsteinium Einsteinium *n*
elapsed-hour counting Betriebsstundenzählung *f*
electric melting furnace Elektrostahlofen *m*
electric separator Elektrofilter *n*
electric storage oven Elektrospeicherofen *m*
electrodeposition Galvanisieren *n*, Galvanikprozeß *m*, Galvanisierung *f*
electrofilter Elektrofilter *n*

electrolysis of fused salts Schmelzflußelektrolyse *f*
electrolysis of water Wasserelektrolyse *f*
electrolyte tank Elektrolyttank *m*
electrolytic separation elektrolytische Trennung *f*
electrolytic slime Elektrolysenschlamm *m*
electromagnetic wave elektromagnetische Welle *f*
electromobile Elektromobil *n*
electronic srap Elektronikschrott *m*
electroplating Galvanisieren *n*
electroplating plant Galvanisieranlage *f*
electroprecipitation Elektroabscheidung *f*
electrosmelting process Elektroschmelzverfahren *n*
electrostatic oil purification elektostatische Ölreinigung *f*
electrothermal elektrothermisch
elemental analysis Elementaranalyse *f*
elementary analysis Elementaranalyse *f*
elementary microanalysis Elementmikroanalyse *f*
elevator bucket Förderbecher *m*, Fördereimer *m*
elimination of acid Entsäuerung *f*
elimination of water Wasserabspaltung *f*
eluant Eluierungsmittel *n*, Eluant *m*
elute/to eluieren, herausspülen, herauslösen
elution Eluierung *f*
elution-stable eluationstabil
elutriate/to abklären, ausschlämmen, reinigen

elutriation Schlämmung *f*, Ausschlämmung *f*, Ausspülung *f*
elutriation process Schlämmverfahren *f*
embankment Uferböschung *f*
embedment Einbettung *f*
embryo mortality Embryosterblichkeit *f*
embryo toxicity Embryotoxizität *f*
embryonic abnormality embryonale Abnormalität *f*
emergency equipment Erste-Hilfe-Einrichtung *f*
emergency generating set Notstromaggregat *n*
emergency lighting Notbeleuchtung *f*
emergency power supply Notstromversorgung *f*
emergency protection Katastrophenschutz *m*
emergency situation Notfallsituation *f*
emery dust Schmirgelstaub *m*
emission Emission *f*, Ausstrahlung *f*, Aussendung *f*, Ausströmung *f*; Abgabe *f*
emission analysis Emissionsanalyse *f*
emission concentration Emissionskonzentration *f*
emission control Emissionsbekämpfung *f*
emission data Emissionsdaten *pl*
emission density Emissionsdichte *f*
emission limit value Emissionsgrenzwert *m*
emission of pollutants Schadstoffemission *f*, Schadstoffabgabe *f*

emission quantity Emissionsmenge *f*
emission register Emissionskataster *m*
emission source Emissionsquelle *f*
emissive ausstrahlend
emit/to emittieren, abstrahlen, aussenden, ausstrahlen
emittance spezifische Ausstrahlung *f*, Ausstrahlungsvermögen *n*
emitted substance emittierter Stoff *m*
emitter Emitter *m*, Aussender *m*
emulsifier Emulgator *m*
emulsion Emulsion *f*
emulsion-loaded sewage emulsionsbelastetes Abwasser *n*
emulsion separation Emulsionsspaltung *f*
encapsulate/to einkapseln, einschließen, einbetten, verkapseln
encapsulation Einkapselung *f*, Einbettung *f*
encapsulation of an old site Einkapselung *f* einer Altlast
encapsulation of special waste Einkapselung *f* von Sonderabfall
endangered species gefährdete Arten *fpl*
endangering of groundwater Grundwassergefährdung *f*
endangering potential Gefährdungspotential *n*
endogeneous endogen
endothermal endotherm
energy Energie *f*
energy-autarkic energieautark
energy balance Energiehaushalt *m*, Energiebilanz *f*
energy conservation Energieerhaltung *f*

energy consumption Energieverbrauch *m*
energy conversion Energieumwandlung *f*
energy crisis Energiekrise *f*
energy-economical energiewirtschaftlich
energy from used tyres Energie *f* aus Altreifen
energy-intensive energieintensiv
energy recovery Energiegewinnung *f*, Energierückgewinnung *f*
energy saving Energieeinsparung *f*
energy source Energiequelle *f*
energy supplier Energielieferant *m*
energy utilization Energienutzung *f*
engineering chemistry chemische Verfahrenstechnik *f*
engineering geology Ingenieurgeologie *f*
enriched uranium angereichertes Uran *n*
enrichment plant Anreicherungsanlage *f*
enthalpy Enthalpie *f*
entropy Entropie *f*
entry of air bubbles Lufteinperlung *f*
entry of pollutants Schadstoffeintrag *m*, Eintrag *m* von Schadstoffen
environment Umwelt *f*
environment of the disposal site Deponieumgebung *f*
environmental umweltbedingt
environmental acceptable umweltgerecht
environmental adviser Umweltberater *m*
environmental analytics Umweltanalytik *f*
environmental angel Umweltengel *m*
environmental aspect Umweltschutzaspekt
environmental charge Umweltabgabe *f*
environmental chemical Umweltchemikalie *f*
environmental compartment Umweltkompartiment *n*
environmental compatibility Umweltverträglichkeit *f*
environmental compatibility test Umweltverträglichkeitsprüfung *f*
environmental compatible test Umweltverträglichkeitsprüfung *f*
environmental crime Umweltkriminalität *f*
environmental crime Umweltkrimineller m
environmental criminal Umweltverbrechen *n*
environmental criminal Umweltverbrecher *m*
environmental damage Umweltschaden *m*
environmental damaging substance Umweltschadstoff *m*
environmental data Umweltdaten pl
environmental effect Umweltauswirkung *f*
environmental endangering umweltgefährdend
environmental engineering Umwelttechnik *f*
environmental gas chromatograph Umwelt-Gaschromatograph *m*
environmental hysteria Umwelthysterie *f*

environmental impact Umweltbelastung *f*
environmental label Umweltengel *m*
environmental legislation Umweltgesetzgebung *f*
environmental matters Umweltfragen *fpl*
environmental media Umweltmedien *npl*
environmental monitoring programme Umweltüberwachungsprogramm *n*
environmental needs Umweltschutzerfordernisse *npl*
environmental policy Umweltpolitik *f*
environmental political competence umweltpolitische Kompetenz *f*
environmental preserving technology umweltschonende Technologie *f*
environmental programme Umweltprogramm *n*
environmental protection Umweltschutz *m*
environmental protection award Umweltschutzpreis *m*
environmental protection law Umweltschutzgesetz *n*
environmental research Umweltforschung *f*
environmental risk Umweltrisiko *n*
environmental sample Umweltprobe *f*
environmental securing Umweltsicherung *f*
environmental sign Umweltzeichen *n*
environmental situation umweltpolitische Lage *f*, Umweltsituation *f*
environmental specimen banking Umweltprobenbank *f*
environmental tax Umweltsteuer *f*
environmental technology Umwelt[schutz]technologie *f*, Umwelt[schutz]technik *f*
environmental tolerance test Umweltverträglichkeitsprüfung *f*
environmentally acceptable umweltverträglich
environmentally acceptable processing umweltverträgliche Aufbereitung *f*
environmentally compatible umweltverträglich, umweltgerecht
environmentally harmful umweltfeindlich
environmentally neutral umweltneutral
environmentally noxious umweltschädlich
enzymatic degradation enzymatischer Abbau *m*
enzymatic sludge processing (treatment) enzymatische Schlammbehandlung *f*
enzyme Enzym *n*
epidemic epidemisch; Epidemie *f*, Seuche *f*, Massenerkrankung *f*
epidemiological epidemiologisch
epidemiological prospective investigation epidemiologische Prospektionsstudie *f*
epidemiological research epidemiologische Erhebung *f*
epidemiological short-term forecast epidemiologische Kurzzeitvorhersage *f*
epidemiology Epidemiologie *f*
epilimnion Epilimnion *n*

epoxy-isolated transformer gießharzisolierter Transformator *m*
equalizing tank Ausgleichstank *m*, Ausgleichsbecken *n*
equilibrium of evaporation Verdampfungsgleichgewicht *n*
equilibrium of forces Gleichgewicht *n* der Kräfte, Kräftegleichgewicht *n*
equipment Ausrüstung *f*
equipment for secondary air Sekundärlufteinrichtung *f*
eradicate/to ausrotten
eradication Ausrottung *f*
erbium Erbium *n*
ergometer Ergometer *n*
ergometry Ergometrie *f*
ergonomics Ergonomie *f*, Ergonomik *f*
erode/to erodieren
erosion Erosion *f*, Abtragung *f*
erosion by wind Winderosion *f*
erosion control Erosionsbekämpfung *f*
erosion protection facility Erosionsschutzeinrichtung *f*
erosion-stable erosionsstabil
erythrocyte Erythrocyt *m*, rotes Blutkörperchen *n*
erythrocytosis Erythrozytose *f*
estate waste Siedlungsabfall *m*
estuary Flußmündung *f*, Meeresarm *m*, Ästuar *n*
etching bath Ätzbad *n*
etching sludge Ätzschlamm *m*
ethics of responsibility Verantwortungsethik *f*
etiology Ätiologie *f*, Lehre von den Krankheitsursachen
europium Europium *n*
eutroph eutroph, nährstoffreich
eutrophic eutroph, nährstoffreich
eutrophication Eutrophierung *f*, Überdüngung *f*
evaluation of areas of suspicion Verdachtsflächenbewertung *f*
evaporation curve Verdampfungskurve *f*
evaporation enthalpy Verdampfungsenthalpie *f*
evaporative cooling tower Verdunstungskühlturm *m*
event preparation Meldungsvorbereitung *f*
evolution of gas Gasentwicklung *f*
example of application Anwendungsbeispiel *n*
excavated material Baggergut *n*, ausgebaggertes Material *n*
excavation and trenching practice Ausgrabungs- und Aushebungspraxis *f*
excavations Erdarbeiten *fpl*
excentric worm pump Exzenterschneckenpumpe *f*
exceptional permission Ausnahmegenehmigung *f*
exceptional recommendation Ausnahmeempfehlung *f*
excess amalgam überschüssiges Amalgam *n*
excess pressure Überdruck *m*
excess sludge Überschußschlamm *m*
excess vapour Überschußdampf *m*
excess water Überstandswasser *n*
exchange chromatography Austauschchromatographie *f*
exchange reaction Austauschreaktion *f*
exchangeability Austauschbarkeit *f*

exchangeable austauschbar, auswechselbar
exchanger Austauscher *m*
excreta disposal Fäkalienbeseitigung *f*
excretion in urine Ausscheidung *f* im Urin
executive data Führungsdaten pl
exemption Ausnahmegenehmigung *f*
exhaust air Abluft *f*
exhaust air purification Abluftreinigung *f*
exhaust air technology Abluftechnik *f*
exhaust gas Abgas *n*
exhaust gas filter Abgasfilter *n*
exhaust gas flow Abluftstrom *m*
exhaust gas purification Abgasreinigung *f*
exhaust gas stack Abluftkamin *m*
exhaust/to absaugen; erschöpfen, verbrauchen
exhausting apparatus Absauggerät *n*
exhausting by suction Absaugung *f*
exhausting plant Absauganlage *f*
exogenic exogen
exogeneous effect exogene Wirkung *f*
expanding agent Blähmittel *n*, Treibmittel *n*
expansion possibility Ausbaumöglichkeit *f*
expert committee Expertengremium *n*
expert opinion Gutachten *n*
expert system Expertensystem *n*
exploitation Abbau *m*, Ausbeutung *f*, Ausnutzung *f*
exploration Erkundung *f*
explore/to erkunden, erforschen, untersuchen
explosion hazard Explosionsgefahr *f*
explosion-pressure-resistant explosionsdruckstoßfest
explosion-proof explosionsgeschützt
explosion-protected explosionsgeschützt
explosion protection Explosionsschutz *m*
explosion risk Explosionsgefahr *f*
explosion wave Explosionswelle *f*
explosive explosiv; Sprengstoff *m*, Explosivstoff *m*, Sprengmittel *n*
exposed exponiert
export restriction Ausfuhrbeschränkung *f*
exposed concrete Sichtbeton *m*
exposition data Expositionsdaten pl
exposition device Expositionseinrichtung *f*
exposition facility Expositionseinrichtung *f*
exposure Exposition *f*
exposure analysis Expositionsanalyse *f*
exposure concentration Konzentration *f* während der Exposition
exposure data Expositionsdaten pl
exposure dose Bestrahlungsdosis *f*
exposure of employees Exposition *f* von Beschäftigten
exposure parameter Expositionsparameter *m*
exposure potential Expositionspotential *n*
exposure time Expositionszeit *f*
exterior durability Außenbeständigkeit *f*, Wetterfestigkeit *f*

exterior paint Außenanstrich *m*, Außenfarbe *f*
external air Außenluft *f*
external exposure externe Exposition *f*
extinction Extinktion *f*
extra work for adaption Anpassungsaufwand *m*
extracellular extrazellulär
extract Extrakt m(n), Auszug *m*
extract/to extrahieren, auslaugen, herauslösen, ausziehen
extractability Extrahierbarkeit *f*
extractable extrahierbar
extractant Extraktionsmittel *n*
extraction Extraktion *f*, Auslaugung *f*, Ausziehung *f*
extraction analysis Extraktionsanalyse *f*
extraction apparatus Extraktionsapparat *m*
extraction under pressure Druckextraktion *f*
extreme value Extremwert *m*

F

face guard Schutzmaske *f*
fact-finding duty Ermittlungspflicht *f*
factorial test Faktortest *m*
factory hygiene Betriebshygiene *f*
factory inspection Betriebskontrolle *f*
faecal coliforms Fäkalkoli *pl*
faecal excretion Kotausscheidung *f*
faecal sludge Fäkalschlamm *m*
faecal substances Fäkalien pl
faecal water Fäkalwasser *n*
faeces Fäkalien *pl*

faeces acceptance station Fäkalienannahmestation *f*
faeces disposal Fäkalienbeseitigung *f*
faeces fouling plant Fäkalienfaulanlage *f*
faeces sludge Kloakenschlamm *m*
fail-safety Störungssicherheit *f*
failure Versagen *n*, Defekt *m*, Fehlschlag *m*, Panne *f*, Störung *f*
failure probability Ausfallwahrscheinlichkeit *f*
failure rate Ausfallquote *f*
fall-out Atomstaub *m*, radioaktiver Niederschlag *m*
falling-film evaporator Fallstromverdampfer *m*
fallow Brachland *n*
fallow land Brachland *n*
fallow recovery Brachlandrückgewinnung *f*
false alarm blinder Alarm *m*
fan station Gebläsestation *f*
fast-breeding reactor Schnellbrutreaktor *m*, schneller Brüter *m*
fastness to washing Waschbeständigkeit *f*
fat cleavage Fettspaltung *f*
fat deficiency Fettmangel *m*
fat deterioration Fettverderb *m*
fat dissolver Fettlösemittel *n*
fat embolism Fettembolie *f*
fat formation Fettbildung *f*
fat product Fettprodukt *n*, Fetterzeugnis *n*
fat sludge Fettschlamm *m*
fatal tödlich, verhängnisvoll
fatigue limit Dauerfestigkeit *f*
fatigue test Ermüdungsversuch *m*, Dauerprüfung *f*

fault Störung *f*
fault detection Störerkennung *f*
fault indication Störmeldung *f*
fault signal Störmeldung *f*
fault signalling and acknowledgement Störmeldung *f* und Quittierung
fauna Tierwelt *f*, Fauna *f*
feculent trüb, unrein, verunreinigt, schlammig
feed back Rückwirkung *f*, Rückführung *f*, Rückkopplung *f*
feed line Speiseleitung *f*
feeding bunker Aufgabebunker *m*
feeding drum Aufgabetrommel *f*
feedwater preheating Speisewasservorwärmung *f*
fen Niedermoor *n*
ferment/to gären, fermentieren, zur Gärung bringen
fermentation Gärungsprozeß *m*, Vergärung *f*, Gärung *f*, Fermentation *f*
fermentation chamber Faulkammer *f*
fermentation fungus Gärungspilz *m*
fermentation gas Biogas *n*
fermented sewage sludge gegorener Klärschlamm *m*
fermenter Fermenter *m*, Gärbottich *m*, Gärtank *m*
fermium Fermium *n*
ferriferous eisenhaltig
ferromagnetic ferromagnetisch
fertilization Düngung *f*; Befruchtung *f*
fertilize/to düngen; befruchten
fertilizer Düngemittel *n*, Dünger *m*
fertilizer combination Düngemittelkombination *f*
fertilizer distribution Düngemittelverteilung *f*
fertilizer factory Düngemittelfabrik *f*
fertilizer production Düngemittelherstellung *f*, Kunstdüngererzeugung *f*
fertilizing agent Düngemittel *n*
fibre dust Faserstaub *m*
fibrogenic fibrogen
fibrous asbestos Faserasbest *m*
fibrous material Faserstoff *m*
field condition Feldbedingung *f*
field diagnosis Felddiagnose *f*
field experiment Feldversuch *m*
field instrumentation Feldinstrumentierung *f*
field investigation Felduntersuchung *f*
field of application Anwendungsbereich *m*, Verwendungsgebiet *n*
field operation Geländearbeit *f*
field pest Ackerschädling *m*
filling and dosing plant Füll- und Dosieranlage *f*
filling device Abfüllvorrichtung *f*
filling level Füllhöhe *f*, Füllstand *m*
filling plan Befüllungsplan *m*
filling station Abfüllstation *f*, Füllstation *f*
filter Filter *n(m)*
filter ash Filterasche *f*
filter bag Filtertasche *f*, Filterbeutel *m*
filter bed Filterschicht *f*
filter cake Filterkuchen *m*
filter cartridge Filterpatrone *f*
filter cleaning Filterreinigung *f*
filter cloth Filtertuch *n*
filter cloth guide Filtertuchführung *f*

filter dust Filterstaub *m*
filter dust residue Filterstaubrest *m*
filter efficiency Filterwirkungsgrad *m*
filter element Filterelement *n*
filter hose Filterschlauch *m*
filter medium Filtermedium *n*
filter plate Filterplatte
filter press Filterpresse *f*
filter salt Filtersalz *n*
filter tissue Filtergewebe *n*
filter/to filtern, filtrieren
filter by suction/to abnutschen
filter by vacuum/to im Vakuum absaugen
filterable filterfähig, filtrierbar
filtercake Filterkuchen *m*
filtering Filtern *n*, Filtrieren *n*, Abklären *n*
filtering basin Klärbecken *n*
filtering hall (shed) Filterhalle *f*
filtering layer Fiterschicht *f*
filtering material Filterstoff *m*
filtering paper Filterpapier *n*
filtering sand Filtriersand *m*
filtering sieve Filtersieb *m*
filtrate Filtrat *n*
filtrate manifold Filtratvorlage *f*
filtrate receiver Filtratvorlage *f*
filtration Filtration *f*, Filtrierung *f*, Filterung *f*
filtration phase Filtrationsphase *f*
filtration plant Filteranlage *f*
filtration residue Filterrückstand *m*, Filtrationsrückstand *m*
filtration speed Filtrationsgeschwindigkeit *f*
final acceptance test Schlußabnahmetest *m*
final check Endkontrolle *f*
final cleaning Nachreinigung *f*

final disposal Endablagerung *f*, Endlager *n*
final disposal plant Endbeseitigungsanlage *f*
final filtering Nachfiltration *f*
final product Endprodukt *n*
final product of the metabolism Stoffwechselendprodukt *n*
final purifier Endreiniger *m*, Nachreiniger *m*
final stage Endstadium *n*
findings on admission Aufnahmebefund *m*
fine chemical Feinchemikalie *f*
fine control Feinregulierung *f*
fine-dust mask Feinstaubmaske *f*
fine filter Feinfilter *n*
fine-grained feinkörnig
fine-preparation Feinaufbereitung *f*
fine rake Feinrechen *m*
fine sieve rake Feinsiebrechen *m*
fine-sieving Feinabsiebung *f*
finely dispersed feindispers, feinverteilt
finely pored feinporig
finely pulverized feinpulverisiert
fire Feuer *n*
fire alarm Feueralarm *m*
fire chamber atmosphere Feuerraumatmosphäre *f*
fire chamber Feuerraum *m*, Heizraum *m*
fire clearing Brandrodung *f*
fire control Feuerführung *f*
fire cultivation Brandrodung *f*
fire damage Brandschaden *m*
fire damp Grubengas *n*, Schlagwetter *n*
fire extinguisher Feuerlöscher *m*
fire-proof paint feuerfeste Farbe *f*

fire regulation Feuerführung *f*
fire retardant Flammschutzmittel *n*
fire risk Brandgefahr *f*
firing Feuerung *f*
firing optimization Feuerungsoptimierung *f*
firing plant Feuerungsanlage *f*
firing technical primary measure feuerungstechnische Primärmaßnahme *f*
firmness Beständigkeit *f*, Festigkeit *f*
first-aid Erste Hilfe *f*
first-aid box (kit) Erste-Hilfe-Kasten *m*, Verbandskasten *m*
first-aid station Unfallstation *f*
fish and wildlife protection Schutz *m* der Fisch und Tierwelt
fish meal Fischmehl *n*
fish mortality Fischsterben *n*
fish poison Fischgift *n*
fish test Fischtest *m*
fish-toxic fischtoxisch
fish toxicity Fischgiftigkeit *f*
fissile material Spaltmaterial *m*
fission chain reaction Kernspaltungsreaktion *f*
fission product Spaltprodukt *n*
fission reactor Kernspaltungsreaktor *m*
fixed bed Schüttschicht *f*, Festbett *n*
fixed-bed adsorption Adsorption *f* im Festbett
fixed-bed circulating reactor Festbettumlaufreaktor *m*
fixed bed reactor Festbettreaktor *m*
fixed connection Festverbindung *f*
fixing bath Fixierbad *n*, Fixierflüssigkeit *f*
flame front Flammenfront *f*
flame ionization detector Flammenionisationsdetektor *m*
flame photometer Flammenphotometer *n*
flame picture Flammenbild *n*
flame-resistant nicht entflammbar
flame retardant Flammschutzmittel *n*
flame shape Flammenbild *n*
flameless flammenlos
flammability Entflammbarkeit *f*, Entzündbarkeit *f*
flammable entflammbar, entzündbar
flash back Flammendurchschlag *m*
flash of lightning Blitz[ein]schlag *m*
flash point Entzündungspunkt *m*, Flammpunkt *m*
flash point tester Flammpunktprüfgerät *n*, Flammpunktprüfer *m*
flashback safeguard Flammenrückschlagsicherung *f*
flexibility of the system Flexibilität *f* des Systems
flexible adaption flexible Anpassung *f*
flocculate/to ausflocken
flocculating agent Ausflockungsmittel *n*, Flockungshilfsmittel *n*, Flokkungsmittel *n*
flocculation Ausflockung *f*, Ausfällung *f*
flocculation agent preparation Flockungsmittelaufbereitung *f*
flocculation filtration Flockungsfiltration *f*
flocculator Flockungsreaktor *m*
flood Flut *f*, Überschwemmung *f*
flood wave Flutwelle *f*
flooding Überflutung *f*
flora Flora *f*, Pflanzenwelt *f*
flotation Flotation *f*, Naßaufbereitung *f*, Schwemmverfahren *n*

flotation cell Flotationszelle *f*
flotation method Flotationsverfahren *n*, Aufschlämmverfahren *n*
flotation plant Flotationsanlage *f*
flotation process Flotationsverfahren *n*, Aufschlämmverfahren *n*
flow aperture Durchflußblende *f*
flow back valve Rückschlagventil *n*
flow behaviour Fließverhalten *n*
flow chart Fließbild *n*, Schaubild *n*
flow direction of groundwater Grundwasserfließrichtung *f*
flow fittings Durchlaufarmatur *f*
flow in open drains Durchfluß *m* in offenen Gerinnen
flow limiter Durchflußbegrenzer *m*
flow orifice Durchflußblende *f*
flowability Fließfähigkeit *f*
flowable fließfähig
fluctuate/to schwanken, fluktuieren, abwechseln
fluctuation Schwankung *f*, Fluktuation *f*
fluctuation of the calorific value Heizwertschwankung *f*
flue Rauchgaskanal *m*
flue ash Flugasche *f*
flue dust Flugstaub *m*
flue gas Rauchgas *n*, Verbrennungsgas *n*, Abgas *n*
flue gas analysis Rauchgasanalyse *f*, Abgasanalyse *f*
flue gas cleaning plant Rauchgasreinigungsanlage *f*
flue gas cleaning system Rauchgasreinigungssystem *n*
flue gas explosion Rauchgasexplosion *f*
flue gas flow Rauchgasströmung *f*
flue gas purification plants Rauchgasreinigungsanlage *f*
flue gas purification residue Rauchgasreinigungsreststoff *m*
flue preheater Rauchgasvorwärmer *m*
fluid Fluid *n*; Flüssigkeit *f*
fluid balance Flüssigkeitshaushalt *m*
fluid bed catalysis Fließbettkatalyse *f*
fluid extract Fluidextrakt *m*
fluid filtration Flüssigkeitsfiltration *f*
fluid intake Flüssigkeitsaufnahme *f*
fluidity Fluidität *f*, Fließfähigkeit *f*
fluidized bed Fließbett *n*, Wirbelschicht *f*
fluidized-bed ash Wirbelschichtasche *f*
fluidized-bed furnace Wirbelschichtofen *m*
fluidized-bed sewage sludge dryer Fließbett-Klärschlammtrockner *m*
fluidized dust Flugstaub *m*
fluidized plant Wirbelschichtanlage *f*
fluidized reactor Wirbelschichtreaktor *m*
fluorescence Fluoreszenz *f*
fluorescence spectroscopy Fluoreszenzspektroskopie *f*
fluorescent screen Röntgenschirm *m*
fluorescent tube Leuchtstoffröhre *f*
fluoridate/to fluoridieren
flouridation Flouridierung *f*
fluorimetry Fluorimetrie *f*
flourinated hydrocarbon flouriertes Kohlenwasserstoff *m*
fluorine Fluor *n*
fluoroscopy Durchleuchtung *f*, Röntgendurchleuchtung *f*

fluorosis Fluorvergiftung *f*
flushing Ausschwemmen *n*
flushing liquor Spüllauge *f*
flushing water Spülwasser *n*
flux Fluß *m*; Flußmittel *n*
fly ash Flugasche *f*
fly ash dump Aufschüttung *f* mit Flugasche
fly ash emission Flugaschenemission *f*
fly ash quality Flugaschequalität *f*
fly ash vitrification Flugascheverglasung *f*
foam Schaum *m*
foam carpet Schaumteppich *m*
foam concrete Schaumbeton *m*
foam extinguisher Schaumlöscher *m*, Schaumfeuerlöscher *m*
foam protection Schaumschutz *m*
foam rubber Schaumgummi *n*
foam trap Schaumfalle *f*
foam/to schäumen
foaming Schäumen *n*
foaming agent Schaummittel *n*, Blähmittel *n*, Treibmittel *n*
foetus Fötus *m*
fog Nebel *m*
foggy neblig, nebelartig
fogging Beschlag *m*
foil sealing Folienabdichtung *f*
food Lebensmittel *n*, Nährmittel *n*
food additive Lebensmittelzusatz *m*, Nahrungsmittelzusatz *m*
food analysis Lebensmitteluntersuchung *f*
food chain Ernährungskette *f*, Nahrungskette *f*, Nahrungsmittelkette *f*
food chemical nahrungsmittelchemisch
food processing Lebensmittelverarbeitung *f*

food uptake Nahrungsaufnahme *f*
food waste Nahrungsmittelabfall *m*
foodstuffs packing Lebensmittelverpackung *f*
forage Viehfutter *n*
forecast result Prognoseergebnis *n*
forecast service Vorhersagedienst *m*
forecast system Prognosesystem *n*
foreign body Fremdkörper *m*
foreign substance Fremdstoff *m*, Fremdsubstanz *f*
forensic chemistry forensische Chemie *f*, gerichtliche Chemie *f*
forensic medicine Gerichtsmedizin *f*
forest Forst *m*, Wald *m*
forest area Waldgebiet *n*
forest border Waldgrenze *f*
forest damage Waldschaden *m*
forest devastation (destruction) Waldverwüstung *f*
forest ecosystem Wald-Ökosystem *n*
forest fire Waldbrand *m*
forest protection agent Forstschutzmittel *n*
forest regeneration Waldregeneration *f*
forest soil Waldboden *m*
forest stand Waldbestand *m*
forest vegetation Waldvegetation *f*
forestry Forstwirtschaft *f*
form of application Anwendungsform *f*
formation of buds Knospenbildung *f*
formation of filter cake Filterkuchenbildung *f*
formation of gas Gasbildung *f*

fossil energy fossile Energie *f*
fossil fuel fossiler Brennstoff *m*
foul sludge Faulschlamm *f*
foul water Abwasser *n*
foul/to faulen, ausfaulen
fouling basin Faulbehälter *m*
fouling chamber Faulkammer *f*
fouling gas Faulgas *n*
fouling gas pipe Faulgasleitung *f*
fouling gas processing Faulgasaufbereitung *f*
fouling gas utilization Faulgasverwertung *f*
fouling plant Faulanlage *f*
fouling sludge degassing Faulschlammentgasung *f*
fouling tank (vessel) Faulbehälter *m*
fouling tower Faulturm *m*
fouling vessel equipment Faulbehälterausrüstung *f*
fouling vessel Faulbehälter *m*
foundry Gießerei *f*, Hüttenwerk *n*
foundry sand Formsand *m*, Gießereisand *m*
fractional distillation fraktionelle Destillation *f*
frame filter Rahmenfilter *n*
francium Francium *n*
free from acid säurefrei
free from odour geruchlos
free from suspended matter schwebstofffrei
free radical freies Radikal *n*
freezing conditioning Gefrierkonditionierung *f*
freezing process Ausfrierverfahren *n*
freezing trap Kühlfalle *f*
frequency converter Frequenzumrichter *m*

fresh-compost Frischkompost *m*
fresh sludge Frischschlamm *m*
freshwater Frischwasser *n*; Süßwasser *n*
freshwater ecosystem Süßwasserökosystem *n*
fringe group Randgruppe *f*
fringe population Randbevölkerung *f*
frontier crossing traffic grenzüberschreitender Verkehr *m*
frost protection Frostschutz *m*
fruit Frucht *f*
fuel Brennstoff *n*, Brennmaterial *m*, Feuerungsmaterial *n*, Heizmaterial *n*; Kraftstoff *m*, Treibstoff *m*
full-face mask Ganzgesichtsmaske *f*
fume Dampf *m*, Dunst *m*, Rauch *m*
fume pipe Abzugsrohr *n*, Gasabzugsrohr *n*
function block Funktionsbaustein *m*
function disorder Funktionsstörung *f*
function test Funktionsprüfung *f*, Funktionstest *m*
functional efficiency Funktionstüchtigkeit *f*
functional environmental definition funktionale Umweltdefinition *f*
functional range Funktionsumfang *m*
functional safety Funktionssicherheit *f*
functional size Funktionsumfang *m*
functionality Funktionalität *f*
fungicidal pilztötend, fungizid
fungicide Fungizid *n*, pilztötendes Mittel *n*, Antimykotikum *n*
fungus Pilz *m*
fungus culture Pilzkultur *f*

furane Furan *n*
furane emission Furanemission *f*
furnace Fabrikofen *m*, Feuerraum *m*, Ofen *m*
furnace exhaust gas Ofenabgas *n*
furnace output Ofenleistung *f*
furnace slag Ofenschlacke *f*
further treatment Weiterbehandlung *f*
further utilization Weiterverwendung *f*
fusible schmelzbar
fusing point Schmelzpunkt *m*
fusion Schmelze *f*, Verschmelzen *n*, Schmelzen *n*

G

gadolinium Gadolinium *n*
gage s. gauge
gallium Gallium *n*
galvanic galvanisch
  galvanic sludge Galvanikschlamm *m*
galvanization Galvanisierung *f*
galvanize/to galvanisieren, verzinken
galvanizing Galvanisieren *n*
  galvanizing company Galvanikbetrieb *m*
  galvanizing plant Verzinkerei *f*
gamma-active gammaaktiv, gammaradioaktiv
gamma counter tube Gammazählrohr *n*
gamma decay Gammazerfall *m*
gamma emitter Gammastrahler *m*
gamma particle Gammateilchen *n*
gamma radiation Gammastrahlung *f*

gamma ray Gammastrahl *m*
garbage Abfall *m*, Müll *m*
garbage collection Müllabfuhr *f*
garbage disposal Müllabfuhr *f*
garbage dump Müllabladeplatz *m*
garbage pit Müllabladeplatz *m*
garbage removal Müllabfuhr *f*
garden refuse Gartenabfälle *mpl*
gas Gas *n*; {*Am*} Benzin *n*
gas analysis Gasanalyse *f*
gas bubble Gasblase *f*
gas burner Gasbrenner *m*
gas burning Gasverbrennung *f*, Gasabfackelung *f*
gas chromatograph Gaschromatograph *m*
gas chromatography Gaschromatographie *f*
gas cleaning plant Gasreinigungsanlage *f*
gas drainage Gasdränage *f*
gas engine Gasmotor *m*
gas evolution Gasentwicklung *f*
gas filter Gasfilter *n*
gas flow Gasstrom *m*
gas hood Gashaube *f*
gas mask Gasmaske *f*
gas motor Gasmotor *m*
gas network Gasnetz *n*
gas-phase reaction Gasphasenreaktion *f*
gas power station for site digestion gas Deponiegaskraftwerk *n*
gas pressure Gasdruck *m*
gas purification Gasreinigung *f*
gas purification technology Abgasreinigungstechnik *f*
gas reservoir Gasspeicher *m*
gas storage Gasspeicherung *f*, Gaslagerung *f*

gaseous

gas tank Gastank *m*
gas-tight gasdicht
gas torch Gasfackel *f*
gas-vapour mixture Gas-Dampf-Gemisch *n*
gaseous gasförmig
gasification Vergasung *f*
gasify/to vergasen
gasifying Vergasen *n*
gasoline Benzin *n*; Gasolin *n*
  gasoline additive Benzinzusatz *m*, Kraftstoffzusatz *m*
  gasoline-air mixture Benzin-Luft-Gemisch *n*
  gasoline level Benzinstand *m*
  gasoline tank Benzintank *m*
  gasoline trap Benzinabscheider *m*
gasometer Gasometer *m*, Gassammler *m*, Gasglocke *f*
gaswork Gaswerk *n*
gate valve Absperrventil *n*
  gate valve position Schieberstellung *f*
gauge Meßgerät *n*, Anzeiger *m*, Anzeigegerät *n*, Eichmaß *n*, Normalmaß *n*
  gauge mark Eichmarke *f*, Eichstrich *m*
  gauge pressure Manometerdruck *m*
  gauge substance Eichsubstanz *f*
gauge/to messen, eichen, kalibrieren, prüfen
gauging Eichen *n*, Kalibrieren *n*, Eichung *f*, Kalibrierung *f*
Geiger counter Geigerzähler *m*
gel Gel *n*, Gallert *n*
  gel chromatography Gelchromatographie *f*
  gel electrophoresis Gelelektrophorese *f*
  gel filtration Gelfiltration *f*

gel formation Gelbildung *f*
gel point Stockpunkt *m*
gel/to gelieren
gelatinate gelieren, gelatinieren
gelatination Gelierung *f*
gelatine Gelatine *f*
gelation Erstarren *n*, Festwerden *n*, Steifwerden *n*
gelling agent Geliermittel *n*
gene Gen *n*, Erbfaktor *m*, Erbeinheit *f*
  gene action Genwirkung *f*
  gene activity Genaktivität *f*
  gene change Genveränderung *f*
  gene conversion Genumwandlung *f*
  gene mutation Genmutation *f*
  gene pool Genpool *m*; Genbank *f*
  gene technology Gentechnik *f*
  gene toxicity Gentoxizität *f*
general metabolism Gesamtstoffwechsel *m*
general plan of the system Anlagenübersicht *f*
generator Generator *m*, Entwickler *m*, Erzeuger *m*
genetechnological gentechnisch
genetechnologically converted substance gentechnisch umgewandelte Substanz *f*
genetic genetisch
  genetic engineering Gentechnologie *f*
  genetic factor Erbfaktor *m*
  genetic trait Erbanlage *f*
geneticist Genetiker *m*
genetics Genetik *f*, Abstammungslehre *f*
genotoxic gentoxisch
genotype Erbmasse *f*, Genotypus *m*
genotypic genotypisch
geobotany Geobotanik *f*

geochemical geochemisch
geochemistry Geochemie *f*
geographic geographisch
geography Geographie *f*, Erdkunde *f*
geological geologisch
geolocical formation geologische Formation *f*
geology Geologie *f*
geometry of the sampler Geometrie *f* der Probenahmepumpe
geophysical geophysikalisch
geophysics Geophysik *f*
geosphere Geosphäre *f*
geotextiles Geotextilien pl
geothermal geothermisch
germ Keim *m*
germ cell Keimzelle *f*
germ filter Entkeimungsfilter *n*
germ-free keimfrei
germ gland Keimdrüse *f*
germ killer Keimtöter *m*
germanium Germanium *n*
germicidal keimtötend
germicidal bath Desinfizierungsbad *n*
germicide keimtötendes Mittel *n*
germinate/to keimen, knospen
germination Keimung *f*, Keimbildung *f*
germinative keimfähig
germless keimfrei
gestation Trächtigkeit *f*, Schwangerschaft *f*
glacier Gletscher *m*
gland Drüse *f*
glandular cell Drüsenzelle *f*
glandular secretion Drüsenabsonderung *f*
glass capillary Glaskapillare *f*
glass collection container Glassammlungscontainer *m*, Glassammelbehälter *m*
glass fibre filter Glasfaserfilter *n*
glass fibre fleece Glasfaservlies *n*
glass powder Glasmehl *n*
glazing Verglasung *f*
global concentration globale (weltumfassende) Konzentration *f*
global environmental monitoring system globales Umweltüberwachungssystem *n*
global radiation Globalstrahlung *f*
gneiss Gneis *m*
goggles Schutzbrille *f*
gold Gold *n*
grade/to sortieren, klassieren, trennen
grader Sortiermaschine *f*; Planierer *m*
gradient Gradient *m*; Neigung *f*, Steigung *f*
grading Klassieren *n*, Sortieren *n*, Trennen *n*
grading screen Klassiersieb *n*
granular activated carbon Aktivkohlegranulat *n*
granulation of used tyres Granulierung *f* von Altreifen
granules of dry sewage sludge Klärschlammtrockengranulat *n*
graphite Graphit *m*
graphite bearing Graphitlager *n*
graphite electrode Graphitelektrode *f*
graphite ferrule Graphitdichtung *f*
graphite layer Graphitschicht *f*
graphite lubricant Graphitschmierstoff *m*
graphite lubrication Graphitschmierung *f*

graphite-moderated reactor graphmoderierter Reaktor *m*
graphitic graphitisch
graphitization Graphitisierung *f*
graphitize/to graphitieren, mit Graphit überziehen, graphitisieren
graphitized carbon graphitisierter Kohlenstoff *m*
grass Gras *n*
grasshopper Heuschrecke *f*
gravel Kies *m*, Geröll *n*
gravel bed Kiesbett *n*
gravel filter Kiesfilter *n*
gravel pit Kiesgrube *f*
gravel screen Kiessieb *n*
gravel soil Kiesboden *m*
gravel treatment plant Kiesaufbereitungsanlage *f*
graveyard Friedhof *m*
gravimetric sorting gravimetrische Sortierung *f*
gravimetry Gravimetrie *f*, Gewichtsanalyse *f*
gravitation Gravitation *f*; Schwerkraft *f*
gravitational deposition Schwerkraftablagerung *f*
gravitational separation Gravitationstrennung *f*, Trennung *f* durch Schwerkraft
gravity Schwerkraft *f*, Schwere *f*
grease Fett *n*, Schmierstoff *m*
grease lubrication Fettschmierung *f*
grease-proof fettdicht, fettundurchlässig
grease separator Fettabscheider *m*
grease solvent Fettlösemittel *n*
grease trap Fettabscheider *m*
grease/to schmieren, einfetten

greasiness Schmierigkeit *f*, Fettigkeit *f*
green alga Grünalge *f*
green bin Grüne Tonne *f*
green dustbin grüne Tonne *f*
green glass Grünglas *m*
green manure Gründünger *m*
green refuse bin grüne Tonne *f*
green wall Pflanzenwand *f*
green waste composting Grünabfallkompostierung *f*
green waste Grünabfall *m*
greenbelt Grüngürtel *m*
greenbelt setting Begrünung *f*
greenfly Blattlaus *f*
greenhouse Treibhaus *n*
greenhouse effect Treibhauseffekt *m*
greenhouse gas Treibhausgas *n*
grinder Brechwerk *n*, Schleifstein *m*, Schleifmaschine *f*
grinding plant Mahlanlage *f*
grinding sludge Schleifschlamm *m*
grit grober Sand *m*, Kies *m*
gross examination Grobuntersuchung *f*, Übersichtsuntersuchung *f*
ground Erdboden *n*, Erde *f*, Grund *m*
ground bacterium Bodenbakterie *f*
ground humidity Bodenfeuchtigkeit *f*
ground rehabilitation (remedial, remediation) Erdreichsanierung *f*
ground tube Masserohr *n*
ground wave Bodenwelle *f*
groundwater Grundwasser *n*
groundwater collection plant Grundwasseranreicherungsanlage *f*
groundwater endangering grundwassergefährdend
groundwater lake Grundwassersee *m*

groundwater level Grundwasserspiegel *m*, Grundwasserstand *m*
groundwater model Grundwassermodell *n*
groundwater pollution Grundwasserverunreinigung *f*
groundwater-protected grundwassergeschützt
groundwater protection Grundwasserschutz *m*
groundwater protection area Grundwasserschutzgebiet *n*
growth Wachstum *n*
growth control Wachstumskontrolle *f*
growth curve Wachstumskurve *f*
growth factor Wachstumsfaktor *m*
growth hormone Wachstumshormon *n*
growth-inhibiting wachstumshemmend, wachstumshindernd
growth inhibitor Wachstumshemmer *m*
growth medium Nährboden *m*, Nährmedium *n*, Nährsubstanz *f*
growth promoter Wuchsstoff *m*
growth promoting wachstumsfördernd
growth regulation Wachstumsregulierung *f*
growth regulator Wachstumsregulator *m*
guiding data Führungsdaten pl, Orientierungsdaten pl, Leitwerte *mpl*
guiding parameter Leitparameter *m*
gully Gully *m*, Abflußrinne *f*, Abflußschacht *m*
gum Gummi *n*, Gummiharz *m*, Kautschuk *m*
guttapercha Guttapercha f(n)
gutter broom Gossenbesen *m*, Rinnsteinbesen *m*
gypseous gipsartig
gypseous earth Gipserde *f*
gypseous marl Gipsmergel *m*
gypsum Gips *m*

H

habit Lebensgewohnheit *f*
habitant Einwohner *m*
habitation Wohnen *n*, Bewohnen *n*
habits of life Lebensgewohnheiten *fpl*
habitual gewohnheitsmäßig
habituate/to gewöhnen
habitude Gewohnheit *f*, Veranlagung *f*
haemanalysis Blutanalyse *f*
haematic substance Blutbestandteil *m*
haematogenesis Blutbildung *f*
haematogenic blutbildend
haemocoagulative blutgerinnend
haemoglobin Hämoglobin *n*, roter Blutfarbstoff *m*
haemogram Hämogramm *n*, Blutbild *n*
haemorrhage Blutung *f*; Blutsturz *m*
haemostatic blutstillend
haemotoxin Blutgift *n*, Hämotoxin *n*
hafnium Hafnium *n*
hail Hagel *m*
half change value Halbwertzeit *f*
half decay period Halbwertsperiode *f*
half-life period Halbwertzeit *f*
half-period Halbperiode *f*, Halbwertzeit *f*

halide Halogenid *n*
halide-free halogenidfrei
halide leak detector Halogenidleckdetektor *m*
hallucinogen Halluzinogen *n*
halogen compound Halogenverbindung *f*
halogen derivative Halogenderivat *n*
halogen Halogen *n*, Salzbildner *m*
halogenatable halogenierbar
halogenate/to halogenieren
halogenated hydrocarbon halogenierter Kohlenwasserstoff *m*
halogenated organic solvent halogeniertes organisches Lösemittel *n*
halogenation Halogenierung *f*
halophyte Halophyt *m*
hammer crusher Hammerbrecher *m*
hammer drill Bohrhammer *m*
hammer mill Hammermühle *f*
hand fire extinghuisher Handfeuerlöscher *f*
hand-held detector tragbarer Detektor *m*
handguard Handschutz *m*
hard coal Steinkohle *f*
hard metal Hartmetall *n*
hard rubber Hartgummi *m*
hardening salt Härtesalz *n*
hardness of water Wasserhärte *f*
hardness removal Enthärtung *f*
hardness tester Härtemesser *m*
harmful schädlich, verletzend, gesundheitsschädlich
harmful effect Schadwirkung *f*
harmful material Gefahrstoff *m*, Schadstoff *m*
harmful waste Gefahrstoffabfall *m*
hashish Haschisch n(m)
hay fever Heuschnupfen *m*

hazard Gefahr *f*, Risiko *n*, Wagnis *n*
hazard identification Gefahrenidentifizierung *f*
hazard potential Gefahrenpotential *n*
hazardous gefährlich, gefahrvoll, riskant
hazardous goods Gefahrgut *n*, gefährliche Güter *npl*
hazardous site Sonderabfalldeponie *f*
hazardous waste Sonderabfall *m*, Sondermüll *m*
hazardous waste management Sonderabfallwirtschaft *f*
hazardous waste transportation Sondermülltransport *m*
hazardous waste treatment process Aufarbeitungsmethode *f* für Gefahrstoffabfall (Sonderabfall)
haze Dunstschleier *m*, feiner Nebel *m*
headwater Oberwasser *n*
healing process Heilvorgang *m*
healing state Heilzustand *m*
health Gesundheit *f*
health effect Gesundheitswirkung *f*
health endangering Gesundheitsgefährdung *f*
health-endangering noise gesundheitsgefährdender Lärm *m*
health food Reformkost *f*
health inspector Sozialarbeiter *m*
health protection Gesundheitsschutz *m*
health service Gesundheitswesen *n*
healthy living gesundes Wohnen *n*
healthy unrisky gesundheitlich unbedenklich
hearing loss Gehörverlust *m*

hearing of experts Expertenanhörung f
hearing protection Gehörschutz m
heat absorption Wärmeaufnahme f, Wärmeabsorption f
heat balance Wärmebilanz f
heat accumulation Wärmestau m
heat-consuming endotherm, wärmeverbrauchend
heat consumption Wärmeverbrauch m
heat content Wärmeinhalt m, Enthalpie f
heat exchanger Wärme[aus]tauscher m
heat insulating effect Wärmedämmwirkung f
heat insulation system Wärmeisolierungssystem n
heat of decomposition Zersetzungswärme f
heat recovery Wärmerückgewinnung f
heat recovery system Wärmerückgewinnungssystem n
heat-resistant hitzebeständig
heath Heide f
heathland Heideland n
heating Heizung f, Befeuerung f, Beheizen n, Erhitzen n
heating bath Heizbad n
heating body Heizkörper m
heating boiler Heizkessel m
heating circuit Heizkreislauf m
heating coil Heizschlange f
heating element Heizelement n, Heizkörper m
heating furnace Heizofen m, Glühofen m
heating pipe Heizleitung f
heating plant Heizungsanlage f

heatproof feuerfest, hitzebeständig
heavy current Starkstrom m
heavy-duty oil Hochleistungsöl n
heavy industry Schwerindustrie f
heavy metal Schwermetall n
heavy metal concentration Schwermetallgehalt m, Schwermetallkonzentration f
heavy metal content Schwermetallgehalt m
heavy metal leaching rate Schwermetallauslaugungsrate f
heavy-water pile Schwerwasserreaktor m
hedge plantation Heckenbepflanzung f
height indicator Höhenanzeiger m, Höhenmesser m
height of bed Schüttungshöhe f, Schütthöhe f
helium Helium n
hemicellulose Hemizellulose f
hepatitis immunization programme Hepatitisimmunisierungsprogramm n
hepatocyte Hepatozyt m
hepatolysis Hepatolyse f
herbicide Unkrautvernichtungsmittel n, Herbizid n
herbicide residue Herbizidrückstand m
hereditary erblich
hereditary anomaly Erbschädigung f
hereditary disease Erbkrankheit f
hereditary factor Erbfaktor m
heredity Vererbung f
hermetic hermetisch, luftdicht
hermetic seal Luftabschluß m
hermetic tight hermetisch dicht

heterogeneous catalysis heterogene Katalyse *f*
heterotrophic heterotroph
heterotrophy Heterotrophie *f*
hibernation Winterschlaf *m*
high-alloyed hochlegiert
high-aromatic hocharomatisch
high-boiling hochsiedend
high-capacity operation Hochlastbetrieb *m*
high-duty hochbeansprucht
high-duty operation Hochlastbetrieb *m*
high-energy energiereich
high frequency Hochfrequenz *f*
high-level tank Hochbehälter *m*
high moor Hochmoor *n*
high-performance filter Hochleistungsfilter *n*
high-performance fouling reactor Hochleistungsfaulreaktor *m*
high-performance liquid chromatography (HPLC) Hochleistungs-Flüssigchromatographie *f*
high-performance UV radiation lamp Hochleistungs-UV-Bestrahlungslampe *f*
high-pressure plant Hochdruckanlage *f*
high-quantity regulation Höchstmengenverordnung *f*
high-tech filter technology hochtechnologische Filtertechnik *f*
high-tech industry High-Tech-Industrie *f*
high-temperature incineration Hochtemperaturverbrennung *f*
high-temperature incineration plant Hochtemperaturverbrennungsanlage *f*

high tide Hochwasser *n*, Fluttide *f*
high water pumping station Hochwasserpumpwerk *n*
high-water level Hochwasserstand *m*
highly active hochaktiv, hochwirksam
highly impacted hochbelastet
histogenesis Histogenese *f*, Gewebebildung *f*
histology Histologie *f*, Gewebelehre *f*
histolysis Gewebezerfall *m*, Histolyse *f*
histopathological histopathologisch
hives Nesselausschlag *m*
hold back basin for rainwater Regenwasserrückhaltebecken *n*
hold tank Speicherbehälter *m*
hold up time Verweilzeit *f*, Aufenthaltszeit *f*
holmium Holmium *n*
homoeopathy Homöopathie *f*
homogeneity Homogenität *f*, Gleichartigkeit *f*
homogeneous homogen, gleichartig
homogeneous catalysis homogene Katalyse *f*
homogenization Homogenisierung *f*
homogenize/to homogenisieren
homogenizer Homogenisiermaschine *f*
homomorphic homomorph, gleichgestaltig
honeycomb catalyst Wabenkatalysator *m*
hood Haube *f*, Abzug *m*, Rauchfang *m*
hookworm Hakenwurm *m*
hormonal action Hormonwirkung *f*

hormone Hormon *n*
hormone balance Hormongleichgewicht *n*
hormone-dependent mammary tumor hormonell bedingter Brusttumor *m*
hormone equilibrium Hormongleichgewicht *n*, hormonelles Gleichgewicht *n*
hormone preparation Hormonpräparat *n*
horn meal Hornmehl *n*
horn shavings Hornabfall *m*, Hornspäne *mpl*
horse hair sieve Roßhaarsieb *n*
hose clip Schlauchschelle *f*
hose coupling Schlauchanschluß *m*
hospital specific waste Kliniksondermüll *m*, krankenhausspezifischer Abfall *m*
host cell Wirtszelle *f*
host organism Wirtsorganismus *m*
host plant Wirtspflanze *f*
hot air Heißluft *f*
hot air drying Heißlufttrocknung *f*
hot air engine Heißluftmaschine *f*
hot air pipe Heißluftleitung *f*
hot cathode Glühkathode *f*
hot-dipped tinning Feuerverzinnung *f*
hot galvanizing Feuerverzinkung *f*
hot gas generator Heißgaserzeuger *m*
house vermin Hausungeziefer *n*
household refrigerating device Haushalts-Kältegerät *n*
household waste Hausmüll *m*
household water Haushaltswasser *n*
housing society Siedlungsstruktur *f*

human carelessness menschliche Sorglosigkeit *f*
human ecology Humanökologie *f*
human error menschliches Versagen *n*
human failure menschliches Versagen *n*
human medicine Humanmedizin *f*
human milk Humanmilch *f*, Muttermilch *f*
human toxicity Humantoxizität *f*
human toxicology Humantoxikologie *f*
humanity Menschlichkeit *f*, Humanität *f*, Menschheit *f*
humanization Humanisierung *f*
humanization of working place Humanisierung *f* des Arbeitsplatzes
humankind Menschheit *f*
humanly menschlich
humic acid Huminsäure *f*
humic substance Huminstoff *m*
humid feucht
humidifier Verdunstungsanlage *f*, Luftbefeuchter *m*
humidifier Verdunstungsanlage *f*, Luftbefeuchter *m*
humidify/to befeuchten
humidity Luftfeuchtigkeit *f*, Luftfeuchte *f*; Nässe *f*, Feuchtigkeit *f*
humidity measurement Feuchtemessung *f*
humous humusartig
humus Humus *m*
humus degradation Humuszersetzung *f*
humus layer Humusschicht *f*
hunting Jagdwesen *n*
hydrate/to hydratisieren, wässern
hydrated hydrathaltig, wasserhaltig

hydration   68

hydration Hydratisierung f, Hydratation f, Wasseranlagerung f
hydraulic hydraulisch
   hydraulic cement Wasserzement m
   hydraulic control system Hydrauliküberwachungssystem n
   hydraulic filter press Plattenpreßfilter n
   hydraulic liquid Hydraulikflüssigkeit f
   hydraulic medium Druckflüssigkeit f
   hydraulic monitoring system Hydrauliküberwachungssystem n
   hydraulic mortar Hydraulikmörtel m, Wassermörtel m
   hydraulic supervision system Hydrauliküberwachungssystem n
   hydraulic system Hydraulikanalage f
hydro-diaphragm pump Hydrodiaphragmapumpe f
hydrocarbon Kohlenwasserstoff m
   hydrocarbon monitor Kohlenwasserstoffmonitor m
   hydrocarbon recovery plant Kohlenwasserstoff-Rückgewinnungsanlage f
hydrocyclone Hydrozyklon m
hydrodynamics Hydrodynamik f, Strömungslehre f
hydroelectric power station Wasserkraftwerk n
hydrogen Wasserstoff m
   hydrogen chloride Salzsäure f
   hydrogen cyanide Blausäure f
   hydrogen peroxide Wasserstoffperoxid n
   hydrogen sulphide concentration Schwefelwasserstoffkonzentration f
hydrogenability Hydrierbarkeit f

hydrogenate/to hydrieren, härten
hydrogenation Hydrierung f
   hydrogenation of coal Kohlehydrierung f
hydrogeological hydrogeologisch
hydrograph curve Gangliniendarstellung f
hydrography Hydrographie f
hydrologic hydrologisch
hydrological cylce Wasserkreislauf m
hydrology Hydrologie f
hydrolysis Hydrolyse f
   hydrolysis reactor Hydrolysereaktor m
hydrolytic hydrolytisch
   hydrolytic degradation hydrolytischer Abbau m
hydrophilic hydrophil, wasserbindend
hydrophobic hydrophob, wasserabstoßend
hydrostatic hydrostatisch
hydrothermal synthesis Hydrothermalsynthese f
hydroxide sludge Hydroxidschlamm m
hygiene Hygiene f
   hygiene regulation Hygienevorschrift f
hygienic hygienisch
   hygienic conditions Hygienebedingungen fpl
hygienics Hygiene f, Gesundheitslehre f
hygrometer Hygrometer n, Feuchtigkeitsmesser m
hygroscopic hygroskopisch
hyperacidity Übersäuerung f
hyperaemia Hyperämie f
hypercalcaemia Hypercalcämie f
hyperfunction Überfunktion f

hypersensitivity Überempfindlichkeit f
hypertension Hypertonie f, Bluthochdruck m
hyperthyroidism Schilddrüsenüberfunktion f
hypertrophic hypertroph
hypertrophy Hypertrophie f
hypoacidity Säuremangel m
hypoalimentation Unterernährung f
hypocalcaemia Kalkmangel m
hypohormonal hormonarm
hypoimmunity Immunschwäche f
hypoinsulinism Insulinmangelkrankheit f
hypolimnion Hypolimnion n
hypolymnic aeration hypolimnische Belüftung f
hypolymnic water Tiefenwasser n
hypotensive blutdrucksenkend
hypothyrosis Schilddrüsenunterfunktion f
hypovitaminosis Vitaminmangel m

I

ice calorimeter Eiskalorimeter n
ice zone Eiszone f
ideal state Idealzustand m
ideality Idealzustand m
identification system Identifizierungssystem n
idioplasm Idioplasma n
igneous electrolysis Schmelzflußelektrolyse f
igneous rock Eruptivgestein n
ignitability Entzündbarkeit f, Zündfähigkeit f
ignitable entflammbar, entzündbar, feuerfangend

ignition Entzündung f, Anzündung f, Zündung f, Zündvorgang m
ignition point Zündpunkt m
ill-smelling übelriechend
illegal barging of spent acid illegale Dünnsäureverklappung f
illuviation Illuviation f
immediate aid Soforthilfe f
immediate measure Sofortmaßnahme f
immersion aerator Tauchbelüfter m
immersion electrode Tauchelektrode f
immersion fitting Eintaucharmatur f
immersion pump Tauchpumpe f
immiscibility Nichtmischbarkeit f, Unvermischbarkeit f
immiscible unmischbar, unvermischbar, nicht mischbar
immission Immission f
immission analysis Immissionsanalyse f
immission-conditioned cadmium deposition immissionsbedingte Cadmiumdeposition f
immission control Immissionskontrolle f
immission distribution Immissionsverteilung f
immission protection Immissionsschutz m
immission protection law Immissionsschutzgesetz n
immission register Immissionskataster n
immission value Immissionswert m
immobilization Immobilisierung f, Festlegung f

immobilization of bacteria Bakterienimmobilisierung f
immobilize/to festlegen, immobilisieren
immune body Immunkörper m
immune defence mechanism Immunabwehrsystem n
immune globuline Immunoglobulin n
immune response Immunoreaktion f
immune system Immunsystem n
immunity Immunität f, Resistenz f
immunization Immunisierung f; Schutzimpfung f
immunize/to immunisieren
immunobiological immunbiologisch
immunobiology Immunbiologie f
immunochemistry Immunchemie f
immunocytochemical immunzytochemisch
immunodetection Immunnachweis m
immunodiagnostics Immundiagnostik f
immunogenetics Immungenetik f
immunologic immunologisch
immunology Immunologie f, Immunitätsforschung f
immunoreaction Immunreaktion f
immunotherapy Immuntherapie f
impact concentration threshold value Belastungskonzentration f
impact mill Prallmühle f
impact of groundwater Grundwasserbelastung f
impact of heavy metals Schwermetalleintrag m
impact of soil Bodenbelastung f
impact of waters Gewässerbelastung f

impact on aquatic systems Gewässerbelastung f
impact on groundwater Grundwasserbelastung f
impact resistance Belastungswiderstand m
impact-resistant casing schlaggeschütztes Gehäuse n
impairment of the efficiency degree Wirkungsgradverschlechterung f
impeller Kreiselmischer m, Kreiselrührer m; Kreiselrad n, Schaufelrad n, Laufrad n
impeller mixer Kreiselmischer m
impermeability Undurchdringbarkeit f, Undurchlässigkeit f
impermeable undurchdringlich, undurchlässig
implementing order Durchführungsverordnung f
impolder/to eindeichen, trockenlegen
impoverishment Verarmung f
impregnated wood getränktes Holz n
impregnating agent Imprägnierungsmittel n
impregnating fluid Imprägnierflüssigkeit f
impregnation Imprägnierung f, Durchdringung f, Durchtränkung f
impressed output current eingeprägter Ausgangsstrom m
impulse counting Impulszählung f
impure unrein, schmutzig, verfälscht, unsauber
impurity Verunreinigung f, Begleitstoff m, Fremdstoff m
impurity atom Fremdatom n
impurity level Verunreinigungsgrad m

impurity water Fremdwasser *n*
imputrescibility Fäulnisbeständigkeit *f*
in vitro in vitro, im Reagenzglas
in vivo in vivo, am lebenden Objekt
inactivation Inaktivierung *f*
incinerate/to verbrennen, einäschern, veraschen
incineration Verbrennung *f*
incinerating process Verbrennungsprozeß *m*
incineration aggregate Verbrennungsaggregat *n*
incineration air Verbrennungsluft *f*
incineration air preheating Verbrennungsluftvorwärmung *f*
incineration dish Veraschungsschälchen *n*
incineration of waste Abfallverbrennung *f*, Müllverbrennung *f*
incineration plant Verbrennungsanlage *f*
incineration residue Verbrennungsrückstand *m*
incineration ship Müllverbrennungsschiff *n*
incineration technique Verbrennungstechnik *f*
incinerator Müllverbrennungsofen *m*, Abfallverbrennungsofen *m*
incinerator system Verbrennungssystem *n*
inclined sorting machine Schrägsortiermaschine *f*
inclusion compound Einschlußverbindung *f*
inclusion method Inklusionsverfahren *n*
incombustibel unverbrennbar, feuerfest

incombustibility Unverbrennbarkeit *f*
incoming denitrification vorgeschaltete Denitrifikation *f*
incorrodible unkorrodierbar, nicht korrodierbar
incorruptible unverderblich
indecomposable unzersetzbar, unzerlegbar
indestructibility Unzerstörbarkeit *f*
indestructible unzerstörbar
index value Richtwert *m*
indication sensitivity Anzeigeempfindlichkeit *f*
indicator Indikator *m*, Anzeiger *m*
indicator bacteria Indikatorbakterien *fpl*
indicator gauge Anzeigegerät *n*
indicator of limit values Grenzwertanzeige *f*
indicator organism Indikatororganismus *m*
indicator paper Indikatorpapier *n*
indicator plant Indikatorpflanze *f*
indicator range Indikatorbereich *m*
indicator reading Zeigerablesung *f*
indium Indium *n*
individual analysis Einzelanalyse *f*
individual introducer Einzeleinleiter *m*
individual protection Personenschutz *m*
indoor air Raumluft *f*, Innenraumluft *f*
indoor air humidity Innenraumluftfeuchte *f*
indoor air pollution Innenraumluftverunreinigung *f*
indoor-fire Wohnungsbrand *m*, Raumbrand *m*

induction air Ansaugluft *f*
induction current Induktionsstrom *m*
induction furnace Induktionsofen *m*
inductive sensor induktiver Aufnehmer *m*
industrial accident Betriebsunfall *m*
industrial chemical Industriechemikalie *f*
industrial chemist Industriechemiker *m*
industrial cleaner Industriereiniger *m*
industrial disease Berufskrankheit *f*
industrial environmental protection Industrie-Umweltschutz *m*
industrial inspection authority Gewerbeaufsichtsbehörde *f*
industrial inspection board Gewerbeaufsichtsamt *n*
industrial location Industriestandort *m*
industrial refuse dump Industrieabfalldeponie *f*
industrial sludge Industrieschlamm *m*
industrial society Industriegesellschaft *f*
industrial waste Gewerbemüll *m*; Industrieabfall *m*
industrial waste water Industrieabwasser *n*
industrialized country Industriestaat *m*, industrialisiertes Land *n*
ineradicable unausrottbar
inert atmosphere Schutzgasatmosphäre *f*
inert gas Inertgas *n*, Schutzgas *n*, Edelgas *n*
inert material Inertmaterial *n*, Ballaststoff *m*, Ballastmaterial *n*
inert residue inerter Rückstand *m*
inertial resistance Trägheitswiderstand *m*
inertialize/to inaktivieren, inertisieren
inertness Reaktionsträgheit *f*, Trägheit *f*
inexplosive explosionssicher
infant mortality Säuglingssterblichkeit *f*
infect/to infizieren, anstecken, verseuchen
infection Infektion *f*, Ansteckung *f*; Verpestung *f*, Vergiftung *f*
infection disease Infektionskrankheit *f*, ansteckende Krankheit *f*
infectious infektiös, ansteckend
infectious waste infektiöser Abfall *m*
infective ansteckend
infective agent Infektionserreger *m*
infertility Unfruchtbarkeit *f*
infestation Befall *m*
infiltrate/to infiltrieren, durchsetzen, tränken, eindringen, durchsickern
infiltration Infiltration *f*, Durchdringung *f*, Eindringung *f*, Einsickerung *f*
inflame/to entflammen, entzünden, entfachen
inflammability Entflammbarkeit *f*, Entzündbarkeit *f*, Brennbarkeit *f*, Feuergefährlichkeit *f*
inflammable entzündbar, brennbar, entflammbar, feuergefährlich
inflammable solvent brennbares Lösemittel *n*
inflammation Entflammung *f*, Entzündung *f*

inflammatory entzündlich, entzündbar
influence of the environment Umwelteinfluß *m*
influence of the surroundings Umgebungseinfluß *m*
influence on the environmemt Einfluß *m* auf die Umwelt
information data-bank for harmful materials Gefahrstoff-Informationsdatenbank *f*
information technical networking informationstechnische Vernetzung *f*
infrared filter Infrarotfilter *n*
infrared radiation Infrarotstrahlung *f*
infrared spectroscopy Infrarotspektroskopie *f*
infrared spectrum Infrarotspektrum *n*
infrastructure Infrastruktur *f*
infructuous unfruchtbar
ingestion Ingestion *f*
inhabitable bewohnbar
inhabitant Bewohner *m*
inhabitant equivalent [value] Einwohnergleichwert *m*
inhalation Inhalation *f*, Einatmung *f*
inhalation uptake Inhalationsaufnahme *f*, Aufnahme *f* durch Inhalation
inhale/to einatmen
inhaler Inhalationsapparat *m*
inhibitor Inhibitor *m*, Hemmstoff *m*
inhibitory effect Hemmwirkung *f*, inhibierende Wirkung *f*
inhomogeneity Inhomogenität *f*
initial concentration Anfangskonzentration *f*, Ausgangskonzentration *f*
initial dose Anfangsdosis *f*
initial material Ausgangsstoff *m*
initial stage Anfangsstadium *n*
injected output current eingeprägter Ausgangsstrom *m*
injection gel Injektionsgel *n*
injurious to health gesundheitsschädlich
inland sea Binnenmeer *n*
inland water Binnengewässer *n*
inlet air Zuluft *f*, Frischluft *f*
innoxious unschädlich
inodorous geruchlos, geruchfrei
inorganic anorganisch
inorganic fertilizer Mineraldünger *m*
input value Eingangswert *m*
insanitary unhygienisch, gesundheitsschädlich
insect control Insektenbekämpfung *f*
insect pest Insektenplage *f*
insecticidal insektizid, insektenvernichtend
insecticide Insektizid *n*, Insektenvernichtungsmittel *n*
insecticide residue Insektizidrückstand *m*
insectivore Insektenfresser *m*
inshore an der Küste, küstennah
inshore fishing Küstenfischerei *f*
in-situ analysis in-situ Analyse *f*
in-situ experiment in-situ Experiment *n*
installation technique Aufbautechnik *f*
instruction-synchronized befehlssynchron
instrumental analysis instrumentelle Analyse *f*
instrumental control instrumentelle Kontrolle *f*
instrumental equipment instrumentelle Ausstattung *f*

instrumental error instrumenteller Fehler *m*
insulating compound Isoliermasse *f*
insulating layer Isolierschicht *f*
insulating material Isolierstoff *m*, Isoliermaterial *n*
insulating pitch Isolierpech *n*
insulation Isolierung *f*, Isolation *f*; Isoliermaterial *n*
insulator Isolator *m*, Isolationsstoff *m*, Isolationsmittel *n*
intensity of evaporation Verdunstungsintensität *f*
intensity of exposure Expositionsintensität *f*
intensity of rainfall Niederschlagsintensität *f*
intensive rot Intensivrotte *f*
interception Interzeption *f*
interference signal Störungssignal *n*
interim regulation Übergangsbestimmung *f*
intermediate clarification Zwischenklärung *f*
intermediate reaction Zwischenreaktion *f*
intermediate storage Zwischenlagerung *f*
intermeshing Vermaschung *f*
intermittent-circuit operation Aussetzschaltung *f*
internal exposure interne Exposition *f*
international environmental commission internationaler Umweltausschuß *m*
intoxicant Rauschmittel *n*, Rauschgift *n*
intoxicate/to berauschen; vergiften
intoxication Rausch *m*, Trunkenheit *f*; Intoxikation *f*, Vergiftung *f*

invasion Invasion *f*
inversion layer Inversionsschicht *f*
inversion weather [situation] Inversionswetterlage *f*
investigation of pipings Untersuchung *f* an Rohrleitungen
investments for environmental protection Umweltschutzinvestitionen *fpl*
iodine Jod *n*, Iod *n*
ion chromatography Ionenchromatographie *f*
ion exchange chromatography Ionenaustauschchromatographie *f*
ion exchange resin Ionenaustauscharz *n*
ion exchanger Ionenaustauscher *m*
ionization Ionisierung *f*, Ionisation *f*
ionization current Ionisationsstrom *m*
ionizing radiation ionisierende Strahlung *f*
ionosphere Ionosphäre *f*
ion selectivity Ionenselektivität *f*
iridium Iridium *n*
irradiation Bestrahlung *f*, Ausstrahlung *f*
irradiator Strahler *m*
irreversible irreversibel
irrigate/to berieseln, künstlich bewässern
irrigation Bewässerung *f*
irrigation basin Bewässerungsbecken *n*
irrigation canal Bewässerungsgraben *m*
irrigation of forests Bewässerung *f* von Wäldern
irrigation technology Bewässerungstechnik *f*
irritability Reizbarkeit *f*

irritable reizbar
irritant Reizstoff *m*
isothermal isotherm
isotope-labelled radiomarkiert
ivory Elfenbein *n*

J

jacket Mantel *m*, Auskleidung *f*, Hülle *f*, Umhüllung *f*, Ummantelung *f*
jacket/to verkleiden, ummanteln, umhüllen
jet engine Düsentriebwerk *n*, Düsenaggregat *n*
jet fuel Düsentreibstoff *m*
jet separator Trenndüse *f*
joint-project Gemeinschaftsprojekt *n*
joule Joule *n*
jungle Dschungel *m*
jute fabric Jutegewebe *n*
jute fibre Jutefaser *f*

K

karst Karst *m*
kerosene Kerosin *n*
key chemical Grundchemikalie *f*, Grundstoff *m*
key register Schlüsselverzeichnis *n*
knackery Abdeckerei *f*
kneader Knetwerk *n*
kneading machine Knetwerk *n*
knock-free klopffrei, nicht klopfend
knock-free inhibitor Antiklopfmittel *n*

L

labelled atom markiertes Atom *n*
laboratory accident Laborunfall *m*
laboratory balance Laborwaage *f*
laboratory data Labordaten pl
laboratory equipment Laborausstattung *f*, Laboreinrichtung *f*
laboratory examination Laboruntersuchung *f*
laboratory scale Labormaßstab *m*
laboratory test Labortest *m*, Laboruntersuchung *f*
labour protection Arbeitsschutz *m*
lacquer waste Lackabfall *m*
lagoon Lagune *f*
lake remediation Seesanierung *f*
lamellar clarifier Lamellenklärer *m*
laminar flow Laminarströmung *f*
land cultivated by man Kulturlandschaft *f*
land office Grundbuchamt *n*
land register Kataster *n*
land saving Flächenschonung *f*
land surveying Landvermessung *f*
land surveyor Geometer *m*, Landvermesser *m*
land utilization map Flächennutzungskarte *f*
landfill Deponietechnik *f*
landfill site Deponiegelände *n*
landscape Landschaft *f*
landscape architecture Landschaftsgestaltung *f*
landscape biology Landschaftsbiologie *f*
landscape diagnosis Landschaftsdiagnose
landscape ecology Landschaftsökologie *f*

landscape preservation Landschaftspflege *f*
landscape protection Landschaftsschutz *m*
landscape spoliation Landschaftszerstörung *f*
landslide Erdrutsch *m*
lanthanum Lanthan *n*
large manufacture Massenherstellung *f*
large-scale composting großtechnische Kompostierung *f*
large-scale lysimeter Großlysimeter *n*
large-scale processing of liquid manure großtechnische Gülleaufbereitung *f*
larvicide Larvizid *n*
laser surveillance system Laser-Überwachungssystem *n*
latency period Inkubationszeit *f*, Latenzzeit *f*, Latenzperiode *f*
latent latent, verborgen
law of consolidation of arable land Flurbereinigungsgesetz *n*
law of environmental planning Raumordnungsgesetz *n*
lawn Rasen *m*
lawrencium Lawrencium *n*
layer of smog Dunstglocke *f*
leach Auslaugung *f*
leach water drainage Sickerwasserdränage *f*
leach/to auslaugen; auswaschen; heraussickern; durchsickern
leach through/to durchsickern, hindurchsickern
leaching Auslaugen *n*, Auslaugung *f*
leaching basin Sickerbecken *n*
leaching concentrate Sickerkonzentrat *n*

leaching resistent slag auslaugungsresistente Schlacke *f*
leaching tank Reinigungstank *m*
leaching test Auslaugtest *m*
leaching water Sickerwasser *n*
leaching water collection Sickerwassersammlung *f*
leaching water disposal Sickerwasserentsorgung *f*
leaching water from waste dumps Deponiesickerwasser *n*
leaching water purification plant Sickerwasserreinigungsanlage *f*
lead Blei *n*
lead accumulator Bleiakkumulator *m*
lead alloy Bleilegierung *f*
lead crystal Bleikristall *n*, Bleiglas *n*
lead dust Bleistaub *m*
lead ore Bleierz *n*
lead poisoning Bleivergiftung *f*
lead/to verbleien
leaf analysis Blattanalyse *f*
leaf damage Blattschädigung *f*
leaf herbicide Blattherbizid *n*
leaf impairment Blattschädigung *f*
leaf mould Lauberde *f*
leaf wax Blachwachs *n*
leak Leck *n*, Undichtigkeit *f*
leak detection Lecksuche *f*
leak detector Lecksucher *m*, Lecksuchgerät *n*
leak/to auslaufen, ausrinnen, durchsickern, entweichen, leck sein
leakage Entweichen *n*; Schwund *m*, Verlust *m*
leakage control system Leckage-Überwachungssystem *n*
leakage indicator Leckanzeigegerät *n*

leakage monitoring system Leckage-Überwachungssystem *n*
leakage tester Lecksuchgerät *n*
leakiness Undichtigkeit *f*, Undichtheit *f*
leaking leck, undicht
leaking oil control Leckölüberwachung *f*
leaking oil control system Leckölüberwachungssystem *n*
leaking oil monitoring Leckölüberwachung *f*
leakproof leckdicht, leckfrei
leaksafe lecksicher
leaky undicht, leck
leather scrap Lederabfall *m*
leather waste Lederabfall *m*
legislator Gesetzgeber *m*
leguminous plant Leguminose *f*
leisure industry Freizeitindustrie *f*
leisure activity Freizeitbeschäftigung *f*, Freizeitaktivität *f*
leisure research Freizeitforschung *f*
lengthwise flow basin längsdurchflossenes Becken *n*
lethal dose Letaldosis *f*, tödliche Dosis *f*, tödliche Menge *f*
lethality Sterblichkeit *f*, Letalität *f*
leucocyte Leukozyt *m*, weißes Blutkörperchen *n*
leucocytosis Leukozytose *f*, Leukozytenzerfall *m*
leucolysis Leukolyse *f*
leucoma Leukom *n*
leucosis Leukämie *f*, Leukose *f*
leukaemia Leukämie *f*, Leukose *f*
leukaemic leukämisch
level limiting value Höhenstandgrenzwert *m*
level monitoring Pegelüberwachung *f*

level of exposure Expositionsgrenzwert *m*
level of plant management Betriebsführungsebene *f*
liability claim Haftpflichtanspruch *m*
liability for compensation Ersatzpflicht *f*
liability insurance Haftpflichtversicherung *f*
life expectancy Lebenserwartung *f*
life space Lebensraum *m*
life span Lebensdauer *f*
lifetime Lebensdauer *f*, Leben *n*
light boiling component niedrigsiedende Komponente *f*
light building material Leichtbaustoff *m*
light distillate fuel Leichtbenzin *n*
light duty detergent Leichtwaschmittel *n*
light fraction of waste Müll-Leichtfraktion *f*
light liquid Leichtflüssigkeit *f*

light-load operation Schwachlastbetrieb *m*
light-load step Schwachlaststufe *f*
light metal Leichtmetall *n*
light-weight concrete Leichtbeton *m*
lightning protection Blitzschutz *m*
lignin Lignin *n*, Holzfaserstoff *m*
lignite Braunkohle *f*
lignite bed Braunkohlenlager *n*
lignite carbonization plant Braunkohleschwelanlage *f*
lignite coke Braunkohlenkoks *m*
lignite pitch Braunkohlenpech *n*
lignite tar Braunkohlenteer *m*

lime Kalk *m*, Kalkerde *f*
lime bin Kalksilo *n*
lime burner Kalkbrenner *m*
lime cast Kalkputz *m*, Kalkverputz *m*
lime cement mortar Kalkzementmörtel *m*
lime concrete Kalkbeton *m*
lime crusher Kalkbrecher *m*
lime milk Kalkmilch *f*
lime powder Kalksteinmehl *n*
lime washer Kalksteinwäscher *m*
lime/to kalken, einkalken
limestone Kalkstein *m*
limestone suspension Kalksteinsuspension *f*
limestone washer Kalksteinwäscher *m*
liming Kalken *n*
liming of forests Kalken *m* der Wälder
limit of absolute safety Unbedenklichkeitsschwelle *f*
limit value Grenzwert *m*
limit value of the level Höhenstandgrenzwert *m*
limit violation Grenzwertverletzung *f*
limiting case Grenzfall *m*
limiting concentration Grenzkonzentration *f*
limiting condition Grenzbedingung *f*
limiting line Grenzlinie *f*
limiting load Grenzbelastung *f*
limiting size Grenzmaß *n*
limy kalkig
line erosion Linienerosion *f*
line pressure Betriebsdruck *m*
lining Auskleidung *f*

linkage valency Bindungswertigkeit *f*, kovalente Wertigkeit *f*
lipocyte Fettzelle *f*, Lipozyt *m*
lipogenesis Lipogenese *f*
lipoid lipoid, fettähnlich, fettartig
lipolysis Lipolyse *f*, Fettspaltung *f*
lipometabolism Fettstoffwechsel *m*, Fetthaushalt *m*
lipophobic lipophob
liposoluble fettlöslich
liquefaction Verflüssigung *f*, Flüssigwerden *n*
liquefaction of coal Kohleverflüssigung *f*
liquefaction of gases Gasverflüssigung *f*
liquefied gas Flüssiggas *n*
liquefier Verflüssigungsapparat *m*
liquefy/to verflüssigen; sich verflüssigen
liquefying plant Verflüssigungsanlage *f*
liquid flüssig, dünnflüssig; Flüssigkeit *f*, flüssiger Körper *m*
liquid air flüssige Luft *f*
liquid chromatography Flüssigchromatographie *f*
liquid circulation Flüssigkeitsumlauf *m*
liquid cooler Flüssigkeitskühler *m*
liquid crystal Flüssigkristall *m*
liquid fertilizing agent Flüssigdüngemittel *n*
liquid gas Flüssiggas *n*
liquid manure Gülle *f*
liquid manure management Güllewirtschaft *f*
liquid phase Flüssigphase *f*
liquid waste Flüssigabfall *m*
liquor Brühe *f*, Flotte *f*; Lauge *f*; Flüssigkeit *f*

lithium Lithium *n*
lithosphere Lithosphäre *f*
litter bin Müllbehälter *m*
litter box Müllbehälter *m*
litter problem Abfallproblem *n*
littorial deposit litorale Ablagerung *f*
littorial transportation litoraler Transport *m*
liver cell Leberzelle *f*
liver dysfunction Leberfunktionsstörung *f*
liver function test Leberfunktionstest *m*
liver metabolism Leberstoffwechsel *m*
liver parenchymal cell Lebergewebszelle *f*
livestock Vieh *n*, Viehbestand *m*
living conditions Lebensbedingungen *fpl*
load Beanspruchung *f*, Belastung *f*; Last *f*, Beladung *f*
load capacity Belastbarkeit *f*
load change Belastungsveränderung *f*
load compensation Belastungsausgleich *m*
load curve Ganglinie *f*
load duration Belastungsdauer *f*
load limit Belastungsgrenze *f*
load/to laden, beladen, aufladen; beschicken, chargieren, speisen
loading Belastung *f*; Aufladung *f*; Beschicken *n*
loading capacity Belastungsvermögen *n*
loading chute Laderutsche *f*
loading device Ladevorrichtung *f*, Beschickungseinrichtung *f*

loading situations Belastungsverhältnisse *npl*
loading test Belastungsprobe *f*
loam Lehm *m*
loam coat Lehmschicht *f*
loamy lehmig, lehmartig
loamy marl Lehmmergel *m*
loamy soil Lehmboden *m*
local disturbance lokale Störung *f*
location restriction Standorteinschränkung *f*
locator Ortungsgerät *n*
loess Löß *m*
logistics of disposal Entsorgungslogistik *f*
long-acting langwirkend
long-chain langkettig
long-path gas cuvette Langweggasküvette *f*
long-term measurement Langzeitmessung *f*
long-term monitoring Langzeitüberwachung *f*
long-term reaction Langzeitreaktion *f*
long-term resistance Langzeitbeständigkeit *f*
long-term stability Langzeitstabilität *f*
loop agitator Schlaufenrührer *m*
lorry driver Lastwagenfahrer *m*
lorry pool Lastwagenpark *m*
loss by erosion Erosionsverlust *m*, Verlust *m* durch Erosion
loss of evaporation Verdampfungsverlust *m*
loudness Lautstärke *f*, Lärm *m*
low-boiling niedrigsiedend
low-boiling halogenated hydrocarbon leichtflüchtiger halogenierter Kohlenwasserstoff *m*

low-carbon kohlenstoffarm
low-grade fuel minderwertiger Brennstoff *m*
low in heavy metals schwermetallarm
low loaded aeration Schwachlastbelebung *f*
low-temperature conversion Niedertemperaturkonvertierung *f*
low tide Niedrigwasser *n*, Ebbe *f*
low-velocity sand filter plant Langsamsandfilteranlage *f*
lowering of the groundwater level Grundwasser[spiegel]absenkung *f*, Grundwasserpegel[ab]senkung *f*
lubricant Schmiermittel *n*, Schmiere *f*, Gleitmittel *n*
lubricant additive Schmierstoffadditiv *n*
lubricant disposal Schmiermittelentsorgung *f*
lubricate/to schmieren, fetten, ölen
lubricating Schmieren *n*, Einfetten *n*, Ölen *n*
lubricating effect Schmiereffekt *m*
lubricating grease Schmierfett *n*
lubricating liquid Schmierflüssigkeit *f*
lubricating oil Schmieröl *n*, Maschinenöl *n*, Motorenöl *n*
lubricating syringe Schmierspritze *f*
lubricator Schmiervorrichtung *f*, Schmierapparat *m*
lubricity Gleitfähigkeit *f*, Schmierfähigkeit *f*
lumber Bauholz *n*, Nutzholz *n*
luminescent lumineszent, leuchtend
lung abscess Lungenabszeß *m*
lutetium Lutetium *n*

lye Lauge *f*, Brühe *f*
lye-proof laugenbeständig, alkalibeständig
lymph node Lymphknoten *m*
lymphadenitis Lymphknotenentzündung *f*, Lymphadenitis *f*
lymphatic gland Lymphdrüse *f*
lymphatic tissue Lymphgewebe *n*
lymphocyte Lymphozyt *m*
lymphocytosis Lymphozytose *f*
lypophilization Gefriertrocknung *f*
lysimeter Lysimeter *n*

M

macroanalysis Makroanalyse *f*
macrobiosis Makrobiose *f*, Langlebigkeit *f*
macroclimate Großklima *n*, Makroklima *n*
macrocosm Makrokosmos *m*
macromolecular makromolekular
macromolecule Makromolekül *n*
macroscopic makroskopisch
macrosomia Großwuchs *m*
macrostructure Makrostruktur *f*, Grobstruktur *f*, Grobgefüge *n*
magnesium Magnesium *n*
magnet Magnet *m*
magnet carrier Magnetträger *m*
magnet crane Magnetkran *m*
magnetic magnetisch
magnetic conveyor belt Magnetbandförderer *m*
magnetic drum Magnettrommel *f*
magnetic-inductive flow measurement magnetisch-induktive Durchflußmessung *f*

magnetic separator Magnetabscheider *m*
magnetic stirrer Magnetrührer *m*
magnetic switch Magnetschalter *m*
magnetic valve Magnetschalter *m*
magnetism Magnetismus *m*
magnetization Magnetisierung *f*
mailing case Versandgefäß *n*
main constituent Hauptbestandteil *m*
main constituent of a solvent Lösemittelhauptkomponente *f*, Lösemittelhauptbestandteil *m*
main inspection Hauptinspektion *f*, große Inspektion *f*
maintenance cost Wartungskosten *f*
maintenance dose Erhaltungsdosis *f*
maintenance of the health of waters Gesunderhaltung *f* von Gewässern
maintenance report Instandhaltungsbericht *m*
maintenance routine work Instandhaltungsarbeit *f*
major component Hauptkomponente *f*, Hauptbestandteil *m*
major metabolite Hauptmetabolit *m*
make acidic/to ansäuern
make alkaline/to alkalisch machen
malaria control Malariabekämpfung *f*
malaria fever Malariafieber *n*
maldevelopment Fehlentwicklung *f*
malformation Mißbildung *f*
malfunction Funktionsstörung *f*
malignancy Bösartigkeit *f*
malignant bösartig; verderblich, schädlich
malignant cell bösartige Zelle *f*
malnutrition Unterernährung *f*

malodorous übelriechend
malodour Übelgeruch *m*, Gestank *m*
maloperation Fehlbedienung *f*
mammal Säugetier *n*
mammary gland Brustdrüse *f*
mammary tissue Brustgewebe *n*
mammary tumour Brusttumor *m*
mammogram Mammogramm *n*
mammography Mammographie *f*
management of hazardous substances Gefahrstoffverwaltung *f*
manganese Mangan *n*
manhole Einstiegsöffnung *f*, Einstiegsschacht *m*
manometer Manometer *n*, Druckmesser *m*, Druckanzeiger *m*
manometer scale Manometerskala *f*
manual work Handarbeit *f*
manually telecontrolled handferngesteuert
manure Dünger *m*, Dung *m*, Mist *m*
manure dose Düngemittelgabe *f*
manure salt Düngesalz *n*
manure vessel Jauchebecken *n*
manure works Düngemittelfabrik *f*
manuring Düngen *n*, Düngung *f*
manuring salt Düngesalz *n*
mariculture Aquakultur *f*
marine climate Seeklima *n*
marine ecosystem marines Ökosystem *n*
marine fouling mariner Bewuchs *m*
marine pollution Seeverschmutzung *f*, Meeresverschmutzung *f*
marketing Vermarktung *f*, Marketing *n*, Absatz *m*
marl Mergel *m*
marl pit Mergelgrube *f*
marrow cell Knochenmarkzelle *f*
marsh Sumpf *m*
marsh gas Sumpfgas *n*

marsh plant Sumpfpflanze f
marshland Marsch f, Marschland n
mass goods Massengüter npl
mass mortality Massensterben n
mass spectrometer Massenspektrometer n
mass tube Masserohr n
material balance Stoffbilanz f
material consumption Materialverbrauch m
material of value Wertstoff m
material stress Materialbeanspruchung f
maximum Maximum n, Höchstgrenze f, Höchstmaß n
maximum concentration Höchstkonzentration f
maximum concentration limit höchste Grenzkonzentration f
maximum content Höchstgehalt m
maximum credible accident (MCA) größter anzunehmender Unfall (GAU)
maximum deviation Höchstabweichung f
maximum dose Höchstdosis f
maximum life Höchstlebensdauer f
maximum permissible dose höchstzulässige Dosis f
maximum quantity Höchstmenge f
maximum speed limit Höchstgeschwindigkeitsgrenze f
meadow Wiese f, Aue f
measurand Meßwert m
measurand pick-up Meßwertaufnahme f
measure of redevelopment Sanierungsmaßnahme f
measured data Meßdaten pl
measured value Meßwert m

measured-value arithmetics Meßwertarithmetik f
measurement of canal depth Kanaltiefenmessung f
measurement of sludge content Schlammgehaltsmessung f
measurement of the sludge level Schlammpegelmessung f
measurement routine Meßvorgang m
measuring and control engineering Meß- und Regelungstechnik f
measuring apparatus Meßapparatur f
measuring arrangement Meßanordnung f
measuring beaker Meßbecher m
measuring cell Meßzelle f
measuring cuvette Meßküvette f
measuring data Meßdaten pl
measuring device of luminous bacteria Leuchtbakterien-Meßgerät n
measuring electrode Meßelektrode f
measuring electronics Meßelektronik f
measuring gas Meßgas n
measuring good Meßgut n
measuring material Meßgut n
measuring medium Meßmedium n
measuring method Meßverfahren n
measuring sensor Sensor m, Meßfühler m
measuring transducer Meßumformer m
meat poisoning Fleischvergiftung f
meat processing Fleischverarbeitung f
mechanical-biological clarification plant mechanisch-biologische Kläranlage f

mechanical shaker Schüttelmaschine *f*
mechanical sorting mechanisches Auslesen *n*
medical environmental control medizinische Umweltkontrolle *f*
medical gymnastics Heilgymnastik *f*
medical prevention medizinische Vorsorge *f*
medicament donation Medikamentenspende *f*
medicinal plant Heilpflanze *f*
medicinal well Heilquelle *f*
medium-sized process control station Prozeßstation *f* mittleren Umfangs
medulla Mark *n*
medullary tumour Knochenmarkstumor *m*
meiosis Meiose *f*
melanoma Melanom *n*
meliorate/to bessern, verbessern, meliorieren
melioration Bodenverbesserung *f*, Melioration *f*
melt Schmelze *f*, Schmelzfluß *m*
melt/to schmelzen, verschmelzen
melt off/to abschmelzen
meltability Schmelzbarkeit *f*
meltable schmelzbar
melting enthalpy Schmelzenthalpie *f*
  melting furnace Schmelzofen *m*
  melting heat Schmelzwärme *f*
  melting index Schmelzindex *m*
  melting loss Schmelzverlust *m*
  melting point Schmelzpunkt *m*
  melting pot Schmelztiegel *m*
  melting process Schmelzprozeß *m*
membrane chamber filter press Membrankammerfilterpresse *f*

membrane diffusion Osmose *f*
membrane permeability Membrandurchlässigkeit *f*
membrane plant purification Membran-Anlagenreinigung *f*
membrane-polarimetric method membranpolarimetrisches Verfahren *n*
membrane pump Membranpumpe *f*
membrane technology Membrantechnik *f*
mendelevium Mendelevium *n*
mensuration analysis Maßanalyse *f*
mercaptane Merkaptan *n*
mercurism Quecksilbervergiftung *f*
mercury Quecksilber *n*
  mercury battery Quecksilberbatterie *f*
  mercury recycling Quecksilberrecycling *n*
  mercury toxicity Quecksilbertoxizität *f*
meristem Meristem *n*, Teilungsgewebe *n*
meshing Analogwert *m*
mesomeric energy Mesomerieenergie *f*
mesomerism Mesomerie *f*, Resonanz *f*
mesophile mesophil
mesotrophic mäßig produktiv
message text Meldungstext *m*
met office Wetteramt *n*
metabiosis Metabiose *f*
metabolic convertible compound metabolisch umsetzbare Verbindung *f*
metabolic disease Stoffwechselkrankheit *f*
metabolic inhibition Stoffwechselblockierung *f*

metabolic pathway Stoffwechselweg *m*
metabolism Metabolismus *m*, Stoffwechsel *m*
metabolite Metabolit *m*, Stoffwechselprodukt *n*
metal chips Metallspäne *mpl*
metal content Metallgehalt *m*
metal detector Metalldetektor *m*, Metallsuchgerät *n*
metal scrap Metallabfall *m*
metal sludge Metallschlamm *m*
metalimnion Metalimnion *n*
metallic colour Metallikfarbe *f*
metallic foil Metallfolie *f*
metallochemistry Metallchemie *f*
metallurgic metallurgisch
metallurgical plant metallurgische Anlage *f*
metamorphosis Metamorphose *f*, Gestaltsveränderung *f*, Umwandlung *f*
metastable metastabil
metastasis Metastase *f*, Tochtergeschwulst *n*
metastasis of cancer Krebsgeschwulst *n*, Krebsmetastase *f*
meteorite Meteorit *m*
meteorological meteorologisch
meteorological map Wetterkarte *f*
meteorology Meteorologie *f*, Wetterkunde *f*
methadone Methadon *n*
methane bacterium Methanbakterium *n*
methane gas Methangas *n*
method combination Verfahrenskombination *f*
method of evaluation Auswertungsmethode *f*

method of hazardous waste processing Aufarbeitungsmethode *f* für Gefahrstoffabfall
method of measurement Meßmethode *f*
micro-ecosystem Mikroökosystem *n*
micro-filtration membrane Mikrofiltrationsmembran *f*
microanalysis Mikroanalyse *f*
microbalance Mikrowaage *f*
microbe Mikrobe *f*, Mikroorganismus *m*
microbial mikrobiell, mikrobisch
microbial culture Mikrobenkultur *f*, Mikrobenzüchtung *f*
microbial slime mikrobischer Schlamm *m*
microbicidal antibiotisch, mikrobentötend
microbiological mikrobiologisch
microbiological reaction mechanism mikrobiologischer Reaktionsmechanismus *m*
microbiologist Mikrobiologe *m*
microbiology Mikrobiologie *f*
microchemical mikrochemisch
microchemistry Mikrochemie *f*
microclimate Mikroklima *n*, Kleinklima *n*
microcolloid Mikrokolloid *n*
microelectronics Mikroelektronik *f*
microflora Mikroflora *f*
microliter Mikroliter *m*
microorganism Mikroorganismus *m*, Klein[st]lebewesen *n*
microporous mikroporös
microprocessor Mikroprozessor *m*
microscope Mikroskop *n*
microscopic mikroskopisch
microscopy Mikroskopie *f*
microsome Mikrosom *n*

microsomia Kleinwuchs *m*
microstrainer system Mikrosieb-Filtersystem *n*
microstructure Mikrostruktur *f*
microwave Mikrowelle *f*
microwave technology Mikrowellentechnik *f*
migration Migration *f*
mildew Moder *m*, Mehltau *m*, Schimmel *m*
mildew-resistant schimmelbeständig, schimmelfest
mildew/to schimmeln, modern
milieu Milieu *n*, Umgebung *f*
milk drying plant Milchtrocknungsanlage *f*
milk powder Trockenmilch *f*, Milchpulver *n*
milk processing Milchverarbeitung *f*
mill dust Mehlstaub *m*
mine Mine *f*, Bergwerk *n*, Zeche *f*
mine disaster Grubenunglück *n*
mine exploitation Bergwerksnutzung *f*, Minennutzung *f*
mine explosion Grubenexplosion *f*
mine gas Grubengas *n*
mine ventilation Grubenbelüftung *f*
miner Bergarbeiter m, Grubenarbeiter *m*
mineral mineralisch; Mineral *n*
mineral acid Mineralsäure *f*, anorganische Säure *f*
mineral analysis Mineralanalyse *f*
mineral coal Steinkohle *f*
mineral dust Gesteinsstaub *m*
mineral fibre industry Mineralfaserindustrie *f*
mineral flax Faserasbest *m*

mineral oil containing sludge mineralölhaltiger Schlamm *m*
mineral oil product Mineralölprodukt *n*
mineral water filling station Mineralwasserabfüllstation *f*
mineralization Mineralisierung *f*
mineralize/to mineralisieren
mineralogical mineralogisch
miniature plant Kleinstanlage *f*
miniaturized ecosystem miniaturisiertes Ökosystem *n*
minimization of pollution Schadstoffminimierung *f*
minimization of residual substances Minimierung *f* von Reststoffen
minimize/to minimieren, verringern
minimum demand Mindestanforderung *f*
minimum effective dose kleinste wirksame Dosis *f*
minimum efficiency Mindestleistung *f*
minimum output Mindestleistung *f*
minimum pressure Mindestdruck *m*
mining Bergbau *m*, Bergwesen *n*, Grubenbau *m*
minister of environmental protection Minister *m* für Umweltschutz
Ministry of Environmental Protection Umweltschutzministerium *n*
Ministry of Health Gesundheitsministerium *n*
miscibility Mischbarkeit *f*
miscible mischbar
misdiagnose Fehldiagnose *f*
misgrowth Mißwuchs *m*, Auswuchs *m*
mist Nebel *m*, Dunst *m;* Sprühregen *m*
mite Milbe *f*

mitosis indirekte Zellkernteilung *f*, Mitose *f*
mitotic mitotisch
mixable mischbar
mixed adhesive Zweikomponentenkleber *m*
mixed asphalt concrete material Asphaltbetonmischgut *n*
mixed bacteria culture bakterielle Mischkultur *f*
mixed culture Mischkultur *f*
mixed forest Mischwald *m*
mixed plastic scrap gemischter Plastikabfall *m*
mixed plastic waste Kunststoffmischabfall *m*
mixed sewage Mischabwasser *n*
mixed waste Mischabfall *m*
mixer Mischer *m*, Mischapparat *m*
mixing apparatus Mischapparat *m*
mixing basin Mischbecken *n*
mixing jet Mischdüse *f*
mixing plant Mischanlage *f*
mixing tank Mischtank *m*
mixing vessel Mischgefäß *n*
mixotrophic mixotroph
mixture Mischung *f*, Gemenge *n*, Gemisch *n*
mixture for freezing Kältemischung *f*
mixture of ash Aschegemisch *n*
mixture of sewage and activated sewage sludge Abwasser-Belebtschlamm-Gemisch *n*
mixture of solids Feststoffmischung *f*
mixture ratio Mischungsverhältnis *n*
mobile solvent Laufmittel *n*, Fließmittel *n*

mobile vessel fahrbares Gefäß *n*
mobilization Mobilisierung *f*
mode of action Wirkungsweise *f*
mode of life Lebensweise *f*
model experiment Modellversuch *m*
model of agricultural structure Agrarstrukturmodell *n*
model simulation Modellsimulation *f*
model substance Modellsubstanz *f*
modelling calculation Modellrechnung *f*
module system Modulsystem *n*, Baukastensystem *n*
moisture analyzer Feuchteanalysator *m*
moisture-attracting feuchtigkeitsanziehend
moisture content Feuchtigkeitsgehalt *m*
moisture control Feuchtigkeitsregelung *f*
moisture-resistant feuchtigkeitsbeständig
molar molar
mole Mol *n*
molecular molekular
molecular biology Molekularbiologie *f*
molecular sieve Molekularsieb *n*
molecular weight Molekulargewicht *n*
molecule Molekül *n*
moloscs Weichtiere *npl*
molluscicide Molluskizid *n*
molybdenum Molybdän *n*
monitor Monitor *m*, Warner *m*, Überwacher *m*
monitor/to überwachen, abhören, mithören

monitoring Überwachung *f*, Kontrolle *f*
monitoring and control Überwachung *f* und Steuerung
monitoring at a workplace Arbeitsplatzüberwachung *f*
monitoring equipment Kontrolleinrichtung *f*
monitoring function Beobachtungsfunktion *f*
monitoring model Modell *n* zur Überwachung
monitoring routine Überwachungsroutine *f*
monitoring station Überwachungsstation *f*
monitoring system for the ambient noise Umweltlärm-Überwachungssystem *n*
monoculture Monokultur *f*
monomer Monomer *n*
monomeric monomer
monthly load curve Monatsganglinie *f*
monthly output curve Monatsganglinie *f*
moor peat Moortorf *m*
morbidity Krankhaftigkeit *f*, Unnatürlichkeit *f*, Morbidität *f*
mordant Beize *f*, Beizmittel *n*
morphogenesis Morphogenese *f*, Gestaltbildung *f*
morphology Morphologie *f*, Formenlehre *f*, Gestaltlehre *f*
mortal tödlich, sterblich
mortality Sterblichkeit *f*, Mortalität *f*
mortality rate Sterblichkeitsrate *f*, Sterblichkeitsziffer *f*
moss Moos *n*
mother liquor Mutterlauge *f*, Stammlauge *f*
mother's milk Muttermilch *f*
motor gasoline Motortreibstoff *m*
motor winding temperature sensor Motorwicklungstemperaturfühler *m*
motorway Autoschnellstraße *f*
mould Schimmel *m*, Moder *m*; Modererde *f*; Schimmelpilz *m*
mould formation Schimmelpilzbildung *f*
mouldering Vermoderung *f*
mouthwash Mundspülwasser *n*
muck-bark compost Torf-Rinden-Kompost *m*
mucosa Schleimhaut *f*
mucous schleimig; schleimabsondernd
mucous membrane Schleimhaut *f*
mucus Schleim *m*
mud Schlamm *m*, Schlick *m*; Schmutz *m*
mud box Schlammkasten *m*
mud filter Schlammfilter *n*
mud flats Wattenmeer *n*
mud trap Schlammabscheider *m*
muddy schlammig; schmutzig, verdreckt
mulch Mulch *m*
mulch/to mulchen
mull Mull *m*
multi-chamber container Mehrkammercontainer *m*
multi-channel analyzer Mehrkanalanalysator *m*
multi-element analysis Multielementanalyse *f*
multi-way valve Mehrwegschieber *m*
multicomponent analysis Mehrstoffanalyse *f*

multicomponent mixture Mehrstoffgemisch *n*
multicomponent system Mehrstoffsystem *n*
multiple-hearth furnace (roaster) Etagenofen *m*
multiple use Mehrfachnutzung *f*, Mehrfachverwendung *f*
multiprocessing Multiprozessing *n*
multiprocessor-capable multiprozessorfähig
multiprocessor concept Multiprozessorkonzept *n*
multistage prepurification mehrstufige Vorreinigung *f*
multistage process mehrstufiges Verfahren *n*
multistage thickening plant mehrstufige Eindickungsanlage *f*
municipal effluent städtisches Abwasser *n*
municipal humus kommunaler Humus *m*
municipal vehicle Kommunalfahrzeug *n*
municipal water supply städtische Wasserversorgung *f*
muscle tissue Muskelgewebe *n*
mushroom poisoning Pilzvergiftung *f*
mutability Veränderlichkeit *f*, Mutationsfähigkeit *f*
mutagen Mutagen *n*
mutagenic mutagen, erbgutverändernd
mutagenic risk mutagenes Risiko *n*
mutagenicity Mutagenität *f*
mutagenesis Mutagenese *f*
mutation Genveränderung *f*, Mutation *f*; Änderung *f*, Umwandlung *f*
mutualism Mutualismus *m*

mycelium Mycel *n*, Pilzgeflecht *n*
mycetism Pilzvergiftung *f*
mycobacterium Mykobakterie *f*
mycology Mykologie *f*, Pilzkunde *f*
mycosis Mykose *f*, Pilzkrankheit *f*
myeloid cell Knochenmarkzelle *f*

N

nanosomia Zwergwuchs *m*
nanous zwergwüchsig
naphta Erdöl *n*, Leuchtöl *n*
naphthalene Naphthalen *n*
narcotic Betäubungsmittel *n*, Narkotikum *n*; Rauschgift *n*
narcotization Narkotisierung *f*
narcotize/to narkotisieren
nasal spray Nasenspray *n*
nascent state Entstehungszustand *m*
national health Volksgesundheit *f*
national park Nationalpark *m*
native nativ, natürlich
native substance Naturstoff *m*
natural balance Naturhaushalt *m*
natural circuit natürlicher Kreislauf *m*
natural colour Naturfarbe *f*, Eigenfarbe *f*
natural disaster Naturkatastrophe *f*
natural drying Lufttrocknung *f*
natural fibre Naturfaser *f*
natural gas Naturgas *n*, Erdgas *n*
natural gas supply system Erdgasnetz *n*
natural product Naturprodukt *n*
natural resources natürliche Ressourcen *fpl*, Naturschätze *mpl*
natural rubber Naturkautschuk *m*
natural ventilation natürliche Belüftung *f*

nature and environmental protection group Natur- und Umweltschutzgruppe f
nature conservation Naturschutz m, Erhaltung f der Natur
nature reserve Naturschutzgebiet n
near-shore strandnah, ufernah
necrobiosis Nekrobiose f
necrose/to absterben, nekrotisieren
necrosis Nekrose f, Absterben n eines Gewebes
needle valve Nadelventil n
nematode Fadenwurm m, Nematode m
neodymium Neodym n
neon Neon n
neoplasm Geschwulstbildung f, Neoplasma n, Gewächs n
nephelometer Nephelometer n, Trübungsmeßgerät n
nerve gas Nervengas n
nerve irritation Nervenreizung f
nervous disease Nervenleiden n
nesting aid Nisthilfe f
net efficiency Nutzleistung f
net load Nutzlast f
net weight Nettogewicht n
network of pipes Rohrleitungsnetz n
network topology Netztopologie f, Netzwerkaufbau m
neurobiology Neurobiologie f
neurochemistry Neurochemie f
neurodermatosis Neurodermatose f
neurotoxin Nervengift n
neutral point Neutralpunkt m
neutral position Nullstellung f
neutral wire Nulleiter m
neutralization Neutralisierung f
neutralization method Neutralisationsverfahren n
neutralization process Neutralisationsvorgang m
neutralize/to neutralisieren
neutron activation analysis Neutronenaktivierungsanalyse f
neutron decay Neutronenzerfall m
neutron emission Neutronenemission f
neutron shield Neutronenschutz m
new ground Neuland n
niche Nische f
nickel Nickel n
nickel bath Nickelbad n
nickel sewage Nickelabwasser n
nicotine Nikotin n
nicotine content Nikotingehalt m
nicotinism Nikotinvergiftung f
niobium Niob n
nitrate Nitrat n
nitrate breathing Nitratatmung f
nitrate-controlled denitrifikation nitratgesteuerte Denitrifikation f
nitrate elimination Nitratelimination f
nitrate leaching Nitratauswaschung f
nitrate-sensitive electrode nitratsensitive Elektrode f
nitrating acid Nitriersäure f
nitration Nitrierung f
nitre Salpeter m
nitrificating clarifier nitrifizierende Kläranlage f
nitrification Nitrifizierung f
nitrify/to nitrieren, nitrifizieren
nitrifying bacteria Nitrifizierungsbakterien fpl
nitrifying bacterium Stickstoffbakterie f
nitrite Nitrit n
nitrite-free nitritfrei

nitro-cellulose

nitro-cellulose Nitrocellulose *f*, Cellulosenitrat *n*
nitrobacteria Nitrobakterien *npl*
nitrogen Stickstoff *m*
nitrogen conversion Stickstoffumsetzung *f*
nitrogen cycle Stickstoffkreislauf *m*
nitrogen cycle in the nature Stickstoffkreislauf *m* in der Natur
nitrogen dioxide Stickstoffdioxid *n*
nitrogen excretion Stickstoffausscheidung *f*
nitrogen problem Stickstoffproblematik *f*
nitrogen release Stickstofffreisetzung *f*
nitrogen removal Stickstoffentfernung *f*
nitrogeneous stickstoffhaltig
nitrosamine Nitrosamin *n*
nobelium Nobelium *n*
noble gas Edelgas *n*
noble metal recycling Edelmetallrecycling *n*
nodular knotig
nodule Knötchen *n*
nodule bacteria Knöllchenbakterien *npl*
nodule forming bacteria Knöllchenbakterien *npl*
noise Geräusch *n*, Lärm *m*; Rauschen *n*
noise ban Lärmverbot *n*
noise control Lärmbekämpfung *f*
noise disturbance Lärmbelästigung *f*
noise emission Lärmemission *f*
noise factor Rauschfaktor *m*
noise level Geräuschpegel *m*

noise nuisance Geräuschbelästigung *f*
noise protecting wall Lärmschutzwand *f*
noise protection Lärmschutz *m*
noise protection wall from plastic waste Lärmschutzwand *f* aus Kunststoffmüll (Kunststoffabfall)
noise protection wall from waste Lärmschutzwand *f* aus Abfall
nominal capacity Nennleistung *f*
nominal load Nennlast *f*
nominal output Nennleistung *f*
nominal value Nennwert *m*, Nominalwert *m*
nominal width Nennweite *f*
non-alkali alkalifrei
non-aqueous nichtwäßrig
non-combustible unverbrennbar
non-conducting nichtleitend
non-contact continuous level measurement berührungslose kontinuierliche Füllstandsmessung *f*
non-corroding nicht korrodierend
non-corrosive korrosionsfrei; rostbeständig; säurefest
non-destructive measurement zerstörungsfreie Messung *f*
non-ferrous metal Buntmetall *n*
non-filterable unfiltrierbar
non-flammable unbrennbar
non-foaming nicht schäumend
non-fusable nicht schmelzbar
non-gasifiable nicht vergasbar
non-knocking klopffrei
non-passable area nichtbegehbarer Bereich *m*
non-returnable bottle Einwegflasche *f*
non-reusable substance Reststoff *m*

non-skid flooring rutschfester Bodenbelag *m*
non-splintering splitterfrei
non-sulphur schwefelfrei
non-systemic nicht systemisch, unsystemisch
non-toxic nicht toxisch, ungiftig, giftfrei
non-volatile nichtflüchtig
normal distribution Normalverteilung *f*
normal output Nennleistung *f*
normal pressure Normaldruck *m*
normal state Normalzustand *m*
normal temperature Normaltemperatur *f*
noxious schädlich, verderblich, ungesund
noxious substance Schadstoff *m*
nozzle dryer Düsentrockner *m*
nuclear nuklear
nuclear age Atomzeitalter *n*
nuclear chain reaction Kernkettenreaktion *f*
nuclear chain reactor Kernreaktor *m*
nuclear charge number Kernladungszahl *f*
nuclear division Kernteilung *f*
nuclear engineering Kerntechnik *f*
nuclear fission Kernspaltung *f*
nuclear power plant Kernkraftwerk *n*
nuclear power supply Kernenergieversorgung *f*, nukleare Energieversorgung *f*
nuclear-weapon-free area atomwaffenfreie Zone *f*
nucleus Kern *m*; Zellkern *m*; Atomkern *m*
nuclide Nuklid *n*, Kernbaustein *m*, Kernteilchen *n*
nut coal Nußkohle *f*
nutrient Nährstoff *m*
nutrient content Nährstoffgehalt *m*
nutrient deficiency Nährstoffmangel *m*
nutrient entry into the sea Nährstoffeinleitung *f* in das Meer
nutrient liquid Nährflüssigkeit *f*
nutrient mass balance Nährstoffhaushalt *m*
nutrient medium Nährboden *m*, Nährmedium *n*
nutrient supply Nährstoffversorgung *f*
nutrient yeast Nährhefe *f*
nutritional disturbance Ernährungsstörung *f*
nutritional physiology Ernährungsphysiologie *f*
nutritious nahrhaft, nährend
nutritiousness Nahrhaftigkeit *f*
nutritive nahrhaft, nährend

O

oasis Oase *f*
obligation of announcement Anzeigepflicht *f*
obligation to take back Rücknahmepflicht *f*
observation balloon Beobachtungsballon *m*
observation function Beobachtungsfunktion *f*
observation hole Schauloch *n*
observation point Beobachtungsstelle *f*, Meßstelle *f*

observation tower Beobachtungswarte f, Beobachtungsturm m
occupational accident Arbeitsunfall m, Berufsunfall m
occupational disease Berufskrankheit f
occupational exposure berufsbedingte (arbeitsbedingte) Exposition f
occupational respiratory disease berufsbedingte Atemwegserkrankung f
occupational safety Arbeitsschutz m; Arbeitssicherheit f
ocean combustion Verbrennung f auf See
ocean dumping Verklappung f
ocean dumping of acid Säureverklappung f
ocean incineration Verbrennung f auf See, Seeverbrennung f
ocean pollution Meeresverschmutzung f
oceanic water Meerwasser n
oceanography Meereskunde f, Ozeanographie f
odorless geruchlos, geruchsfrei
odorous wohlriechend
odorous intensive geruchsintensiv
odour-active geruchsaktiv
odour emission Geruchsemission f
odour-intensive geruchsintensiv
odour nuisance Geruchsbelästigung f
oedema Ödem n, Wassersucht f
off-shore küstennah, ablandig
offer palette Angebotspalette f
oil Öl
oil absorption Ölabsorption f
oil aerosol separator Ölnebelabscheider m
oil basin Ölbehälter m
oil burner Ölbrenner m
oil catch pan Ölauffangpfanne f
oil change Ölwechsel m
oil chemistry Erdölchemie f
oil-containing ölhaltig
oil-containing waste water ölhaltiges Abwasser n
oil field Ölfeld n
oil film Ölfilm m
oil-polluted waste water ölverschmutztes Abwasser n
oil pollution Ölverschmutzung f
oil scum Ölschaum m
oil slick Ölteppich m
oil trap Ölfang m
oil vacuum pump Ölvakuumpumpe f
oil vapour Öldampf m
oil well fire Ölquellenbrand m
old-deposit Altablagerung f
old deposition Altablagerung f
old equipment Altanlage f
old page Altseite f
old plant Altanlage f
old site Altlast f
old-site redevelopment Altlastensanierung f
old site treatment Altlastenbehandlung f
olefine Olefin n
oleiferous ölhaltig
oligotrophic nährstoffarm, oligotroph
on-off control Zweipunktregelung f
on-site investigation Vorortuntersuchung f
on-site redevelopment On-Site-Sanierung f
oncogenesis Tumorbildung f
oncogenic tumorbildend

one-component system Einstoffsystem *n*
one-layer filter Einfachfilter *n*
one-way container Einwegcontainer *m*, Einwegbehälter *m*
one-way dishes Einweggeschirr *n*
one-way method Einwegverfahren *n*
one-way packing Einwegverpackung *f*
open-air test Freilandversuch *m*
open-cast working Tagebaugewinnung *f*
open circuit monitoring Drahtbruchüberwachung *f*
open land area Freilandfläche *f*
open working Tagebau *m*
operator Bediener *m*
operating and monitoring panel Bedien- und Überwachungstafel *f*
operating condition Betriebszustand *m*
operating data Betriebsdaten pl
operating function Bedienfunktion *f*
operating mode Betriebsart *f*
operating time system Laufzeitsystem *n*
operation set Befehlsvorrat *m*
operational condition Verarbeitungsbedingung *f*; Betriebsbedingung *f*
operational diary Betriebstagebuch *n*
operational parameter Betriebsparameter *m*
operational safety Betriebssicherheit *f*
operations log Betriebstagebuch *n*
operator command Bedienanweisung *f*
operator communication channel Bedienkanal *m*
operator course Bedienablauf *m*
operator environment Bedieneroberfläche *f*
operator error Bedienungsfehler *m*
operator function Bedienfunktion *f*
operator interface Bedieneroberfläche *f*
operator pathway Bedienpfad *m*
operator station Bedien- und Beobachtungsstation *f*
operator's console Bedienungspult *n*
optical activity optische Aktivität *f*
optical brightener optischer Aufheller *m*
optical refraction Lichtbrechung *f*
optical warning and report system optisches Warnmeldesystem *n*
optimum heating rate optimale Heizrate *f*
oral vaccine Schluckimpfstoff *m*
ore Erz *n*
ore benefication Erzaufbereitung *f*
ore crusher Erzbrecher *m*
ore deposit Erzlager *n*, Erzlagerstätte *f*
ore roasting Erzröstung *f*
ore slag Erzschlacke *f*
ore slime Erzschlamm *m*
ore sludge Erzschlamm *m*
ore vein Erzader *f*
organic organisch
organic chemistry organische Chemie *f*
organic compound organische Verbindung *f*
organic degradation product organisches Abbauprodukt *n*

organic domestic waste organischer Hausmüll *m*
organic dust organischer Staub *m*
organic pollutant organischer Schadstoff *m*
organic solvent organisches Lösemittel *n*
organization for environmental protection Umweltschutzorganisation *f*
organochlorine compound Organochlorverbindung *f*
organohalogen compound Organohalogenverbindung *f*
organologic organologisch
organology Organologie *f*
organometallic metallorganisch
original equipment Erstausstattung *f*, Erstausrüstung *f*
original material Ausgangsmaterial *n*, Ausgangsstoff *m*
original packing Originalverpackung *f*
original product Ausgangsprodukt *n*
osmium Osmium *n*
osmosis Osmose *f*
osmotic osmotisch
out-of-limit condition Grenzwertverletzung *f*
outbreak Ausbruch *m*
outbreak of a disease Krankheitsausbruch *m*
outburst Ausbruch *m*
outdoor durability Außenbeständigkeit *f*, Witterungsbeständigkeit *f*
outdoor paint Außenfarbe *f*
outflow Ausfluß *m*, Auslaß *m*; Ausströmen *n*
outflow control Abflußsteuerung *f*, Ablaufsteuerung *f*
outflow quality Ablaufqualität *f*

outgas/to entgasen
outgrowth Auswuchs *m*
outlet Auslauf *m*, Ablauf *m*, Abfluß *m*
outlet cock Auslaßhahn *m*
outlet controller Ausflußkontrolle *f*
outlet pipe Ausflußrohr *n*
outlet valve Auslaßventil *n*
output capacity Durchsatzleistung *f*
output current Ausgangsstrom *m*
output curve Ganglinie *f*
output curve representation Ganglinieniendarstellung *f*
output maximum Leistungsmaximum *n*
output module Ausgabebaugruppe *f*
oven soot Ofenruß *m*
overacidification Übersäuerung *f*
overacidify/to übersäuern
overall yield Gesamtausbeute *f*
overcrowded area Ballungsgebiet *n*, Ballungsraum *m*
overdosage Überdosierung *f*
overdose/to überdosieren
overexploitation Raubbau *m*
overfeeding Überfütterung *f*; Überernährung *f*
overflow dam Überlaufwehr *n*
overflow drain Überlaufkanal *m*
overhead line Freileitung *f*
overhead reservoir Hochbehälter *m*
overhead tank Hochbehälter *m*
overhead wiring Oberleitung *f*
overload protection Überlastungsschutz *m*
overpopulation Überbevölkerung *f*
oversaturation Übersättigung *f*
overstress Überbeanspruchung *f*

overweight Übergewicht *n*
oxidant Oxidant *m*, Oxidationsmittel *n*
oxidation Oxidation *f*
oxidation catalyst Oxidationskatalysator *m*
oxidation plant Oxidationsanlage *f*
oxide Oxid *n*
oxidize/to oxidieren
oxygen Sauerstoff *m*
oxygen activated sludge process Sauerstoff-Belebtschlammverfahren *n*
oxygen bleach Sauerstoffbleiche *f*
oxygen breathing Sauerstoffatmung *f*
oxygen concentration Sauerstoffkonzentration *f*
oxygen consumption Sauerstoffverbrauch *m*
oxygen control loop Sauerstoffregelkreis *m*
oxygen control system Sauerstoffregelkreis *m*
oxygen cycle Sauerstoffkreislauf *m*
oxygen deficiency Sauerstoffmangel *m*
oxygen demand Sauerstoffbedarf *m*
oxygen-enriched sauerstoffangereichert
oxygen entry capacity Sauerstoffeintragsleistung *f*
oxygen entry control Sauerstoffeintragsregelung *f*
oxygen input capacity Staffelung *f* der Sauerstoffeintragsleistung
oxygen-poor water sauerstoffarmes Wasser *n*
oxygen saturation value Sauerstoffsättigungswert *m*

oxygen supplier Sauerstoffversorger *m*
oxygen supply Sauerstoffversorgung *f*
oxygen ventilation Sauerstoffbelüftung *f*
oxygenation Oxigenierung *f*, Sauerstoffanreicherung *f*
ozonator Ozonerzeuger *m*
ozone Ozon *n*
ozone concentration Ozonkonzentration *f*, Ozongehalt *m*
ozone contact Ozonkontakt *m*
ozone content Ozongehalt *m*
ozone damage Ozonschaden *m*
ozone-damaging ozonschädlich
ozone deficiency Ozonmangel *m*
ozone effect Ozonwirkung *f*
ozone hole Ozonloch *n*
ozone layer Ozonschicht *f*
ozonic ozonisch, ozonhaltig
ozonization Ozonisierung *f*, Behandlung *f* mit Ozon
ozonization chamber Ozonkammer *f*
ozonization facility Ozonungsanlage *f*
ozonolysis Ozonolyse *f*, Ozonspaltung *f*
ozonometer Ozonmesser *m*, Ozonometer *n*
ozonometry Ozonmessung *f*

P

packaging cardboard Dichtungspappe *f*
packaging case Versandschachtel *f*
packaging density Packungsdichte *f*

packaging machine Verpackungsmaschine *f*
packaging material Verpackungsmaterial *n*, Packmaterial *n*, Verpackungsstoff *m*
packaging paper Packpapier *n*
packaging plant Abfüllbetrieb *m*
packaging press Bündelpresse *f*
packed height Füllkörperhöhe *f*
packed tower Füllkörperturm *m*
packed tube Füllkörperrohr *n*
packing Verpackung *f*
packing expenditure Verpackungsaufwand *m*
packing output Verpackungsaufkommen *n*
paddle dryer Schaufeltrockner *m*
paddle mixer Paddelrührer *m*, Schaufelrührer *m*
paddle stirrer Paddelrührer *m*, Schaufelrührer *m*
paddle wheel Schaufelrad *n*
paint Farbe *f*; Lack *m*; Anstrich *m*
paint factory Farbenfabrik *f*
paint remover Abbeizmittel *n*
palladium Palladium *n*
pallet Palette *f*
pallet truck Gabelhubwagen *m*
pampas Pampa *f*
pancreas gland Bauchspeicheldrüse *f*
panic Panik *f*
panic reaction Kurzschlußhandlung *f*
paper Papier *n*
paper bag Papiertüte *f*, Papiertragetasche *f*
paper coating Papierbeschichtung *f*
paper consumption Papierverbrauch *m*
paper cup Papierbecher *m*

paper filler Papierfüller *m*
paper intercepting (trapping) fence Papierfang *m*, Papierfangzaun *m*
paper mill sewage Papierfabrikabwasser *n*
paper packing Papierverpackung *f*, Verpackung *f* aus Papier
paper recyclate Papierrecyclat *n*
paper sack Papiersack *m*
paper trap Papierfang *m*
paraffin Paraffin *n*
paraformaldehyde Paraformaldehyd *m*
parallel experiment Parallelversuch *m*
paralyzer Katalysatorgift *n*, Kontaktgift *n*
parameter Parameter *m*
parameter-dependent control parameterabhängige Steuerung *f*
parameter entry mask Parametriermaske *f*
parameter of valuation Bewertungsmaßstab *m*
parasite Parasit *m*, Schmarotzer *m*
parasitic parasitisch, schmarotzend, schmarotzerhaft
paratonic wachstumshemmend, paratonisch
parc Park *m*
parenchyma Parenchym *n*, Grundgewebe *n*
partial combustion Teilverbrennung *f*, unvollständige Verbrennung *f*
partial flow of flue gas Rauchgasteilstrom *m*
partial load Teilladung *f*
partial pressure Partialdruck *m*
partial solubility Teillöslichkeit *f*
particle Partikel *n*, Teilchen *n*, Stückchen *n*

particle accelerator Teilchenbeschleuniger *m*
particle bombardment Teilchenbeschuß *m*
particle diameter Teilchendurchmesser *m*
particle emission Partikelemission *f*
particle filter Partikelfilter *n*
particle free sample flow partikelfreier Probenstrom *m*
particle residue Partikelrückstand *m*
particle shape Partikelform *f*, Teilchenform *f*
particulate emission Partikelemission *f*
particulate matter partikelförmiger Stoff *m*
parting agent Trennmittel *n*
partition coefficient Verteilungskoeffizient *m*
passive sampler Passivprobenehmer *m*, Diffusionssammler *m*
passive smoking Passivrauchen *n*
pasture Weide *f*
patch test Pflasterprobe *f*, Epikutantest *m*
pathobiological pathobiologisch
pathobiology Pathobiologie *f*
pathogen Krankheitserreger *m*, Pathogen *n*
pathogenic pathogen, krankheitserregend, krankmachend
pathological pathologisch, krankhaft
pathology Pathologie *f*, Krankheitslehre *f*
pathway of flue gas Rauchgasweg *m*
peak load Spitzenbelastung *f*, Höchstbelastung *f*

peat Torf *m*
peat bath Moorbad *n*
peat briquet Torfbrikett *n*
peat coal Lignit *m*
peat digging Torfgewinnung *f*
peat gas Torfgas *n*
peat plate Torfplatte *f*
peat water Torfwasser *n*
peaty soil Torferde *f*, Moorerde *f*
pebble Kies *m*, Kiesel *m*, Kieselstein *m*, Grobkies *m*
pebble filter Kiesfilter *n*
pedology Bodenkunde *f*
pellet Pellet *n*, Granulat *n*; Kügelchen *n*; Tablette *f*
pellet/to pelletieren; tablettieren
pelletizer Granuliermaschine *f*
peltry Pelzwerk *n*, Pelzwaren *fpl*
penetration depth Eindringtiefe *f*
penetrative effect Tiefenwirkung *f*
percentage purity Reinheitsgrad *m*
perchlorate Perchlorat *n*
percolate/to durchseihen, durchsickern, läutern
percolating amount Versickerungsmenge *f*
percolating area Versickerungsbereich *m*
percolating basin Sickerbecken *n*
percolating oil Sickeröl *n*
percolating range Sickerungsstrecke *f*
percolating water Sickerwasser *n*
percolating water of a refuse dump Deponiesickerwasser *n*
percolation Perkolation *f*, Durchsickern *n*, Filtration *f*
percolator Perkolator *m*, Filtertuch *n*, Filtersack *m*
perforated brick Lochziegel *m*
period Periode *f*

period of fine weather Gutwetterperiode *f*
period of half change Halbwertszeit *f*
period of poor weather Schlechtwetterperiode *f*
period of redevelopment Sanierungszeitraum *m*
period of retention Retentionszeit *f*
periodic periodisch
periodic system Periodensystem *n*
permanent immunity bleibende Immunität *f*
permanent load Dauerbelastung *f*, Dauerbeanspruchung *f*
permanent stress Dauerbeanspruchung *f*
permeability Permeabilität *f*, Durchlässigkeit *f*
permeable permeabel, durchlässig, durchdringbar
permeate/to durchdringen, durchsetzen
permeation Permeation *f*, Durchdringung *f*
permeation resistance Permeationswiderstand *m*
permeation tube Permeationsrohr *n*
permissible tolerance zulässige Abweichung *f*
permitted load level zulässige Belastung *f*
peroxide Peroxid *n*
persist/to beharren, fortbestehen, bleiben, fortdauern
persistence Beharrung *f*, Fortdauer *f*, Wirkungsdauer *f*, Persistenz *f*
persistent ausdauernd, beharrlich, persistent
personal monitoring device Personenüberwachungsgerät *n*

personal protection Personenschutz *m*
pest Schädling *m*; Ungeziefer *n*
pest control Schädlingsbekämpfung *f*
pesticidal schädlingsbekämpfend, pestizid
pesticide Pestizid *n*, Schädlingsbekämpfungsmittel *n*; Pflanzenschutzmittel *n*
pesticide analytics Pestizidanalytik *f*
pestilence Seuche *f*, Pest *f*
petrochemical Petrochemikalie *f*
petrochemical plant Petrochemieanlage *f*
petrochemistry Petrochemie *f*
petrol Benzin *n*, Kraftstoff *m*, Treibstoff *m*
petrol pipe Benzinleitung *f*
petrol separator Benzinabscheider *m*
petrol station Tankstelle *f*
petrol substitute Benzinersatz *m*
petrol trap Benzinabscheider *m*
petrol vapour disposal and recovery Benzindampfentsorgung *f* und -rückgewinnung
petrol vapour recovery plant Benzindampfrückgewinnungsanlage *f*
petroleum Petroleum *n*, Erdöl *n*, Mineralöl *n*
petroleum processing Erdölverarbeitung *f*
petroleum product Erdölprodukt *n*
petroleum refinery Erdölraffinerie *f*
pH determination pH-Bestimmung *f*
pH electrode pH-Elektrode *f*
pH value pH-Wert *m*

pharmaceutical chemistry Pharmakochemie *f*, Heilmittelchemie *f*
pharmaceutical industry Pharmaindustrie *f*
pharmaceutical product Pharmaprodukt *n*, pharmazeutisches Erzeugnis *n*
pharmacochemistry Pharmakochemie *f*
pharmacokinetical pharmakokinetisch
pharmacokinetics Pharmakokinetik *f*
pharmacological pharmakologisch
pharmacological research Arzneimittelforschung *f*
pharmacology Pharmakologie *f*
pharmacy Apotheke *f*; Arzneimittelkunde *f*, Pharmazie *f*
pharyngitis Rachenkatarrh *m*, Pharyngitis *f*
phenol Phenol *n*
phenol formaldehyde resin Phenolformaldehydharz *n*
phenylurethane Phenylurethan *n*
phon Phon *n* {Lautstärkeeinheit}
phonometer Phonmesser *m*, Schallmeßgerät *n*
phosphate Phosphat *n*
phosphate elimination Phophatelimination *f*, Phosphateliminierung *f*
phosphate elimination plant Phosphateliminierungsanlage *f*
phosphate fertilizer Phosphatdüngemittel *n*
phosphate load Phosphatbelastung *f*
phosphate precipitation Phosphatfällung *f*
phosphate/to phosphatieren
phosphorescence Phosphoreszenz *f*

phosphorus Phosphor *m*
photo-assimilatory oxygen production photoassimilatorische Sauerstoffproduktion *f*
photoabsorption Photoabsorption *f*
photoactive photosensitiv
photoassimilation Photoassimilation *f*
photocatalysis Photokatalyse *f*
photochemical photochemisch
photodegradation Photoabbau *m*
photolysis Photolyse *f*
photometer Photometer *n*, Belichtungsmesser *m*
photomorphogenesis Photomorphogenese *f*
photorespiration Photorespiration *f*
physical chemistry physikalische Chemie *f*
physical condition Gesundheitszustand *m*
physicochemical physikochemisch
physics Physik *f*
physiological response physiologische Reaktion *f*
physiology Physiologie *f*
physiotherapy Physiotherapie *f*, Heilgymnastik *f*, physikalische Therapie *f*
phytocenosis Phytozönose *f*
phytochemistry Phytochemie *f*, Pfanzenchemie *f*
phytohormone Phytohormon *n*, Pflanzenhormon *n*
phytology Pflanzenkunde *f*, Phytologie *f*
phytopathology Phytopathologie *f*
phytoparasite Pflanzenparasit *m*, Phytoparasit *m*
phytoplankton Phytoplankton *n*
phytoplasm Phytoplasma *n*

phytoprotein Pflanzenprotein n, Pflanzeneiweiß n
phytotoxic phytotoxisch, pflanzentoxisch
phytotoxin Phytotoxin n, Pflanzengift n
pick-up Meßwertaufnehmer m
piezo-resistant detector (sensor) piezo-resistiver Aufnehmer m
pig-and-scrap process Roheisen-Schrott-Verfahren n
pigment Pigment n, Farbstoff m
pigment bacterium Pigmentbakterie f
pigment cell Pigmentzelle f
pile coking Meilerverkokung f
pilot model Versuchsmodell n
pilot plant Pilotanlage f, Versuchsanlage f
pilot plant scale Großversuchsmaßstab m
pipe burst Rohrbruch m
pipe cleaning Rohrreinigung f
pipe renewal Rohrerneuerung f
pipeline network Rohrleitungsnetz n, Kanalnetz n
pipework system Rohrleitungssystem n
piping Rohrleitung f, Rohrnetz n, Röhrenwerk n
piston engine Kolbenmotor m
pit closure Zechenstillegung f
pit coal Steinkohle f, Grubenkohle f
pit fire Grubenbrand m
pit ventilation Grubenlüftung f
pitch Pech n
pitch-like pechartig
pitch oil Pechöl n
placenta Plazenta f, Mutterkuchen m
plague Pest f, Seuche f

plan of land utilization Flächennutzungsplan m
planetary agitator Planetenrührer m
planetary mill Planetenmühle f
planetary mixer Planetenmischer m
planetary stirrer Planetenrührer m
plankton Plankton n
plankton algae Planktonalgen fpl
plant Anlage f; Pflanze f
plant chemistry Pflanzenchemie f
plant disease Pflanzenkrankheit f
plant documentation Anlagendokumentation f
plant ecology Pflanzenökologie f
plant experiment Pflanzenversuch m
plant extract Pflanzenextrakt m
plant flow diagram Anlagenfließbild n
plant flow sheet Anlagenfließbild n
plant for nitrogen removal Entstickungsanlage f
plant management Anlagenmanagement n
plant mimic Anlagenfließbild n
plant nutrient Pflanzennährstoff m
plant pest Pflanzenschädling m
plant physiology Pflanzenphysiologie f
plant protection Pflanzenschutz m
plant protection agent Pflanzenschutzmittel n
plant protection law Pflanzenschutzgesetz n
plant protein pflanzliches Protein n
plant sector Anlagenabschnitt m
plant-specific data anlagenspezifische Daten pl
plant status Anlagenzustand m

plant survey Anlagenübersicht *f*
plant-tolerable pflanzenverträglich
plant wall Pflanzenwand *f*
plantation Bepflanzung *f*
plasmolysis Plasmolyse *f*, Zellschrumpfung *f*
plastic Plastik *n*, Kunststoff *m*
plastic bottle Kunststoffflasche *f*
plastic chips Kunststoffschnitzel *npl*
plastic-coated kunststoffbeschichtet
plastic coating Kunststoffbeschichtung *f*
plastic container Plastikbehälter *m*
plastic material Kunststoffmaterial *n*
plastic sludge Kunststoffschlamm *m*
plastic utilization plant Kunststoffverwertungsanlage *f*
plastic waste Kunststoffabfall *m*, Plastikabfall *m*
plastic waste recycling Plastikabfallwiederaufbereitung *f*, Plastikrecycling *n*
plasticizer Weichmacher *m*
plate column Bodenkolonne *f*
plate condenser Plattenkondensator *m*
plate evaporator Plattenverdampfer *m*
plate exchanger Plattenaustauscher *m*
plate heat exchanger Plattenwärmeaustauscher *m*, Plattenwärmetauscher *m*
plate number Bodenzahl *f*
plate scrap Blechschrott *m*
platinum Platin *n*

plough Pflug *m*
ploughed land Ackerland *n*
plough/to pflügen
plug valve Hahn *m*, Hahnventil *n*
plumbic bleihaltig
plumbous bleihaltig
plunge cell Tauchzelle *f*
plunger pump Tauchpumpe *f*
plutonium Plutonium *n*
plutonium breeder Plutoniumbrutreaktor *m*
pneumatic cleaning Luftwäsche *f*, Luftaufbereitung *f*
pneumatic conveyance pneumatische Förderung *f*
pneumatic drill Preßluftbohrer *m*
pneumatic hammer Preßlufthammer *m*
pneumatic pressure Luftdruck *m*
pneumatic pump Luftpumpe *f*
pneumococcosis Pneumokokkeninfektion *f*
pneumonia Lungenentzündung *f*, Pneumonie *f*
point of disturbance Störstelle *f*
pointshaped punktförmig
poison Gift *n*, Giftstoffe *mpl*
poison gas Giftgas *n*, Kampfgas *n*
poison/to vergiften
poisoned wheat Giftweizen *m*
poisoning Vergiftung *f*
poisoning of the catalysator Katalysatorvergiftung *f*
poisonous giftig, toxisch
poisonous action Giftwirkung *f*
poisonous effect Giftwirkung *f*
poisonous gas Giftgas *n*
poisonous matter Giftstoff *m*
polar region Polarregion *f*, Polarkreis *m*
polarimetry Polarimetrie *f*

polarograph Polarograph *f*
polarography Polarographie *f*
pollen Pollen *m*, Blütenstaub *m*
pollen analysis Pollenanalyse *f*
pollen catarrh Heuschnupfen *m*
pollination Bestäubung *f*
pollutant Schadstoff *m*, Verunreinigung *f*
pollutant-containing schadstoffhaltig
pollutant control Schadstoffüberwachung *f*
pollutant disposal Schadstoffbeseitigung *f*, Schadstoffentsorgung *f*
pollutant emission Schadstoffemission *f*
pollutant flow analysis Schadstofffließanalyse *f*
pollutant freight Schadstofffracht *f*
pollutant-impacted ground schadstoffbelastetes Erdreich *n*
pollutant migration Schadstoffwanderung *f*
pollute/to verunreinigen, beschmutzen, beflecken, besudeln, verschmutzen
polluted area Altlast *f*, Altlasten *fpl*
polluted area registration Altlasterfassung *f*
polluted verunreinigt, verschmutzt
polluter Umweltverschmutzer *m*
pollution Verschmutzung *f*, Verunreinigung *f*, Beschmutzung *f*, Beflekkung *f*
pollution control Schadstoffkontrolle *f*, Reinhaltung *f*; Umweltschutz *m*
pollution level Verschmutzungsgrad *m*

pollution unit of direct introducers Schadeinheit *f* von Direkteinleitern
poly-electrolyte Polyelektrolyt *m*
polychlorinated biphenyl polychloriertes Biphenyl *n*
polycyclic aromatic hydrocarbon polycyclischer aromatischer Kohlenwasserstoff *m*
polyelectrolyte Polyelektrolyt *m*
polyester Polyester *m*
polyhalogenated aromatic compound polyhalogenierte aromatische Verbindung *f*
polymer Polymer *n*
polymerization Polymerisation *f*
polymerize/to polymerisieren
polymorphic polymorph
polymorphism Polymorphismus *m*
polytrophic übermäßig nährstoffreich
polyurethane Polyurethan *n*
pond Teich *m*, Becken *n*, Weiher *m*, Tümpel *m*
pond aeration Teichbelüftung *f*
poor in waste abfallarm
population cycle Populationszyklus *m*
population density Bevölkerungsdichte *f*
population dynamics Populationsdynamik *f*
population ecology Populationsökologie *f*
population explosion Bevölkerungsexplosion *f*
population genetics Humangenetik *f*
population growth Bevölkerungswachstum *n*

population statistics Populationsstatistik f
pore Pore f
pore fluid Porenflüssigkeit f
pore size Porengröße f
pore space Porenraum m
poriferous porig, porös
porosity Porosität f, Durchlässigkeit f; Porenweite f
porous porös, porig; durchlässig; schwammig
porous concrete Schaumbeton m
portable extraction analyzer tragbarer Extraktionsanalysator m
portable fire extinguisher Handfeuerlöscher m, Handfeuerlöschgerät n
possibility of manual intervention Handeingriffsmöglichkeit f
post-exposure period Nachexpositionszeit f
post-mortem examination Untersuchung f nach dem Tode
post-term birth Spätgeburt f
postnatal postnatal, nach der Geburt
postnatal mortality Sterblichkeit f nach der Geburt
pot experiment Gefäßversuch m, Topfversuch m
potable trinkbar
potable water Trinkwasser n
potassium Kalium n
potential danger potentielle Gefahr f, potentielles Risiko n
potentiometer Potentiometer n
potentiometry Potentiometrie f
poudrette Fäkaldünger m, Mistpulver n
pour point Stockpunkt m
pourability Gießbarkeit f

pourable gießbar
powder fire extinguisher Trockenfeuerlöscher m, Pulverlöscher m
powdered coal Pulverkohle f
powdered manure Düngepulver n
powdered milk Milchpulver n
powdered stone Gesteinsmehl n
power balance Energiebilanz f
power consumer Stromverbraucher m
power consumption Energieverbrauch m
power engineering equipment energietechnische Ausrüstung f
power-heat combination Kraft-Wärme-Kopplung f
practice of approval Genehmigungspraxis f
practise alarm Probealarm m
praseodym Praseodym n
pre-thickening of sludge Schlammvoreindickung f
precast concrete Fertigbeton m
precipitate Präzipitat n, Ausfällung f, Niederschlag m, Bodensatz m, Fällungsprodukt n
precipitate/to ausfällen, fällen, abscheiden, niederschlagen, präzipitieren
precipitating agent Fällungschemikalie f
precipitation Fällung f, Ausfällung f, Ausfällen n; Niederschlag m
precipitation electrode Niederschlagselektrode f
precipitation product Fällungsprodukt n
precipitation tank Absetzbecken n, Absetzbehälter m, Klärtank m
precise balance Analysenwaage f

precision adjusting valve Feinregulierventil *n*
preclarification Vorklärung *f*
preclarifier Vorfluter *m*
  preclarifier basin Vorklärbecken *n*
preclarify/to vorklären
precleaner Vorreiniger *m*
preclinical sign of poisoning vorklinische Vergiftungserscheinung *f*
precoat Anschwemmschicht *f*
  precoat filter Anschwemmfilter *n*
precombustion Vorverbrennung *f*
preconditioning Vorbehandeln *n*
precool/to vorkühlen
precrushing Vorzerkleinerung *f*
  precrushing plant Vorbrecheranlage *f*
precursor Vorläufer *m*
predryer Vortrockner *m*
predrying Vortrocknung *f*
prefilter Vorfilter *n*
prefuse/to vorschmelzen
pregnancy Schwangerschaft *f*
  pregnancy test Schwangerschaftstest *m*
pregnant schwanger
preheat/to vorheizen
preliminary cleaning Vorreinigung *f*
  preliminary drying Vortrocknung *f*
  preliminary experiment Vorversuch *m*
  preliminary filtering Vorfilterung *f*
  preliminary treatment Vorbehandlung *f*
premature birth Frühgeburt *f*
premix/to vormischen
prenatal pränatal, vor der Geburt
preoperative präoperativ
preozonation Vorozonisierung *f*
prepacked abgepackt, vorgepackt

preparation method Aufbereitungsmethode *f*
preparation of waste water Abwasseraufbereitung *f*
preparation plant Aufbereitungsanlage *f*
prepurified sewage vorgereinigtes Abwasser *n*
prepurified waste water vorgereinigtes Abwasser *n*
prepurify/to vorreinigen
prerot Vorrotte *f*
  prerot area Vorrottefläche *f*
  prerot drum Vorrottetrommel *f*
  prerot unit Vorrotte-Einheit *f*
prerot/to vorrotten
prerotting Vorrottung *f*
preservation Aufbewahrung *f*; Erhaltung *f*; Haltbarmachung *f*, Konservierung *f*
  preservation of food Lebensmittelkonservierung *f*
  preservation of species Arterhaltung *f*, Artenschutz *m*
  preservation of the oceans Erhaltung *f* der Meere
preservative Konservierungsmittel *n*; Schutzmittel *n*
press for rake good Rechengutpresse *f*
pressed cake Preßkuchen *m*
pressure aeration system Druckbelüftungssystem *n*
  pressure bottle Druckflasche *f*
  pressure controller Druckregler *m*
  pressure difference Druckdifferenz *f*
  pressure dispenser Druckzerstäuber *m*
  pressure drop Druckabfall *m*

pressure equalization Druckausgleich *m*
pressure filter plant Druckfilteranlage *f*
pressure filtration Druckfiltration *f*
pressure probe Drucksonde *f*
pressure-proportional level druckproportionaler Höhenstand *m*
pressure release flotation Druckentspannungsflotation *f*
pressure transducer Druckmeßumformer *m*
pressure vessel Druckbehälter *m*
prestressed concrete Spannbeton *m*
pretreatment Vorbehandlung *f*
pretreat/to vorbehandeln
prevention of accidents Unfallverhütung *f*
prevention of air pollution Luftreinhaltung *f*
prevention of water pollution Gewässerschutz *m*
preventive präventiv
preventive action strategy Vorbeugungsstrategie *f*
preventive medical checkup Vorsorgeuntersuchung *f*
primary crusher Grobbrecher *m*, Grobzerkleinerer *m*
primary energy from waste Primärenergie *f* aus Müll
primary filter Vorfilter *n*
primary industry Grundstoffindustrie *f*
primary sludge Primärschlamm *m*
primeval forest Urwald *m*
printing ink Druckfarbe *f*
priority of disturbances Störungspriorität *f*
proband Proband *m*, Versuchsperson *f*
probe Meßkopf *m*, Sonde *f*

problem of old-sites Altlastenproblematik *f*
problem waste Problemabfall *m*
process analysis Prozeßanalyse *f*, Verfahrensanalyse *f*
process automation Prozeßautomatisierung *f*
process automation task Prozeßautomatisierungsaufgabe *f*
process control Prozeßführung *f*
process control system Prozeßleitsystem *n*
process data Prozeßdaten pl
process data acquisition Prozeßdatenerfassung *f*
process data communication Prozeßdatenübermittlung *f*
process disturbance Prozeßstörung *f*
process element Prozeßelement *n*
process engineering Verfahrenstechnik *f*, Prozeßtechnik *f*
process flow sheet Prozeßfließbild *n*
process fuel Prozeßbrennstoff *m*
process-integrated prozeßintegriert, verfahrensintegriert
process line Prozeßlinie *f*
process malfunction Prozeßstörung *f*
process management Prozeßführung *f*
process observation Prozeßbeobachtung *f*
process of combustion Verbrennungsvorgang *m*
process of decomposition Zersetzungsvorgang *m*
process overview Prozeßüberblick *m*

process signal Betriebsmeldung *f*
process stability Prozeßstabilität *f*
process technology Prozeßtechnologie *f*
process visualization Prozeßvisualisierung *f*
process water Prozeßwasser *n*
processed slag aufbereitete Schlacke *f*
processibility Verarbeitbarkeit *f*
processing Verarbeitung *f*
  processing centre Behandlungszentrum *n*
  processing centre for recyclable solid waste materials Aufarbeitungszentrum *n* für wiederverwertbare feste Abfallmaterialien
  processing method Aufbereitungsverfahren *n*
  processing of an old site Altlastaufarbeitung *f*, Aufarbeitung *f* von Altlasten
  processing of batch data Chargendatenverwaltung *f*
  processing of liquid manure Gülleaufbereitung *f*
  processing of used (spent) oil Altölaufbereitung *f*
  processing of waste Abfallaufbereitung *f*
  processing plant Behandlungsanlage *f*, Aufbereitungsanlage *f*
procuring right Beschaffungsrecht *n*
producer of hazardous waste Sonderabfallerzeuger *m*
product safety Produktsicherheit *f*
  product-specific remaining material produktspezifischer Reststoff *m*
production plant Produktionsanlage *f*
production specific industrial residue produktionsspezifischer Industrierückstand *m*
profile analysis Profilanalyse *f*
progesterone Progesteron *n*, Gelbkörperhormon *n*
prognosis result Prognoseergebnis *n*
prognosis system Prognosesystem *n*
programme for the purification of exhaust air Abluftreinigungsprogramm *n*
project execution Projektabwicklung *f*
proliferate/to proliferieren, wuchern, sich stark vermehren
proliferation Proliferation *f*, Wachstum *n*, Wucherung *f*
promethium Promethium *n*
proof-decree for waste Abfall-Nachweisverordnung *f*
proof-regulation for waste Abfall-Nachweisverordnung *f*
propagation of pollutant Schadstoffausbreitung *f*
propellant Treibmittel *n*, Treibgas *n*
propellant gas Treibgas *n*
propeller mixer Propellermischer *m*, Propellerrührer *m*
propeller pump Rotationspumpe *f*
prophylactic prophylaktisch, vorbeugend; Prophylaktikum *n*, Abwehrmittel *n*
prophylaxis Prophylaxe *f*, Vorbeugung *f*
prosthesis Prothese *f*, Ersatz *m*
protactinium Protactinium *n*
  protecting cover Schutzhaube *f*
  protecting mask Schutzmaske *f*
protection Schutz *m*; Absicherung *f*, Abschirmung *f*

protection coating Schutzbeschichtung *f*
protection layer Schutzschicht *f*
protection of groundwater Grundwasserschutz *m*
protection of resources Ressourcenschonung *f*
protection of species Artenschutz *m*
protection of the countryside Landschaftsschutz *m*
protection of prehensile birds Greifvogelschutz *m*
protection regulation Schutzbestimmung *f*
protective schützend
protective bandage Schutzverband *m*
protective clothes Schutzkleidung *f*
protective clothing Schutzkleidung *f*
protective coating Schutzanstrich *m*
protective effect Schutzwirkung *f*
protective equipment Schutzausrüstung *f*
protective insulation Schutzisolierung *f*
protective jacket Schutzmantel *m*
protective requirement Schutzanforderung *f*
protein analysis Proteinanalyse *f*
protein complex Proteinkomplex *m*
protein deficiency Proteinmangel *m*, Eiweißmangel *m*
protein degradation Eiweißabbau *m*
protein diet Eiweißdiät *f*
proteolysis Proteolyse *f*, Eiweißabbau *m*
proton Proton *n*, Wasserstoffkern *m*
proton accelerator Protonenbeschleuniger *m*
proton bombardment Protonenbeschuß *m*
protoplasm Protoplasma *n*
provisional measure Übergangsmaßnahme *f*
provisional solution Übergangslösung *f*
proximate analysis Schnellanalyse *f*, Kurzanalyse *f*
pseudocroup Pseudokrupp *m*
psycho-drug Psychopharmakon *n*
psychochemical Psychopharmakon *n*
psychomotor test psychomotorischer Test *m*
psychosomatic psychosomatisch
public data network öffentliches Datennetz *n*
public health service Gesundheitswesen *n*
public mains öffentliches Stromnetz *n*
pulmonary emphysema Lungenemphysem *n*
pulmonary function test Lungenfunktionsprüfung *f*
pulmonary oedema Lungenödem *n*
pumice Bims[stein] *m*
pumis concrete Bimsbeton *m*
pump shaft Pumpenschacht *m*
pump sump Pumpensumpf *m*
pure culture Reinkultur *f*
pure water Reinwasser *n*
purge device Spülvorrichtung *f*
purge system Spülsystem *n*
purge/to reinigen, säubern; spülen; klären, läutern
purging programme Spülprogramm *n*
purification Reinigung *f*

purification plant Kläranlage *f*
purification process Reinigungsprozeß *m*, Reinigungsvorgang *m*
purification stage Reinigungsstufe *f*
purifier Reiniger *m*, Reinigungsmittel *n*; Reinigungsapparat *m*
purify/to reinigen, säubern
purity Reinheit *f*, Echtheit *f*
putrefaction Fäulnis *f*, Faulen *n*, Verfall *m*, Verwesung *f*
putrefaction bacterium Fäulnisbakterie *f*
putrefactive fäulnisserregend
putrefactive bacteria Fäulnisbakterien *fpl*
putrefy/to verfaulen, faulen, verwesen
putrescible substance verrottbarer Stoff *m*
pyretic fiebererzeugend
pyrexia Fieberzustand *m*
pyrolysis Pyrolyse *f*
pyrolysis gas Pyrolysegas *n*
pyrolysis incineration plant Pyrolyseverbrennungsanlage *f*
pyrolysis method Pyrolyseverfahren *n*
pyrolysis reactor Pyrolysereaktor *m*
pyrolytic pyrolytisch
pyrolyze/to pyrolisieren
pyrometer Pyrometer *n*, Hitzemesser *m*
pyrophoric pyrophor, luftentzündlich

Q

quagmire Sumpf *m*, Morast *m*
qualitative analysis qualitative Analyse *f*
qualitative prognosis method qualitatives Prognoseverfahren *n*
quality of groundwater Grundwasserqualität *f*
quality of waters Gewässergüte *f*
quantitative analysis quantitative Analyse *f*
quantitative yield quantitative Ausbeute *f*
quantity minimization Betragsminimierung *f*
quantity to be measured zu bestimmende Größe *f*, Analyt *m*
quarantine Quarantäne *f*
quarrying Tagebauförderung *f*, Tagebaubetrieb *m*
quartz Quarz *m*
quartz glass Quarzglas *n*
quasi ecological convention quasi-ökologische Konvention *f*
quenched charcoal Löschkohle *f*
quenching Löschen *n*; Abschrecken *n*, Ablöschen *n*
quenching agent Ablöschmittel *n*, Abschreckmittel *n*
quick ash Flugasche *f*
quick assay Schnelltest *m*
quick freezing Schnellgefrieren *n*
quick setting cement Schnellbinder *m*, schnellbindender Zement *m*
quick test Vorprobe *f*
quicklime gebrannter Kalk *m*
quicksand Schwimmsand *m*, Treibsand *m*
quickset hedge [lebende] Hecke *f*

quickset noise protection wall lebende Lärmschutzwand f

R

rabid tollwütig
rabies Tollwut f
racemate Razemat n
racemic razemisch
radiant strahlend
  radiant flux density Strahlungsdichte f
  radiant heat Strahlungswärme f
  radiant intensity Strahlungsintensität f
radiate/to ausstrahlen, strahlen, Strahlen aussenden, leuchten
radiation Strahlung f, Ausstrahlung f, Abstrahlung f, Strahlenemission f
  radiation behaviour Strahlungsverhalten n
  radiation belt Strahlungsgürtel m
  radiation biology Strahlenbiologie f
  radiation chemistry Strahlenchemie f
  radiation genetics Strahlengenetik f
radiative strahlend, Strahlungs-
radiator lining Heizkörperverkleidung f
radical Radikal n
  radical scavenger Radikalfänger m
radioactive radioaktiv
  radioactive chemical radioaktive Chemikalie f
  radioactive constant Zerfallskonstante f
  radioactive contamination radioaktive Verseuchung f
  radioactive decay radioaktiver Zerfall m
  radioactive decontamination radioaktive Entseuchung f
  radioactive waste radioaktiver Abfall m
  radioactive waste disposal radioaktive Abfallbeseitigung f
  radioactive waste water radioaktives Abwasser n
radioactivity Radioaktivität f
radiobiological radiobiologisch
radiobiology Radiobiologie f
radioecology Radioökologie f
radiogram Röntgenbild n
radiolabelled compound radiomarkierte Verbindung f
radiological radiologisch, röntgenologisch
  radiological result Röntgenbefund m
radiology Radiologie f, Röntgenkunde f, Strahlenkunde f
radiolucent strahlendurchlässig
radionuclide Radionuklid n
radiopaque strahlenundurchlässig
radioprobe Radiosonde f
radiosensitive strahlenempfindlich
radiotherapy Strahlentherapie f, Röntgentherapie f
radiotoxicology Radiotoxikologie f
radium Radium n
  radium-contaminated radiumverseucht
  radium therapy Radiumtherapie f
radon Radon n
  radon level Radonpegel m
  radon spring Radonquelle f
rain Regen m

rainfall 110

rain basin Regenbecken *n*
rain forest Regenwald *m*
rain gauge Regenmesser *m*
rain spillway basin (reservoir) Regenüberlaufbecken *n*
rain storage reservoir Regenrückhaltebecken *n*
rain water Regenwasser *n*
rainfall Regen *m*, Niederschlag *m*, Schauer *m*
rainfall basin Regenbecken *n*
rainfall level Regenhöhe *f*
rainfall measurement Regenmessung *f*
rainfall rate Niederschlagsmenge *f*, Regenhöhe *f*
rainproof regenfest
raintight regendicht
rainwater canal Regenwasserkanal *m*
rainwater overflow basin Regenüberlaufbecken *n*
rainy season Regenzeit *f*
rake Rechen *m*, Harke *f*
rake good Rechengut *n*
rake grid Rechengitter *n*
rake material Rechengut *n*, Rechenmaterial *n*
rake material press Rechengutpresse *f*
rake plant Rechenanlage *f*
rake screenings Rechengut *n*
rake system Rechensystem *n*
random error Zufallsfehler *m*
random fluctuation zufällige Schwankung *f*
random sample Stichprobe *f*
random series Zufallsfolge *f*
random test Stichprobentest *m*
range of measurement Meßbereich *m*

rapid analysis Schnellanalyse *f*
rapid charge Schnellaufladung *f*
rapid cooling system Schnellkühlsystem *n*
rapid filtration plant Schnellfilteranlage *f*
rapid method Schnellverfahren *n*
rapid testing method Schnelltestverfahren *n*
rare earth metal Seltenerdmetall *n*
rare gas Edelgas *n*
rat Ratte *f*
rat poison Rattengift *n*
rat test Rattentest *m*, Versuch *m* mit Ratten
rate of discharge Ausströmgeschwindigkeit *f*, Entladegeschwindigkeit *f*
rate of heat transfer Wärmeübergangsrate *f*
rate of mineralization Mineralisationsrate *f*
rating Auslegung *f*
ratty rattenverseucht; rattenartig
ravage Verwüstung *f*
raw density Rohdichte *f*
raw gas Rohgas *n*
raw material Rohstoff *m*, Ausgangsmaterial *n*
raw material recovery Rohstoffrückgewinnung *f*, Rohstoffwiedergewinnung *f*
raw material shortage Rohstoffmangel *m*
raw sewage Rohabwasser *n*
raw water Brauchwasser *n*
raw water resource Rohwasserressource *f*
ray filter Strahlenfilter *n*
ray quantum Strahlenquant *m*
reabsorb/to resorbieren

reabsorbable resorbierbar
reabsorption Resorption *f*
react/to reagieren
reactant Reaktant *m*, Reaktionspartner *m*, Reaktionsteilnehmer *m*
reaction Reaktion *f*; Einwirkung *f*; Gegenwirkung *f*, Rückwirkung *f*; Umwandlung *f*
reaction air Reaktionsluft *f*
reaction chemical Reaktionschemikalie *f*
reaction course Reaktionsverlauf *m*
reaction gas Reaktionsgas *n*
reaction product Reaktionsprodukt *n*, Umwandlungsprodukt *n*
reaction rate Reaktionsrate *f*, Umsatzrate *f*, Umsatzgeschwindigkeit *f*
reactivate/to reaktivieren
reactivation Reaktivierung *f*, Reaktivieren *n*, Wiederbelebung *f*
reactive reaktiv, rückwirkend, gegenwirkend
reactivity Reaktivität *f*
reactor Reaktor *m*
reactor control Reaktorsteuerung *f*
reactor core Reaktorkern *m*
reactor engineering Reaktortechnik *f*
reactor volume Reaktorvolumen *n*
readiness for announcement of the population Anzeigebereitschaft *f* der Bevölkerung
reading error Ablesefehler *m*
reading uncertainty Ablesefehler *m*
readjust/to nachregeln, wieder anpassen, korrigieren
readjustment Wiederanpassung *f*, Neuordnung *f*, Korrektur *f*
ready-made compost Fertigkompost *m*

readyness for operation Funktionsbereitschaft *f*
reafforest/to wiederaufforsten
reafforestation Wiederaufforstung *f*
reagent Reagens *n*, chemisches Nachweismittel *n*, Nachweisreagens *n*
reagent solution Reagenslösung *f*
real-time behaviour Realzeitverhalten *n*
real-time data collection Echtzeitdatenerfassung *f*
reanimation Wiederbelebung *f*
recap/to runderneuern
recast/to umschmelzen, umgießen
recasting Wiedereinschmelzen *n*
receiver basin Vorlagebecken *n*
recessive rezessiv, zurückgehend
recharge/to wieder aufladen; wieder beschicken; wiederbeladen
recipe system Rezeptsystem *n*
recirculation Rezirkulation *f*, Rückleitung *f*
recirculation pump Umwälzpumpe *f*
reclaimed rubber Regeneratgummi *n*
reclaiming Regeneration *f*, Wiedergewinnung *f*, Rückgewinnung *f*
recombination Rekombination *f*, Wiedervereinigung *f*
recombination reaction Rekombinationsreaktion *f*
recombine/to rekombinieren, wiedervereinigen
recondition/to rekonditionieren, wieder instandsetzen
reconditioning Wiederherstellung *f*, Instandsetzung *f*
reconstruct/to umbauen; rekonstru-

reconstruction 112

ieren, wieder aufbauen, wieder herstellen
reconstruction Wiederaufbau *m*, Wiederherstellung *f*; Umbau *m*
reconvert/to zurückverwandeln, zurückbilden; sich zurückverwandeln
recording instrument Registriergerät *n*, Aufzeichnungsgerät *n*
recording paper Registrierpapier *n*
recoverable wiedererlangbar; wiederherstellbar; regenerierbar, zurückgewinnbar, wiedergewinnbar
recoverable waste Wertstoff *m*
recovered acid Abfallsäure *f*
recover/to wiedergewinnen, zurükkerhalten, [zu]rückgewinnen; genesen, gesunden, bergen
recovery Gewinnung *f*, Rückgewinnung *f*, Wiedergewinnung *f*; Bergung *f*; Erholung *f*, Genesung *f*, Gesundung *f*
recovery of chemicals Chemikalienrückgewinnung *f*
recovery of heat Wärmerückgewinnung *f*
recovery of soil Wiedergewinnung *f* von Boden
recovery of solvents Lösemittelrückgewinnung *f*
recovery process Rückgewinnungsverfahren *n*
recrystallization Rekristallisation *f*
rectification Rektifikation *f*, Rektifizierung *f*
rectifier Destillationsapparat *m*, Rektifaktor *m*; Gleichrichter *m*; Rektifizierapparat *m*
recultivation Rekultivierung *f*
recultivation layer Rekultivierungsschicht *f*

recyclability Rückführbarkeit *f*, Wiederverwertbarkeit *f*
recyclable rückführbar, wiederverwendbar; wiederaufarbeitungsfähig
recyclable waste wiederverwertbarer Abfall *m*
recyclate Recyclat *n*, Wiedergewinnungsprodukt *n*, Wertstoff *m*
recyclate method Wertstoffverfahren *n*
recycled paper Umweltschutzpapier *n*
recycle/to rückführen, zurückführen, wiederverwerten, wiederverwenden, recyclieren
recycling Rückgewinnung *f*, Rückführung *f*, Rezyklierung *f*, Wertstoffrückgewinnung *f*, Recycling *n*, Wiederaufarbeitung *f*, Zurückgewinnung *f*, Abfallverwertung *f*
recycling method Wiederverwertungsverfahren *n*
recycling model Recyclingmodell *n*
recycling of debris Wiederverwertung *f* von Bauschutt
recycling of recyclates of plastics Wertstoffrecycling *n* von Kunststoffen
recycling paper Umweltschutzpapier *n*
recycling plant Abfallverwertungsanlage *f*, Rückgewinnungsanlage *f*, Recyclinganlage *f*
recycling priority Verwertungsvorrang *m*
recycling programme Abfallverwertungsprogramm *n*
recycling rate Recyclingrate *f*
recycling residue Recyclingrückstand *m*

recycling technique Recyclingverfahren *n*
red data book Rote Liste *f*
redevelop/to sanieren
redevelopment Sanierung *f*
redevelopment costs Sanierungskosten *pl*
redevelopment instruction Sanierungsvorschrift *f*
redevelopment of old-sites Altlastensanierung *f*
redox reaction Redoxreaktion *f*
reducing agent Reduktionsmittel *n*
reef Klippe *f*, Riff *n*
reference input Sollwert *m*
reference level Vergleichspegel *m*
refill Nachfüllpackung *f*
refill/to wiederauffüllen, nachfüllen, auffüllen
refillable nachfüllbar
refilling Nachfüllung *f*
refine/to raffinieren; aufbereiten, frischen; läutern
refined raffiniert; aufbereitet, gefrischt; geläutert
refinery Raffinerie *f*, Raffinationsanlage *f*; Frischerei *f*
refinery sewage Raffinerieabwasser *n*
refinery waste Raffinationsabfall *m*
refining Raffinierung *f*; Frischen *n*; Läuterung *f*, Reinigen *n*
reflux Rückfluß *m*, Rücklauf *m*, Zurückfließen *n*, Zurückfluten *n*
reflux condenser Rückflußkühler *m*
reforestation Wiederaufforstung *f*
refraction Brechung *f*, Brechungsvermögen *n*, Refraktion *f*
refraction index Brechungsindex *m*
refractive lichtbrechend

refractometer Brechungsmesser *m*, Refraktometer *n*
refractory lining feuerfeste Auskleidung *f*
refrigerant Kühlmittel *n*, Kältemittel *n*
refrigerate/to kühlen
refrigerating Kühlen *n*, Kühlung *f*, Kälteerzeugung *f*
refrigerating agent Kühlmittel *n*
refrigerating aggregate Kühlmaschine *f*, Kühlaggregat *n*
refrigerating chamber Kühlraum *m*
refrigerating engine Kühlmaschine *f*
refrigerating machine Kühlmaschine *f*
refrigerator Kühlschrank *m*
refrigerator truck Kühlwagen *m*
refuge Refugium *n*
refuse Abfall *m*, Abraum *m*, Müll *m*
refuse bin Müllbehälter *m*, Mülltonne *f*
refuse collecting plant Abfallsammelanlage *f*
refuse collection lorry Müllsammelfahrzeug *n*
refuse collection Müllabfuhr *f*
refuse composition Müllzusammensetzung *f*
refuse compactor Müllverdichter *m*
refuse container Müllbehälter *m*, Müllcontainer *m*
refuse container lid with integrated handle Müllcontainerdeckel *m* mit integriertem Griff
refuse depot Abfalldepot *n*
refuse density Mülldichte *f*
refuse sack Müllsack *m*

refuse transfer station Müllumladestation *f*, Müllumschlagstation *f*
refuse vessel Müllbehälter *m*
regain/to wiedergewinnen
regaining Rückgewinnung *f*, Wiedergewinnung *f*
regenerate Regenerat *n*
regenerate/to neu schaffen, neu beleben, neu bilden, umgestalten, erneuern; regenerieren, wiedergewinnen
regenerated material Regenerat *n*
regeneration Regenerierung *f*, Regeneration *f*, Rückgewinnung *f*; Reformierung *f*, Neuschaffung *f*, Neubildung *f*, Wiederherstellung *f*
regeneration station Regenerierungsstation *f*
regeneration time (period) Regenerationszeit *f*
regenerative regenerativ
regenerative energy regenerative Energie *f*
regenerative heat recovery regenerative Wärmerückgewinnung *f*
regenerative plant Rückgewinnungsanlage *f*
regional environmental committee regionaler Umweltausschuß *m*
register Kataster *n*
register of hazardous substances Gefahrstoffkataster *m*
registration system for hazardous substances Gefahrstoff-Registriersystem *n*
regresssion Regression *f*, Rückbildung *f*, Rückentwicklung *f*
regulation for large-scale incineration plants Großfeuerungsanlagenverordnung *f*
regulation for limit values Grenzwertregelung *f*

regulation for the disposal Rahmenbestimmung *f* für die Entsorgung, Verordnung *f* für die Entsorgung
regulation for the transportation of hazardous goods Gefahrgutvorschrift *f*
regulation in case of disturbance (accident) Störfallverordnung *f*
regulation of quantity Mengenregelung *f*
rehabilitation of the waters Gewässersanierung *f*
reimportation Wiedereinfuhr *f*
reinforced concrete Stahlbeton *m*
reject Ausschuß *m*
relative humidity relative Luftfeuchte *f*
relay and contactor technique Relais- und Schütztechnik *f*
relay technique Relaistechnik *f*
release of a pollutant Schadstoffaustritt *m*
release prohibition Freisetzungsverbot *n*
relict Relikt *n*
reloading plant Umladeanlage *f*
remainder Rest *m*, Rückstand *m*, Überrest *m*
remaining material (matter, substance) Reststoff *m*
remedial Sanierung *f*
remedial action Sanierungsaktion *f*
remedial of old disposal sites Sanierung *f* von Altdeponien
remedial of old sites Altlastensanierung *f*
remedial of the waters Gewässersanierung *f*
remedial with long-term effect Sanierung *f* mit Langzeiteffekt

remediation Sanierung *f*
remediation method Sanierungsmethode *f*
remedy Arznei *f*, Heilmittel *n*; Gegenmittel *n*
remelt/to umschmelzen, wieder einschmelzen
remote action Fernwirkung *f*
remote control Fernbedienung *f*, Fernlenkung *f*
remote effect Fernwirkung *f*
remote monitoring Fernüberwachung *f*
removal of iron Enteisenung *f*
removal of manganese Entmanganung *f*
renaturation Renaturierung *f*
renovate/to renovieren, restaurieren
renovation phase Renovierungsphase *f*
reparation Wiedergutmachung *f*, Wiederherstellung *f*; Reparatur *f*
repeated determination Doppelbestimmung *f*
repeated test Doppelbestimmung *f*
replacement reaction Verdrängungsreaktion *f*, Substitutionsreaktion *f*
replacement surgery Ersatzchirurgie *f*
replant/to neu pflanzen, verpflanzen, umpflanzen
replantation Wiedereinpflanzung *f*, Reimplantation *f*
replicate test Wiederholungsbestimmung *f*, Mehrfachbestimmung *f*
repopulation Wiederbesiedlung *f*
report printer Berichtsdrucker *m*
repository condition Endlagerungsbedingung *f*
representative for landscape protection Naturschutzbeauftragter *m*

reprocess/to wiederaufarbeiten
reprocessing plant Wiederaufarbeitungsanlage *f*
reproducibility Reproduzierbarkeit *f*
reproduction Reproduktion *f*, Fortpflanzung *f*
reproduction rate Reproduktionsrate *f*
repurchase obligation Rücknahmepflicht *f*, Rücknahmeverpflichtung *f*
rescue Rettung *f*, Bergung *f*
rescue party Rettungsmannschaft *f*, Bergungstruppe *f*
reservation Reservat *n*, Naturschutzgebiet *n*
reservoir Reservoir *n*, Vorrat *m*; Speicher *m*, Sammelbecken *n*, Staubecken *n*
residence time Verweilzeit *f*
residual activity Restaktivität *f*
residual amount Restmenge *f*
residual material Reststoff *m*
residual material from power plants Kraftwerkreststoff *m*
residual matter Reststoff *m*
residual moisture Restfeuchte *f*
residual product (substance) Reststoff *m*
residual waste Restmüll *m*
residual waste substance Abfallreststoff *m*
residue Rückstand *m*, Rest *m*, Reststoff *m*
residue analysis Rückstandsanalyse *f*
residue determination Rückstandsbestimmung *f*
residue recycling Rückstandsrecycling *n*
residues of plant protection agents Pflanzenschutzmittelrückstände *mpl*

resin Harz *n*
resistance thermometer Widerstandsthermometer *n*
resistance to abrasion Abriebfestigkeit *f*
resistance to acid Säurebeständigkeit *f*
resistance to light Lichtbeständigkeit *f*
resistant to chemicals chemikalienbeständig
resistant to fracture bruchfest
resistant to frost frostbeständig
resistant to weathering wetterfest, wetterbeständig
resolve/to auflösen, trennen, zerlegen
resorption Resorption *f*
resource Ressource *f*, Hilfsquelle *f*
resource exploitation Ressourcenausnutzung *f*
resource inventory Bestandsaufnahme *f*
respirable dust lungengängiger Staub *m*
respirator Respirator *m*, Beatmungsgerät *n*
respiratory respiratorisch; Atem-, Atmungs-
respiratory disease Erkrankung *f* der Atemwege
respiratory function test Atemfunktionsprüfung *f*
respiratory poison Atemgift *n*
respiratory protection apparatus Atemschutzgerät *n*
respiratory tract Atemweg *m*, Atemtrakt *m*
response Antwort *f*, Response *f*
response area betroffenes Gebiet *n*

response time behaviour Antwortzeitverhalten *n*
responsibility of approval Genehmigungszuständigkeit *f*
restorable wiederherstellbar
restoration Restaurierung *f*, Rekonstruktion *f*, Wiederherstellung *f*
restoration of health Genesung *f*
restore/to restaurieren, erneuern, wiederherstellen, instandsetzen
restricted combustion unvollständige Verbrennung *f*
retardant Verzögerungsmittel *n*
retardation Verzögerung *f*, Verlangsamung *f*, Entwicklungshemmung *f*
retention Beibehaltung *f*, Zurückhaltung *f*
retention factor Rf-Wert *m*, Verzögerungsfaktor *m*, Rückhaltefaktor *m*
retention time Verweilzeit *f*
retort graphite Retortengraphit *m*
retread/to runderneuern
retreat/to nachbehandeln, nochmals behandeln
retreatment Nachbehandlung *f*
return condenser Rückflußkühler *m*
return of sludge Schlammrückführung *f*
return sludge Rücklaufschlamm *m*
returnable bottle Mehrwegflasche *f*
returnable packing Mehrwegverpackung *f*
returnable plastic bottle Kunststoffpfandflasche *f*
reusable wiederverwertbar
reuse Wiederverwendung *f*, Wiederverwertung *f*
reuse system Wiederverwendungssystem *n*
reuse/to wiederverwenden, wiederverwerten

reutilizable wiederverwertbar
reutilizable substance wiederverwertbarer Stoff *m*
reutilization Wiederverwendung *f*
reutilization system Wiederverwendungssystem *n*
reverse osmosis Umkehrosmose *f*
reverse reaction Rückreaktion *f*, Gegenreaktion *f*
reversed osmosis Umkehrosmose *f*
reversibility Reversibilität *f*, Umkehrbarkeit *f*
reversible umkehrbar, reversibel
revision shaft Revisionsschacht *m*
revitalization Revitalisierung *f*, Wiederbelebung *f*
revival Wiederbelebung *f*, Sanierung *f*, Wiederaufgreifen *n*, Wiederaufleben *n*
revive/to wiederbeleben, wiederherstellen
revivify/to regenerieren, wiederbeleben, reaktivieren
rewash/to nachwaschen, nachwässern
rework/to umarbeiten
rhenium Rhenium *n*
rheological rheologisch
rheology Rheologie *f*, Strömungslehre *f*
rhinitis Rhinitis *f*, Katarrh *m*, Schnupfen *m*
rhodium Rhodium *n*
rich coal Fettkohle *f*
rich gas Reichgas *n*
right for provision Beschaffungsrecht *n*
right of approval (permission) Genehmigungsrecht *n*
rinse/to abspülen, spülen, waschen, ausspülen

rinsing Abspülen *n*, Spülen *n*, Spülung *f*
rinsing liquid Spülflüssigkeit *f*
rising generation Nachwuchsgeneration *f*
risk analysis Gefährdungsanalyse *f*, Risikoanalyse *f*
risk assessment Risikoabschätzung *f*, Risikofeststellung *f*
risk evaluation Risikobewertung *f*
risk of cancer Krebsrisiko *n*
risk of infection Ansteckungsgefahr *f*
risk of liability Haftungsrisiko *n*
risk potential Gefährdungspotential *n*, Risikopotential *n*, Gefahrenpotential *n*
risk valuation Risikobeurteilung *f*
river and lake protection Gewässerschutz *m*
river bank Flußufer *n*
river basin Flußbecken *n*
river bed Flußbett *n*
river dam Staudamm *m*, Talsperre *f*
river police Wasserschutzpolizei *f*
river water Flußwasser *n*
road bed Bahnkörper *m*
road construction Straßenbau *m*
road construction material Straßenbaumaterial *n*
road damage Straßenschaden *m*
road safety Straßenverkehrssicherheit *f*
roasted material Röstgut *n*
roasted ore Abbrand *m*
roasting gas Röstgas *n*
roasting plant Röstanlage *f*
rock Fels[en] *m*, Gestein *n*
rock salt mine Salzbergwerk *n*
rock wool Mineralwolle *f*
rodent Nagetier *n*

rodenticide Rodentizid *n*
roentgenogram Röntgenaufnahme *f*, Röntgenbild *n*
roentgenography Röntgenographie *f*
room heating Raumheizung *f*
room temperature Raumtemperatur *f*
rot Rotte *f*, Fäule *f*, Fäulnis *f*, Verwesung *f*
rot cell Rottezelle *f*
rot chamber Rottezelle *f*
rot compost Rottekompost *m*
rot drum Rottetrommel *f*
rot period Rottedauer *f*
rot plate Rotteplatte *f*
rot-preventing fäulnisverhindernd
rot procedure Rottevorgang *m*
rot warming Rotteerwärmung *f*
rot [away]/to faulen, verfaulen, vermodern, verrotten, verwesen, verwittern, modern
rotary atomizer Rotationszerstäuber *m*
rotary drum filter Trommeldrehfilter *n*
rotary dryer Trockentrommel *f*, Drehtrockner *m*
rotary evaporator Rotationsverdampfer *m*
rotary furnace Drehofen *m*, Drehrohrofen *m*
rotary sprayer (sprinkler) Drehsprenger *m*
rotary tube furnace Drehrohrofen *m*
rotary tube furnace plant Drehrohrofenanlage *f*
rotation of crops Fruchtfolge *f*
rotational diffusion coefficient Rotationsdiffusionskoeffizient *m*

rotational dryer Rotationstrockner *m*
rotproof fäulnisbeständig, unverrottbar
rotproofness Fäulnisbeständigkeit *f*
rotten odour Fäulnisgeruch *m*, Modergeruch *m*
rottenness Fäule *f*, Fäulnis *f*, Moder *m*
round flask Rundkolben *m*
routine analysis Routineanalyse *f*
rubber abrasion Gummiabrieb *m*
rubber adhesive Gummikleber *m*
rubber apron Gummischürze *f*
rubber glove Gummihandschuh *m*
rubber hose Gummischlauch *m*
rubber stopper Gummistopfen *m*
rubble Schotter *m*, Steinschutt *m*, Schutt *m*; Bauschutt *m*
rubble dump Schutthalde *f*, Bauschuttkippe *f*
rubble recycling Bauschuttrecycling *n*
rubble recycling plant Baustoffrecyclinganlage *f*
rubble refuse dump Bauschuttdeponie *f*
rubble treatment plant Bauschuttaufbereitungsanlage *f*
rubidium Rubidium *n*
rumen Pansen *m*
ruminant Wiederkäuer *m*
run a blank/to eine Blindprobe machen
run wild/to außer Kontrolle geraten
running time system Laufzeitsystem *n*
rural area ländliche Gegend *f*
rural development ländliche Entwicklung *f*

rural population Landbevölkerung *f*
rust Rost *m*
rust-free rostfrei
rust-preventing rostverhütend
rust-preventive paint Rostschutzfarbe *f*
rust protection Korrosionsschutz *m*, Rostschutz *m*
rust removal Entrostung *f*
rust remover Rostentferner *m*
rust-resistant rostbeständig
rust slag Rostschlacke *f*
rust/to rosten, verrosten, einrosten
rusted eingerostet, verrostet
rustproof nichtrostend, rostgeschützt, rostfrei, rostsicher
rustproof steel rostfreier Stahl *m*
rusty rostig, verrostet
ruthenium Ruthenium *n*

S

safety Sicherheit *f*, Gefahrlosigkeit *f*, Verläßlichkeit *f*, Zuverlässigkeit *f*, Schutz *m*
safety and environmental aspect Sicherheits- und Umweltaspekt *m*
safety area Sicherheitsbereich *m*, Sicherheitsgebiet *n*
safety barrier Sicherheitsbarriere *f*
safety bottle Sicherheitsflasche *f*
safety cap Sicherheitskappe *f*
safety concept Sicherheitskonzept *n*
safety device Sicherheitseinrichtung *f*
safety disposal site Sicherheitsdeponie *f*
safety equipment Sicherheitsausrüstung *f*
safety harness Sicherheitsuniform *f*, Sicherheitskluft *f*
safety investigation Sicherheitsuntersuchung *f*
safety landfill site Sicherheitsdeponie *f*
safety measure Sicherheitsmaßnahme *f*, Sicherungsmaßnahme *f*
safety refuse dump Sicherheitsdeponie *f*
safety-related equipment Sicherheitsausrüstung *f*
safety requirement Sicherheitsanforderung *f*
safety standard Sicherheitsstandard *m*
saliferous salzhaltig
salification Salzbildung *f*
salify/to Salz bilden, in ein Salz überführen
salina Salzsee *m*, Salzsumpf *m*; Salzquelle *f*
saline salzig
saline deposits Abraumsalze *npl*
saline soil Salzboden *m*
salinity Salinität *f*
salinization of the soil Versalzung *f* des Bodens
salmon culture Lachszucht *f*
salt Salz *n*, Kochsalz *n*
salt aerosol Salzaerosol *n*
salt bed Salzlager *n*
salt dome Salzstock *m*
salt equilibrium Salzhaushalt *m*
salt freight Salzbelastung *f*, Salzfracht *f*
salt lake Salzsee *m*
salt melt Salzschmelze *f*
salt mine Salzbergwerk *n*

salt pit Salzgrube *f*
salvage Rettung *f*, Bergung *f*; Wiedergewinnung *f*
salvage vessel Bergungsfahrzeug *n*
salvage work Aufräumungsarbeiten *fpl*
salvage/to rückgewinnen, wiedergewinnen; bergen
samarium Samarium *n*
sample bottle Probeflasche *f*
sample box Probenbox *f*
sample composition Probenzusammensetzung *f*
sample flow conveyance Probenstromförderung *f*
sample material Probematerial *n*, Probegut *n*
sample storage Probenaufbewahrung *f*, Probenlagerung *f*
sampler Probenahmegerät *n*; Probenahmepumpe *f*
sampling Probe[ent]nahme *f*, Probeziehen *n*
sampling device Probenahmeeinrichtung *f*
sampling from a ship Probenahme *f* vom Schiff
sampling material Probematerial *n*
sampling point Probeentnahmestelle *f*
sampling rate Probenahmerate *f*, Probenahmegeschwindigkeit *f*
sampling system Probenahmesystem *n*
sand bag Sandsack *m*
sand bath Sandbad *n*
sand classifier Sandklassierer *m*
sand filter Sandfilter *n*
sand filtering system Sandfiltersystem *n*
sand-lime-brick Kalksandstein *m*

sand-oil sludge Sand-Öl-Schlamm *m*
sand trap Sandfang *m*
sand trapping blower Sandfanggebläse *n*
sand trapping equipment Sandfangausrüstung *f*
sand trapping (intercepting) fence Sandfang *m*, Sandfangzaun *m*
sand trapping system Sandfangsystem *n*
sandblast Sandstrahl *m*, Sandstrahlgebläse *n*
sandstone Sandstein *m*
sandy sandig
sandy marl Sandmergel *m*
sandy shale Sandschiefer *m*
sandy soil Sandboden *m*
sandy stratum Sandschicht *f*
sanitary sanitär, gesundheitlich, hygienisch
sanitary engineering Sanitärtechnik *f*
sanitary police Gesundheitspolizei *f*
sanitize/to sterilisieren, keimfrei machen
sanitizer Desinfektionsmittel *n*
saprobiotic system Saprobiensystem *n*
satellite earth station Erdefunkstelle *f*
satellite picture Satellitenbild *n*
satellite town Trabantenstadt *f*
saturate/to sättigen, saturieren; durchtränken, durchsetzen, tränken
saturation Sättigung *f*; Durchdringung *f*, Durchtränkung *f*
saturation concentration Sättigungskonzentration *f*

saturation humidity Sättigungsfeuchte *f*
saturation limit Sättigungsgrenze *f*
saturation temperature Sättigungstemperatur *f*
scale formation Kesselsteinbildung *f*, Krustenbildung *f*
scandium Skandium *n*
scanning electron microscope Rasterelektronenmikroskop *n*
scheme of measuring (locations) points Meßstellenschema *n*
scour valve Spülventil *n*
scrap Ausschuß *m*, Abfall *m*, Rest *m*; Schrott *m*
scrap dealer Schrotthändler *m*
scrap heap Schrotthaufen *m*
scrap iron Alteisen *n*, Eisenschrott *m*
scrap metal Altmetall *n*, Schrott *m*
scrap of processing Bearbeitungsabfälle *mpl*
scrap preparation plant Schrottaufbereitungsanlage *f*
scrap/to verschrotten, abwracken
scrapping of used tyres Granulierung *f* von Altreifen
screen belt dryer Laufbandtrockner *m*
screen rejects Siebgut *n*
screening belt press Siebbandpresse *f*
screening effect Siebeffekt *m*
screening facility Rechenanlage *f*
screening residue Rechenrückstand *m*
screening of solids Feststoffabsiebung *f*
screw conveyor Förderschnecke *f*
screw dryer Schneckentrockner *m*
screw mill Schraubenmühle *f*

scrubber Skrubber *m*, Wäscher *m*, Turmwäscher *m*
scrubbing process Waschvorgang *m*
scrubbing tower Berieselungsturm *m*
scum Schaum *m*, Abschaum *m*; Schlamm *m*
sea water Meerwasser *n*
sea disaster Seekatastrophe *f*
sea pollution Seeverschmutzung *f*
seal mortality Robbensterben *n*
seal Schutzanstrich *m*; Siegel *n*, Verschluß *m*
seal/to versiegeln, abdichten; plombieren
sealant Dichtungsmittel *n*
sealed tube Einschmelzrohr *n*
sealing compound Abdichtmasse *f*, Vergußmasse *f*
sealing of a landfill Deponieabdichtung *f*
sealing of the disposal (waste) site Deponieabdichtung *f*
sealing of the refuse dump Deponieabdichtung *f*
sealing packing Dichtmanschette *f*
sealing wall Dichtwand *f*
seaquake Seebeben *n*
seat belt Sicherheitsgurt *m*
secondary action Nebenwirkung *f*
secondary air Sekundärluft *f*
secondary component Nebenbestandteil *m*
secondary effect Nebeneffekt *m*
secondary injury Folgeschaden *m*
secondary raw material Sekundärrohstoff *m*
secondary reaction Nebenreaktion *f*
secondary treatment Nachbehandlung *f*

secretion Sekretion f, Absonderung f, Ausscheidung f
section of the clarification plant Klärwerksabschnitt m
securing of old sites Altlastensicherung f
security adviser Sicherheitsberater m
security check Sicherheitsüberprüfung f
security clearance Unbedenklichkeitsbescheinigung f
security consulting Sicherheitsberatung f
security control Sicherheitskontrolle f
security device Sicherheitseinrichtung f
security testing Sicherheitsüberprüfung f
sedative poison lähmendes Gift n
sediment Sediment n, Ablagerung f, Bodensatz m, Satz m
sediment/to sedimentieren, sich absetzen
sedimentary sedimentär
sedimentation Sedimentation f, Sedimentierung f, Bodensatzbildung f; Schlämmverfahren n
sedimentation analysis Sedimentationsanalyse f
sedimentation basin Absetzbecken n, Absetztank m, Absetzgefäß n, Klärbecken n
sedimentation behaviour Absetzungsverhalten n
sedimentation sludge Sedimentationsschlamm m, abgesetzter (sedimentierter) Schlamm m
sedimented dust sedimentierter Staub m

seeding sludge Impfschlamm m
seeds processing plant Saatgutaufbereitungsanlage f
segregate/to absondern, entmischen, trennen
seismic seismisch; Erdbeben-
seismic wave Erdbebenwelle f
seismogram Seismogramm n, Erdbebenkurve f
seismograph Seismograph m, Erdbebenmesser m
selection Auswählen n, Selektion f; Ansteuerung f, Wahl f
selective filter Selektivfilter n
selenium Selen n
self catalysis Autokatalyse f
self-cleaning Selbstreinigung f
self-cleaning capability Selbstreinigungsvermögen n
self-cleaning device Selbstreinigungseinrichtung f
self-cleaning power Selbstreinigungskraft f
self-cleaning sieve drum selbstreinigende Siebtrommel f
self regulation Sebstregulation f
self-sufficient autark
self-supporting autark
semi-annual halbjährlich
semi-mobile washing equipment semi-mobile Waschanlage f
semi-volatile mittelflüchtig
semiindustrial halbtechnisch
semimechanical halbautomatisch
semipermeable halbdurchlässig, semipermeabel
sensitivity measurement Sensitivitätsanalyse f
sensitivity to acid Säureempfindlichkeit f

sensitivity to light Lichtempfindlichkeit *f*
sensitization Sensibilisierung *f*
sensor Sensor *m*, Fühler *m*
sensor cable Sensorkabel *n*
sensory sensorisch
separatability Trennbarkeit *f*
separate collection Getrenntsammlung *f*
separating by weight Gewichtsauslese *f*
separating column Trennsäule *f*
separating funnel Scheidetrichter *m*
separating layer Trennungsschicht *f*
separating line Trennlinie *f*
separating plant Sortieranlage *f*, Sichtungsanlage *f*
separating stage Trennstufe *f*, Trennschritt *m*
separation Abscheidung *f*, Abtrennung *f*
separation of dust Staubabscheidung *f*
separation of fats Fettabtrennung *f*
separation process Trennverfahren *n*
separator Abscheider *m*
separator for light-liquids Leichtflüssigkeitsabscheider *m*
septic septisch, faulend
septum Septum *n*, Scheidewand *f*
sequence control operation (system) Ablaufsteuerung *f*
sequential function chart Funktionsplan *m*
series-connected denitrification vorgeschaltete Denitrifikation *f*
series of measurements Meßreihe *f*

service industry Dienstleistungsgewerbe *n*
service life Dauerhaltbarkeit *f*, Lebensdauer *f*
service routine Wartungsprogramm *n*
service water Gebrauchswasser *n*
setpoint Fixpunkt *m*, Sollwert *m*
setpoint and limit value selection Soll- und Grenzwertvorgabe *f*
setpoint operation Sollwertbedienung *f*
settle/to absetzen, klären
settlement Absetzen *n*, Klären *n*
settlement area Siedlungsfläche *f*
settlement of groundwater Grundwasserabsenkung *f*
settling basin Absetzbassin *n*
settling filter Anschwemmfilter *n*
settling tank Absetztank *m*, Klärtank *m*
settling velocity Absetzgeschwindigkeit *f*, Absinkgeschwindigkeit *f*
sewage Abwasser *n*; Sickerwasser *n*
sewage biology Abwasserbiologie *f*
sewage chemistry Abwasserchemie *f*
sewage clarification Abwasserklärung *f*
sewage collection system Abwassersammlungssystem *n*
sewage composition Abwasserzusammensetzung *f*
sewage conveyance Abwasserförderung *f*
sewage discharge Abwasserabfluß *m*
sewage disinfection Abwasserdesinfektion *f*, Abwasserentkeimung *f*
sewage disposal Abwasserbeseitigung *f*

sewage disposal system Abwasserbeseitigungssystem *n*
sewage farm Rieselfeld *n*
sewage flow Abwasserzufluß *m*
sewage immersion pump Abwassertauchpumpe *f*
sewage inflow Abwassereinlauf *m*
sewage inlet Abwassereinlauf *m*
sewage-intensive abwasserintensiv
sewage irrigation Abwasserverregnung *f*; Rieselanlage *f*
sewage load Abwasserfracht *f*
sewage pipe Abwasserleitung *f*
sewage pipe system Abwasserkanalsystem *n*
sewage pond Abwasserteich *m*
sewage problem Abwasserproblem *n*
sewage pumping station Abwasserpumpstation *f*
sewage purification Abwasserreinigung *f*
sewage quantity Abwassermenge *f*
sewage regulation Abwasserverordnung *f*
sewage removal system Abwasserbeseitigungssystem *n*
sewage sludge Klärschlamm *m*
sewage sludge drying Klärschlammtrocknung *f*
sewage sludge granules Klärschlammgranulat *n*
sewage sludge incineration Klärschlammverbrennung *f*
sewage sludge powder Klärschlammpulver *n*
sewage sludge processing Klärschlammaufbereitung *f*
sewage sludge volume Klärschlammvolumen *n*

sewage standard value Abwassernorm *f*
sewage tank Abwassertank *m*
sewage-technological abwassertechnisch
sewage technology Abwassertechnologie *f*, Abwassertechnik *f*
sewage transport Abwassertransport *m*
sewage treatment Abwasserbehandlung *f*
sewage water discharge Abwasserausfluß *m*
sewer Abwasserkanal *m*, Abwassergraben *m*, Kloake *f*
sewer analysis system Kanal-Analyse-System *n*
sewer gas Faulschlammgas *n*, Klärgas *n*
sewer gas plant Klärgasanlage *f*
sewer gas utilization Klärgasnutzung *f*
sewer monitoring Kanalnetzüberwachung *f*
sewer network Abwasserleitungsnetz *n*, Abwassertransportnetz *n*
sewer remediation Kanalsanierung *f*
sewer slime Sielhaut *f*
sewer system Abwasserkanalsystem *n*, Abwasserkanalisation *f*, Kanalisationssystem *n*
sewerage Kanalisation *f*
shaft Schacht *m*
shaft trickling dryer Schachtrieseltrockner *m*
shaker screen Schüttelsieb *n*
shale Schiefer *m*
sheet filter Schichtfilter *n*
sheet zinc Zinkblech *n*
shelf life Haltbarkeit *f*

shelf time Haltbarkeit f, Lagerbeständigkeit f, Lagerfähigkeit f, Umschlagszeit f
shield Schild n, Schutzschild n, Schutzschirm m, Schutz m
shield/to abschirmen, schützen
shielding Abschirmung f, Schutzhülle f
shielding of the workplace Arbeitsplatzabschirmung f
shifting sand Treibsand m, Flugsand m
shipwreck Schiffbruch m; Schiffswrack n
shock cooling Schockabkühlung f
shore Küste f, Ufer n, Strand m
shore erosion Küstenerosion f
shore protection Küstenschutz m
shore recession Uferrückgang m
short-lived kurzlebig
short-term analysis automat Kurzzeitanalysenautomat m
short-term measurement Kurzzeitmessung f
short-term stability Kurzzeitstabilität f
short test Kurztest m
shredder Zerfaserer m, Zerkleinerer m, Reißwolf m, Schredder m
shredder classification plant Schredder-Sortieranlage f
shredder technology Schreddertechnik f, Schreddertechnologie f
shredder waste Schredderabfall m
shredding machine Reißwolf m
shred/to zerfasern, zerkleinern, zerschnitzeln
shrink film Schrumpffolie f
shut-off valve Abschaltventil n
shutdown Außerbetriebsetzung f, Betriebsstillegung f; Betriebsstörung f
shutdown relay Abschaltrelais n
shutter Spund m, Verschluß m, Verschlußblende f
siccative Trockenmittel n
side area Nebenfläche f
side drain Nebenkanal m
side effect Nebenwirkung f
side finding Nebenbefund m
side reaction Nebenreaktion f
side wall of the fire chamber Feuerraumseitenwand f
sieve Sieb n
sieve analysis Siebanalyse f
sieve cross section Siebquerschnitt m
sieve drum Siebtrommel f
sieve dryer Siebtrockner m
sieve plate Siebboden m
sieve rake Siebrechen m
sieve residue Siebrückstand m
sieve tray Siebboden m
sieve/to sieben
sift/to sieben, durchsieben
sifting Siebung f
signal alarm Alarmeinrichtung f
signal effect Signalwirkung f
signal fire Blinkfeuer n
signal lamp Kontrollampe f, Signallampe f
signal paint Markierungsfarbe f
silage Silage f
silage covering Silageabdeckung f
silage fodder Silofutter n, Gärfutter n
silencing Schalldämpfung f
silicagel Kieselgel n, Silikagel n
silicagel adsorption Silikageladsorption f
silicate Silikat n

silicify/to verkieseln
silicon Silizium *n*
silicon-containing dust siliziumhaltiger Staub *m*
silicon oil Silikonöl *n*
silicosis Silikose *f*
silo Silo *m*, Rotteturm *m*
  silo composting Mietenkompostierung *f*
silt Schlamm *m*, Schlick *m*
silver Silber *n*
silver-containing sludge silberhaltiger Schlamm *m*
simulation model Simulationsmodell *n*
simultaneous denitrification simultane Denitrifikation *f*
simultaneous precipitation Simultanfällung *f*
single connection charge einmalige Anschlußgebühr *f*
single determination Einzelbestimmung *f*
single result Einzelergebnis *n*
sink Ausguß *m*, Abfluß *m*, Abflußrohr *n*, Gießrinne *f*
  sink-and-float process Sinkscheideverfahren *n*
sinter Sinter *m*
sinter/to sintern
siphon/to abhebern, aushebern
siphonage Abhebern *n*, Aushebern *n*
site body Deponiekörper *m*
  site capacity Deponiekapazität *f*
  site component Deponiekomponente *f*
  site digestion gas Deponiegas *n*
  site digestion gas collection Deponiegassammlung *f*
  site digestion gas quality Deponiegasqualität *f*

  site gas Deponiegas *n*
  site location Deponiestandort *m*
site of a power plant Kraftwerksstandort *m*
site register Deponiekataster *m*
site-selection Standortwahl *f*
situation of danger Gefahrensituation *f*
situation of exposure Expositionssituation *f*
situation with respect to harmful materials Gefahrstoffsituation *f*
skull guard Schutzhelm *m*
slag Schlacke *f*
  slag concrete Schlackenbeton *m*
  slag crusher Schlackenbrecher *m*
  slag from waste incineration plants Müllverbrennungsschlacke *f*
  slag heap Halde *f*
  slag mill Schlackenmühle *f*
  slag pile Schlackenhaufen *m*
  slag smelting behaviour Schlackeschmelzverhalten *n*
slaughterhouse Schlachthof *m*
slaughterhouse odour Schlachthofgeruch *m*
slaughtering waste Schlachtabfall *m*
slick Ölfleck *m*, Ölfläche *f* auf Gewässer
slide valve Absperrschieber *m*, Schieber *m*
  slide valve position Schieberstellung *f*
  slide valve station Schieberstation *f*
slime Schleim *m*; Schlamm *m*
  slime-forming schleimbildend
slimy schleimig; schlammig
slope sealing Böschungsabdichtung *f*
slope protection Böschungsschutz *m*

slow poison schleichendes Gift *n*
sludge Schlamm *m*, Aufschlämmung *f*, Bodensatz *m*; Faulschlamm *m*
sludge activation Schlammbelebung *f*
sludge activation method Schlammbelebungsverfahren *n*
sludge aftertreatment Schlammnachbehandlung *f*
sludge age Schlammalter *n*
sludge analysis Schlammanalyse *f*
sludge barging Schlammverklappung *f*
sludge bed Schlammbett *n*
sludge cake Schlammkuchen *m*
sludge cake pelletizing Schlammkuchenpelletierung *f*
sludge collecting zone Schlammsammelzone *f*
sludge decomposition efficiency Schlammabbauleistung *f*
sludge deposition Schlammablagerung *f*
sludge dewatering Schlammentwässerung *f*
sludge disinfection Schlammdesinfizierung *f*, Schlammentkeimung *f*
sludge drying Schlammtrocknung *f*
sludge formation Schlammbildung *f*
sludge incineration Schlammverbrennung *f*
sludge removal Schlammentfernung *f*
sludge screening Schlammsiebung *f*
sludge thickening Schlammeindickung *f*
sludge thickening method Schlammeindickungsmethode *f*
sludge transportation pipe Schlammtransportleitung *f*
sludge trap Schlammabscheider *m*
sludge utilization Schlammverwertung *f*
sludge water discharge Schlammwasserableitung *f*
sludgy schlammig
sluice Schleuse *f*, Absperrglied *n*; Schleusenkanal *m*, Abflußkanal *m*
slurry Schlamm *m*, Aufschlämmung *f*, Brei *m*
slurry reactor Schlammreaktor *m*
slurry thickener Schlammwassereindicker *m*
slurrying Aufschlämmen *n*
slush Schlamm *m*, Schmiere *f*
small game Niederwild *n*
small-scale experiment Kleinversuch *m*
smelting Verhüttung *f*, Verhütten *n*, Ausschmelzung *f*, Schmelzen *n*
smelting flux Schmelzfluß *m*
smelting flux electrolysis Schmelzflußelektrolyse *f*
smelting plant Schmelzanlage *f*
smog Smog *m*, rauchdurchsetzter Nebel *m*
smog alarm Smogalarm *m*
smog disaster Smogkatastrophe *f*
smog-forming smogbildend
smoke Rauch *m*
smoke alarm Rauchmelder *m*
smoke bomb Nebelbombe *f*, Rauchbombe *f*
smoke density Rauchdichte *f*, Rauchstärke *f*
smoke density meter Rauchgasdichtemesser *m*
smoke formation Rauchbildung *f*
smoking chamber Räucherkammer *f*

smolder/to schwelen, glimmen
smoldering Schwelen *n*, Glimmen *n*; Verschwelung *f*
  smoldering fire Schwelfeuer *n*, Schwelbrand *m*
  smoldering zone Schwelzone *f*
  smoldering gas Schwelgas *n*
sneeze gas Reizgas *n*
snow water Schneewasser *n*
snowfall Schneefall *m*
soak/to einweichen, durchnässen, vollsaugen; wässern
soak in/to einsickern, einziehen
soakage Aufsaugen *n*, Einsaugen *n*, Durchsickern *n*; Sickerwasser *n*
soakaway Sickergrube *f*
soap bubble flow meter Seifenblasenzähler *m*
soap suds Seifenlauge *f*
soda Soda *f*, Natriumcarbonat *n*
sodium Natrium *n*
softening agent Weichmacher *m*
soil Boden *m*, Erde *f*, Land *n*, Erdboden *m*
  soil aeration Bodenbelüftung *f*
  soil analysis Bodenanalyse *f*
  soil atmosphere Bodenluft *f*
  soil atmosphere concentration Bodenluftkonzentration *f*
  soil atmosphere exhaustion Bodenluftabsaugung *f*
  soil atmosphere sampling method Bodenluft-Probenahmeverfahren *n*
  soil bacteria Bodenbakterien *npl*
  soil chemistry Bodenchemie *f*
  soil column Bodensäule *f*
  soil composting in boxes Boxenkompostierung *f*
  soil conditions Bodenbedingungen *fpl*
  soil conservation Bodenerhaltung *f*
  soil decontamination Bodenentseuchung *f*
  soil erosion control Bodenerosionsbekämpfung *f*
  soil excavation Bodenaushub *m*
  soil fauna Bodenfauna *f*
  soil filter Bodenfilter *n*
  soil flora Bodenflora *f*
  soil investigation Bodenuntersuchung *f*
  soil layer Bodenschicht *f*
  soil map Bodenkarte *f*
  soil mechanics Bodenmechanik *f*
  soil melioration agent Bodenverbesserungsmittel *n*
  soil microorganism Bodenmikroorganismus *m*
  soil morphology Bodenmorphology *f*
  soil organism Bodenorganismus *m*
  soil poisoning Bodenvergiftung *f*
  soil pollution control Bodenreinhaltung *f*
  soil protection Bodenschutz *m*
  soil purification method Bodenreinigungsverfahren *n*
  soil remediation Bodensanierung *f*
  soil respiration Bodenatmung *f*
  soil sampling Bodenprobenahme *f*
  soil science Bodenkunde *f*
  soil-specific bodenspezifisch
  soil structure Bodengefüge *n*
  soil treatment [mikrobiologische] Bodenbehandlung *f*
  soil type Bodenart *f*
  soil washing plant Bodenwaschanlage *f*
sol Sol *n* {*Salzwasser*}
solar solar; Sonnen-
  solar cell Solarzelle *f*
  solar energy Solarenergie *f*

solar heat Sonnenwärme *f*
solar power Sonnenenergie *f*
solar power plant Sonnenenergieanlage *f*
solar radiation Solarstrahlung *f*, Sonneneinstrahlung *f*, Sonnenstrahlung *f*
solid Feststoff *m*, Festkörper *m*
solid constituent Trockenbestandteil *m*
solid-liquid separation Fest-Flüssig-Trennung *f*
solid matter Feststoff *m*; Feststoffgehalt *m*
solid phase Bodenkörper *m*, Festphase *f*
solid state fester Zustand *m*
solid-state chemistry Festkörperchemie *f*
solid waste Festabfall *m*
solidification Erstarren *n*, Erstarrung *f*, Festwerden *n*, Verfestigung *f*
solidify/to verfestigen, erstarren lassen, fest werden lassen
solubility Löslichkeit *f*, Lösbarkeit *f*
solubility product Löslichkeitsprodukt *n*
soluble löslich, lösbar, auflösbar
solution Auflösung *f*, Lösung *f*
solution enthalphy Lösungsenthalpie *f*
solvable auflösbar, löslich
solvate/to anlösen, solvatisieren
solve/to lösen
solvent Lösungsmittel *n*, Lösemittel *n*, Solvens *n*
solvent-containing lösemittelhaltig
solvent-containing exhaust air lösemittelhaltige Abluft *f*
solvent-containing sewage lösemittelhaltiges Abwasser *n*
solvent emission Lösemittelemission *f*
solvent evaporation Lösemittelverdampfung *f*
solvent-free paint lösemittelfreie Farbe *f*
solvent mixture Lösemittelgemisch *n*
solvent recovery Lösemittelrückgewinnung *f*
solvent wastes Lösemittelabfälle *mpl*
soot Ruß *m*
soot-adsorbed pollutant rußadsorbierter Schadstoff *m*
soot/to rußen, verrußen
sooting Verrußen *n*
sootless nichtrußend
sooty rußig
sooty coal Rußkohle *f*
sorbent Sorbens *n*
sorption Sorption *f*
sorption efficiency Sorptionseffektivität *f*
sorption residue Sorptionsrückstand *m*
sorting Sortieren *n*, Auslesen *n*, Klassieren *n*
sorting band Leseband *n*
sorting by weight Gewichtsauslese *f*
sorting equipment Sortieranlage *f*
sorting machine Sortiermaschine *f*
sorting of materials of value Wertstoffsortierung *f*
sorting of recyclates Wertstoffsortierung *f*
sorting of recycling materials Wertstoffsortierung *f*
sorting system Sortiersystem *n*
sound Schall *m*; Klang *m*, Ton *m*

sound attenuation Schalldämpfung *f*
sound insulation Schallisolierung *f*, Schallschutz *m*
sound level analyzer Schallpegelanalysator *m*
sound level meter Geräuschpegelmesser *m*, Lautstärkemeßgerät *n*
source of danger Gefahrenherd *m*, Gefahrenquelle *f*
source of emission Emissionsquelle *f*
source of energy Energiequelle *f*
source of nitrogen Stickstoffquelle *f*
source of raw material Rohstoffquelle *f*
sow down/to begrünen
space chemistry Stereochemie *f*, Kosmochemie *f*
space formula Raumformel *f*, Stereoformel *f*, Konfigurationsformel *f*
space lattice Kristallgitter *n*, Raumgitter *n*
special compost Spezialkompost *m*
special container Spezialcontainer *m*
special situation of burden besondere Belastungssituation *f*
special waste Sonderabfall *m*, Sondermüll *m*
specimen Probe *f*, Probenkörper *m*
specimen banking Probenbank *f*
spectral colour Spektralfarbe *f*
spectral distribution spektrale Verteilung *f*, Spektralverteilung *f*
spectral line Spektrallinie *f*
spectrometry Spektrometrie *f*
spectroscopy Spektroskopie *f*
spectrum Spektrum *n*

spectrum analysis Spektralanalyse *f*
spectrum evaluation by computer Spektrumauswertung *f* durch Computer
spectrum recording Spektrumaufnahme *f*
spent acid Abfallsäure *f*
spent-air cleaning Abluftreinigung *f*
spent catalyst verbrauchter Katalysator *m*
spent hardening salts verbrauchte Härtesalze *npl*
spent material Altstoff *m*
spent oil Altöl
spent oil collection Altölsammlung *f*
spent plant Altanlage *f*
spermicidal samentötend
spill control Überlaufkontrolle *f*
spillway Überlauf *m*
spoil Aushub *f*
spoil/to beschädigen, verderben, ruinieren; faulen
spoilage Verderb *m*
spontaneous evolution of heat Selbsterwärmung *f*
spontaneous heating Selbsterwärmung *f*
spontaneous ignition Selbstentzündung *f*
spontaneous ignition temperature Selbstentzündungstemperatur *f*
spontaneous inflammation Selbstentzündung *f*
spore Spore *f*
spore formation Sporenbildung *f*
spot sample Stichprobe *f*
spray Spray *n*, Sprühmittel *n*; Sprühdose *f*

spray absorption Sprühabsorption *f*
spray absorption method Sprühabsorptionsverfahren *n*
spray aerosol Sprühaerosol *n*
spray arm Sprüharm *m*
spray booth Spritzkabine *f*
spray can Sprühdose *f*
spray can shredder Spraydosenschredder *m*
spray cooler Berieselungskühler *m*
spray/to aufsprühen, aufstäuben, berieseln, besprühen, sprühen, verstäuben
spraying Besprühen *n*, Aufsprühen *n*, Aufspritzen *n*
spread Ausbreitung *f*, Verbreitung *f*
spreading pathway Ausbreitungspfad *m*
spring water Quellwasser *n*
sprinkler Berieselungsapparat *m*, Sprinkler *m*, Spritzapparat *m*
sprinkler irrigation Beregnung *f*
sprinkler system Sprinkleranlage *f*
sprinkle/to benetzen, besprengen, bespritzen, sprühen
square cascade Stufenkaskade *f*
stabilization grid Stabilisierungsgitter *n*
stabilization of sewage sludge Klärschlammstabilisierung *f*
stabilization period Stabilisierungszeitraum *m*
stabilization time Stabilisierungszeit *f*
stabilizer Stabilisator *m*
stack Kamin *m*, Schornstein *m*; Stapel *m*
stack gas Abgas *n*
stagnant stagnierend, stehend, stillstehend

stagnant volume Totvolumen *n*
stagnation Stagnation *f*
stain remover Fleckentferner *m*, Fleckentfernungsmittel *n*, Fleckenreinigungsmittel *n*
stainless fleckenfrei; korrosionsbeständig, korrosionsfrei, rostbeständig
stainless steel Edelstahl *m*, rostfreier Stahl *m*
stand-alone system autarkes System *n*
stand-alone version autarke Version *f*
stand-by interference Entstörbereitschaft *f*
standard acid Maßlösung *f* einer Säure, Säuremaßlösung *f*
standard deviation Standardabweichung *f*
standard mass framwork Standardmengengerüst *n*
standard of living Lebensstandard *m*
standard of performance Leistungsstandard *m*
standardization Standardisierung *f*, Normung *f*, Eichung *f*, Normalisierung *f*, Normenaufstellung *f*, Vereinheitlichung *f*
standardization committee Normenausschuß *m*
standardization measure Eichmaß *n*, Kalibriermaß *n*
standardization method Standardverfahren *n*
standardize/to standardisieren, normieren
standby function Funktionsbereitschaft *f*
standby terminal Bereitschaftsterminal *n*

start-up Inbetriebnahme *f*, Inbetriebsetzung *f*, Ingangsetzung *f*
start-up phase Inbetriebnahmephase *f*
starting bath Ansatzbad *n*
starting material Ausgangsmaterial *n*
starting point Ausgangspunkt *m*, Ansatzpunkt *m*, Ausgangsbasis *f*
starvation Hungern *n*, Aushungern *n*
starvation ration Hungerration *f*
starve/to hungern; verhungern lassen; verhungern
state of analogue output Analogwertzustand *m*
state of design Ausführungsstand *m*
state of equilibrium Gleichgewichtszustand *m*
state of health Gesundheitszustand *m*
static air layer ruhende Luftschicht *f*
static bed ruhendes Feststoffbett *n*
static load Ruhebelastung *f*
station control unit Stationsleitgerät *n*
stationary waste press stationäre Müllpresse *f*
statistical statistisch
statistical certainty statistische Sicherheit *f*
statistical distribution Häufigkeitsverteilung *f*
statistical quality index statistische Güteziffer *f*
steam Dampf *m*; Wasserdampf *m*
steam blower Dampfgebläse *n*
steam boiler Dampfkessel *m*
steam boiler feed water Dampfkesselspeisewasser *n*
steam cooler Dampfkühler *m*

steam discharge Dampfabführung *f*, Dampfableitung *f*
steam generation Dampferzeugung *f*
steam-volatile dampfflüchtig, wasserdampfflüchtig
steam/to dämpfen, dünsten; dampfen
steaming Dämpfen *n*, Eindampfen *n*
steaming jet Dampfstrahl *m*
steaming jet sprayer Dampfstrahlzerstäuber *m*
steaming sterilizer Dampftopf *m*
steaming trap Dampfentwässerer *m*
steel processing Stahlherstellung *f*
steel scrap Stahlschrott *m*
steel sheet piling Stahlspundwände *fpl*
steep/to einweichen; wässern; tränken, quellen
step of attrition Attritionsstufe *f*
step reaction Stufenreaktion *f*
steppe Steppe *f*
steppization Versteppung *f*
sterile filtration Sterilfiltration *f*
sterilization Sterilisation *f*, Unfruchtbarmachung *f*
sterilize/to sterilisieren, unfruchtbar machen; keimfrei machen, entkeimen
sterilizer Sterilisator *m*, Sterilisiergerät *n*
sterilizing keimtötend
stillbirth rate Totgeburtenrate *f*
stimulating substance Reizsubstanz *f*, Reizstoff *m*
stir/to rühren, durchrühren, umrühren
stirrer Rührer *m*, Rührapparat *m*, Rührwerk *n*; Rührstab *m*
stirring apparatus Rührwerk *n*
stirring motor Rührmotor *m*
stirring vessel Rührkessel *m*

stock farming Viehzucht *f*
stoppered flask Stöpselflasche *f*
storage Lagerung *f*, Aufbewahrung *f*, Speicherung *f*
storage basin Vorrratsgefäß *n*, Tank *m*
storage battery Akkumulator *m*, Speicherbatterie *f*
storage capacity Speicherkapazität *f*, Speichervermögen *n*
storage reservoir Speicherbecken *n*
store/to lagern, einlagern, ablagern, aufbewahren, speichern
storm water retention basin (tank) Regenrückhaltebecken *n*
storm water spillway basin Regenüberlaufbecken *n*
stove Brennofen *m*; Einbrennofen *m*; Trockenkammer *f*
stove/to erwärmen, erhitzen, heiß machen; warmhalten
strategy of substitute values Ersatzwertstrategie *f*
stratification Aufschichten *n*, Schichtenbildung *f*, Aufschichtung *f*
stratosphere Stratosphäre *f*
stratum Schicht *f*, Flöz *n*
straw clay Strohlehm *m*
straw combustion Strohverbrennung *f*
straw light-clay Strohleichtlehm *m*
straw utilization Strohverwertung *f*
stream pollution Flußverunreinigung *f*
strength limit Bruchgrenze *f*
strength test Festigkeitsprüfung *f*
streptococcus Streptokokkus *m*
stress Streß *m*
stripping Abtreiben *n*, Austreiben *n*, Strippen *n*, Strippung *f*

stripping column Abtreibkolonne *f*, Austreibekolonne *f*, Stripper *m*
stroma Grundgewebe *n*, Stroma *n*
strontium Strontium *n*
structural chemistry Strukturchemie *f*
structural element Strukturelement *n*
structural technique Aufbautechnik *f*
structural unit Struktureinheit *f*
structure Struktur *f*, Gefüge *n*, Gliederung *f*
structure-function relationship Struktur-Funktions-Beziehung *f*
structure of the disposal site Deponieaufbau *m*
structure-sensitive strukturempfindlich
structureless strukturlos, gefügelos
structurization Strukturierung *f*
structurize/to strukturieren
stunted growth Zwergwuchs *m*
styrene Styrol *n*
subacid schwach sauer
subacute toxicity subakute Toxizität *f*
subalkaline schwach alkalisch, schwach basisch
subchronic effect subchronische Wirkung *f*
subchronical subchronisch
subcooling Unterkühlung *f*
subcritical unterkritisch
subculture Tochterkultur *f*
subcutaneous subkutan, unter der Haut
subfamily Unterfamilie *f*
subgroup Untergruppe *f*
sublethal subletal
sublethal dose subletale Dosis *f*

sublethal toxicity subletale Toxizität *f*
sublimate/to sublimieren
sublimation Sublimation *f*, Verflüchtigung *f*
submerge/to untertauchen, versenken
submerged bearing Unterwasserlager *n*
submerged pump Tauchpumpe *f*
subnutrition Unterernährung *f*, Fehlernährung *f*
subsequent utilization Weiterverwendung *f*
substance group Stoffgruppe *f*
substance of value Wertstoff *m*
substance-related effect substanzbezogener Effekt *m*, substanzbezogene Wirkung *f*
substance-specific risks stoffspezifische Risiken *npl*
substitute Ersatzstoff *m*
substitute material Ersatzstoff *m*
substitute/to substituieren, austauschen, ersetzen; vertreten
substitution Substitution *f*, Einsetzung *f*, Ersetzung *f*, Ersatz *m*, Austausch *m*
substitution product Ersatzprodukt *n*
substrate Substrat *n*; Schichtträger *m*
substrate inhibition Substrathemmung *f*
subterranean unterirdisch
subtropical subtropisch
succession Folge *f*, Reihe *f*, Aufeinanderfolge *f*, Sukzession *f*
successive aufeinanderfolgend
successive experiment Reihenversuch *m*, Serienversuch *m*

suction aerator Ansaugbelüfter *m*
suction bottle Saugflasche *f*
suction device Saugvorrichtung *f*, Saugeinrichtung *f*
suction effect Saugwirkung *f*
suction facility Absauganlage *f*
suction filtration Saugfiltration *f*
suction force Saugkraft *f*
suction hood Absaughaube *f*
suction line Saugleitung *f*
suction plant Absauganlage *f*, Sauganlage *f*
suction power Saugleistung *f*
suitable for disposal entsorgungsfreundlich, entsorgungsgerecht
sulphate Sulfat *n*
sulphide Sulfid *n*
sulphite Sulfit *n*
sulphur Schwefel *m*
sulphur dioxide Schwefeldioxid *n*
sulphur/to schwefeln, ausschwefeln
sum parameter Summenparameter *m*
sump Sumpf *m*, Kolonnensumpf *m*
sun simulation plant Sonnensimulationsanlage *f*
superacidic sehr stark sauer
superacidity Übersäuerung *f*
superalimentation Überernährung *f*
superalkaline sehr stark alkalisch
superannuation Überalterung *f*
superconducting supraleitend, supraleitfähig
supercritical fluid überkritische Flüssigkeit *f*
supercritical pressure überkritischer Druck *m*
supersaturate/to übersättigen
supersaturation Übersättigung *f*
supervision routine Überwachungsroutine *f*

supervisory authority Aufsichtsbehörde *f*
supervisory board Aufsichtsbehörde *f*
supply monitoring Bezugsüberwachung *f*
supply network Versorgungsnetz *n*
supported catalyst Trägerkatalysator *m*
suppressor Suppressor *m*
supraconduction Supraleitung *f*
supraconductor Supraleiter *m*
surface Oberfläche *f*
surface action Oberflächenwirkung *f*
surface-active oberflächenaktiv
surface discharge Oberflächenentladung *f*
surface distribution Oberflächenverteilung *f*
surface filtration Oberflächenfiltration *f*
surface finish Oberflächenbeschaffenheit *f*
surface of bulky goods Schüttgutoberfäche *f*
surface reaction Oberflächenreaktion *f*
surface sealing Oberflächenabdichtung *f*
surface soil Oberboden *m*
surface water Oberflächenwasser *n*
surface water level Oberwasserpegel *m*
surfactant oberflächenaktive Substanz *f*
surveillance Überwachung *f*, Observation *f*
surveillance function Beobachtungsfunktion *f*

surveillance obligation Überwachungspflicht *f*
surveillance office Überwachungsbehörde *f*
survival Überleben *n*
survival time Überlebenszeit *f*
survivor Überlebender *m*
suspend/to aufschlämmen, suspendieren
suspended feinverteilt
suspended matter Schwebstoff *m*
suspended particulate matter Schwebestaubmasse *f*
suspending Aufschwemmen *n*
suspensibility Schwebefähigkeit *f*
suspension Suspension *f*, Aufschlämmung *f*, Aufschwemmung *f*
sustained loading Langzeitbeanspruchung *f*
swamp Sumpf *m*, Morast *m*
swampland Sumpfland *n*
swampy area Moorgebiet *n*
sweeper Kehrmaschine *f*
swelling behaviour Quellverhalten *n*
swelling capacity Quellvermögen *n*
swelling power Quellvermögen *n*
switch clock Schaltuhr *f*
symptom Symptom *n*, Anzeichen *n*, Merkmal *n*; Krankheitszeichen *n*
symptomatic symptomatisch
syncrude synthetisches Rohöl *n*
synecosystem Synökosystem *n*
synergism Synergismus *m*, Zusammenwirken *n*
synergist Synergist *m*
synoptic model synoptisches Modell *n*
system analysis Systemanalyse *f*
system documentation Anlagendokumentation *f*
system failure Systemausfall *m*

T

tabletting Tablettierung *f*
tag/to markieren
tagged atom markiertes Atom *n*
tail flap Schmutzfänger *m*
talc Talkum *n*, Talk *n*
tan Gerbmittel *n*, Eichenlohe *f*
tan/to gerben; beizen
tank Tank *m*, [geschlossen] Behälter *m*; Bottich *m*; Zisternentank *m*
tank biology Tankbiologie *f*
tank coating Tankanstrich *m*
tank container Tankcontainer *m*
tank truck Tankfahrzeug *n*
tank waggon Tankwagen *m*
tankage Fassungsvermögen *n*
tanker Tanker *m*, Tankschiff *n*
tannery sewage Gebereiabwasser *n*
tanning Gerben *n*, Gerbung *f*
tanning drum Gerbfaß *n*
tantalum Tantal *n*
tap Hahn *m*, Anstich *m*, Auslaufventil *n*
tap water Leitungswasser *n*
tap/to abstechen, abzapfen, anzapfen, zapfen
taphole Abstichloch *n*
tapping Abzapfen *n*, Anzapfen *n*, Abstich *m*
tar Teer *m*
  tar asphalt Teerasphalt *m*
  tar collector Teersammler *m*
  tar content Teergehalt *m*
  tar extractor Teerabscheider *m*
  tar formation Teerbildung *f*
  tar pit Teergrube *f*
  tar pitch Teerpech *n*
  tar residue Teerrückstand *m*
  tar separation Teerabtrennung *f*
tar/to teeren

tarred board Teerpappe *f*
tarry teerartig
task of surveillance Überwachungsaufgabe *f*
tear gas Reizgas *n*, Tränengas *n*
technical development technische Entwicklung *f*
technical impurity technische Verunreinigung *f*
technical reliability technische Zuverlässigkeit *f*
technical welfare organization technisches Hilfswerk *n*
technochemistry Industriechemie *f*
technological environmental protection Studium *n* des technischen Umweltschutzes
technology Technologie *f*
technology of redevelopment Sanierungstechnologie *f*
technology transfer Technologietransfer *m*
tectonic movement Gebirgsbewegung *f*
telecommunication Fernmeldeverkehr *m*, Telekommunikation *f*
telecontrol Fernsteuerung *f*, Fernlenkung *f*
telecontrol information Fernwirkinformation *f*
telecontrol master station Fernwirkzentrale *f*
telecontrol substation Fernwirkunterstation *f*
telecontrol system Fernwirksystem *n*
telecontrol unit Fernwirkeinheit *f*
telemetry system Fernmeldesystem *n*
telemonitoring Fernüberwachung *f*
tellurium Tellur *n*

temperature gradient Temperaturgradient *m*
temporary hardness temporäre Härte *f*
teratogenic teratogen, Mißbildungen erzeugend
terbium Terbium *n*
termite Termite *f*
terrace Terrasse *f*, Geländestufe *f*
terrain Terrain *n*, Gelände *n*
terrestrial terrestrisch, irdisch
terrestrial field system terrestisches Feldsystem *n*
terrestric model ecosystem terrestrisches Modellökosystem *n*
territorial waters Territorialgewässer *npl*
test Test *m*, Versuch *m*, Untersuchung *f*, Probe *f*, Experiment *n*; Nachweis *m*
test aerosol Prüfaerosol *n*
test animal Versuchstier *n*
test practice Versuchsdurchführung *f*, Versuchspraxis *f*, Testpraxis *f*
test probe Versuchssonde *f*
test protocol Versuchsprotokoll *n*
test reaction Nachweisreaktion *f*, Identifikationsreaktion *f*
test report Prüfbericht *m*, Versuchsbericht *m*
test result Untersuchungsergebnis *n*
test run Probelauf *m*
test strip Prüfstreifen *m*, Teststreifen *m*, Probestreifen *m*
test system Testsystem *n*, Versuchssystem *n*
test time Laufzeit *f*, Versuchsdauer *f*
test tube Reagenzglas *n*
test value Meßwert *m*

test/to ausprobieren, erproben, probieren, testen, untersuchen
testable prüfbar, untersuchbar
tester Prüfer *m*, Tester *m*
testing Prüfen *n*; Prüfwesen *n*
testing apparatus Prüfapparat *m*
testing arrangement Prüfvorrichtung *f*
testing chamber Prüfkammer *f*
testing equipment Testeinrichtung *f*
testing institute Prüfanstalt *f*
testing material Prüfmaterial *n*
testing plant Versuchsanlage *f*
textile industrial waste water textilindustrielles Abwasser *n*
textile waste Textilabfall *m*
texture Textur *f*, Faserung *f*, Gefüge *n*, Beschaffenheit *f*
thallium Thallium *n*
thallophytes Thallophyten *mpl*
thaw/to auftauen
therapy Therapie *f*, Behandlung *f*, Heilverfahren *n*, heilende Maßnahme *f*
thermal thermisch; Wärme-, Hitze-
thermal absorption Wärmeabsorption *f*, Wärmeaufnahme *f*
thermal analysis Thermoanalyse *f*, thermische Analyse *f*
thermal analyzer Thermoanalyseapparatur *f*
thermal balance Wärmebilanz *f*
thermal capacity Wärmekapazität *f*
thermal conduction Wärmeleitung *f*
thermal conductivity Wärmeleitfähigkeit *f*
thermal decomposition thermischer Abbau *m*, thermische Zersetzung *f*

thermal degradation thermischer Abbau *m*
thermal desorption thermische Desorption *f*
thermal diffusion thermische Diffusion *f*, Thermodiffusion *f*
thermal disposal thermische Entsorgung *f*
thermal efficiency Wärmewirkungsgrad *m*
thermal expansion thermische Ausdehnung *f*, Wärmeausdehnung *f*
thermal flow Wärmefluß *m*
thermal insulating panel Wärmedämmplatte *f*
thermal insulation Wärmeisolierung *f*
thermal power Heizwert *m*
thermal power station Wärmekraftanlage *f*
thermal process thermisches Verfahren *n*
thermal property thermische Eigenschaft *f*, Temperaturverhalten *n*
thermal purification of exhaust air thermische Abluftreinigung *f*
thermal radiation Wärmestrahlung *f*
thermal resistance Wärmedämmwert *m*
thermal spring Thermalquelle *f*, Therme *f*
thermal sulphur bath Thermalschwefelbad *n*
thermal treatment thermische Behandlung *f*, Wärmebehandlung *f*
thermal treatment of waste thermische Abfallbehandlung *f*
thermal unit Wärmeeinheit *f*
thermal utilization thermische Verwertung *f*
thermal vitrification thermische Verglasung *f*
thermal waste treatment plant thermische Abfallbehandlungsanlage *f*
thermal waste utilization thermische Abfallverwertung *f*
thermally stable thermisch stabil
thermistor Thermistor *m*
thermobiotic thermobiotisch
thermochemical thermochemisch
thermochemistry Thermochemie *f*
thermocouple Thermoelement *n*
thermodiffusion Thermodiffusion *f*
thermodynamics Thermodynamik *f*
thermophilic fouling (digestion) thermophile Faulung *f*
thermoresistant thermostabil, thermisch stabil
thicken/to eindicken, verdicken; trüben; entwässern
thickening Dickwerden *n*, Eindikken *n*
thickening agent Verdickungsmittel *n*
thickening by flotation Eindikkung *f* durch Flotation
thickening by gravity Eindikkung *f* durch Schwerkraft
thickening device Eindickungsvorrichtung *f*
thin-emulsion film Dünnschichtfilm *m*
thin-layer chromatography Dünnschichtchromatographie *f*
thin-layer electrophoresis Dünnschichtelektrophorese *f*
thin-sludge Dünnschlamm *m*
thinner Verdünner *m*, Verdünnungsmittel *n*

thinning of woodlands Durchforstung f
three-phase diesel aggregate Dieseldrehstromaggregat n
three-way catalyst Dreiwegekatalysator m
three-way valve Dreiwegeventil n
threshold concentration Auslösekonzentration f
threshold detector Schwellendetektor m
threshold dose Toleranzdosis f, kritische Dosis f
threshold limit Auslöseschwelle f, auslösender Wert m, Grenzschwelle f
threshold measurement Schwellenwertmessung f
threshold value Schwellenwert m
throttle valve Drosselventil n, Drosselklappe f, Luftklappe f
throw-away bottle Einwegflasche f
thyroid activity Schilddrüsenfunktion f
tide Tide f
tillable bebaubar, bestellbar
tillage Bodenbestellung f, Ackerbau m; Ackerland n
timber forest Hochwald m
timber line Baumgrenze f
time of repair Instandsetzungszeit f
tin Zinn n
tin-plate/to verzinnen
tin/to verzinnen; eindosen, in Dosen verpacken
tin plate can Weißblechdose f
tinplate recycling Weißblechrecycling n
tissue analysis sample Gewebeanalysenprobe f
tissue culture Gewebekultur f
tissue filter Gewebefilter n

tissue fluid Gewebeflüssigkeit f
tissue metabolism Gewebestoffwechsel m
titanium Titan n
titrant Titrans n, Titersubstanz f
titrate/to titrieren
titration Titration f, Maßanalyse f, Titrierung f
titre Titer m
titre value Titerwert m
titrimetric volumetrisch; Titrier-, Maß-
titrimetric analysis Titrimetrie f, Maßanalyse f, Volumetrie f
titrimetric standard Urtitersubstanz f
titrimetry Titrimetrie f, Maßanalyse f
tolerable tolerabel, zulässig; erträglich, leidlich
tolerance Toleranz f, Fehlergrenze f, Maßabweichung f, Spielraum m
tolerance dose Toleranzdosis f
tolerance test Verträglichkeitsprüfung f, Belastungsprobe f
tolerance value Toleranzwert m
toluene Toluol n
top fermentation Obergärung f
topography Topographie f, Geländekunde f
topsoil Mutterboden m
torch Brenner m
torch equipment Abfackelanlage f
total appearance of waste Gesamtabfallaufkommen n
total body burden Ganzkörperbelastung f
total concentration Gesamtkonzentration f
total efficiency Gesamtwirkungsgrad m

total load Gesamtfracht f
total metabolism Gesamtstoffwechsel m
total molecular formula Bruttoformel f, Summenformel f
total quantum number Hauptquantenzahl f
total reserve Gesamtreserve f, Gesamtvorrat m
total sludge appearance Gesamtschlammanfall m
total yield Gesamtausbeute f
tower acid Turmsäure f
town-centre renewal Altstadtsanierung f
town gas Stadtgas n
toxic giftig, toxisch; Gift
toxic agent Giftstoff m
toxic degradation product toxisches Abbauprodukt n
toxic effect Giftwirkung f, toxische Wirkung f
toxic hazard monitor Gefahrstoffüberwachungsgerät n
toxic reaction toxische Reaktion f
toxic threshold Toxizitätsgrenze f
toxicant toxisch wirkender Stoff m, Giftstoff m, Toxikum n
toxicity Toxizität f, Giftigkeit f
toxicity measurement Toxizitätsmesssung f
toxicity testing Toxizitätsprüfung f
toxicogenic gifterzeugend
toxicokinetics Toxikokinetik f
toxicological toxikologisch
toxicology Toxikologie f, Giftkunde f
toxify/to vergiften
toxin Toxin n, Giftstoff m, Gift n
trace Spur f, kleine Menge f
trace detection Spurennachweis m
trace element Spurenelement n
trace gas Spurengas n
trace metal Spurenmetall n
trace substance Spurenstoff m
trace substance inertization Spurenstoffinertisierung f
tracer chemistry Spurenchemie f
trade Gewerbe n
trade waste Gewerbemüll m
traffic Verkehr m
traffic accident Verkehrsunfall m
traffic chaos Verkehrschaos n
traffic jam Verkehrsstauung f
traffic noise Verkehrslärm m
traffic queue Fahrzeugschlange f
traffic regulations Verkehrsordnung f
traffic safety Verkehrssicherheit f
traffic sign Verkehrszeichen n
transducer Meßumformer m
tranducer design Meßumformerausführung f
tranquillizer Psychosedativum n, Tranquillizer m, Beruhigungsmittel n
transformation constant Zerfallskonstante f
transformation process Umwandlungsprozeß m
transformation product Umwandlungsprodukt n
transformer oil Transformatorenöl n
transition metal Übergangsmetall n
transition period Übergangszeit f
transition range Übergangsbereich m
transition state Übergangszustand m
transition temperature Umwandlungstemperatur f, Übergangstemperatur f, Sprungtemperatur f
transmutation Transmutation f

transmute/to umwandeln, verwandeln
transportable transportabel, transportfähig
transportation of dangerous substances Gefahrstofftransport *m*, Transport *m* gefährlicher Stoffe
transportation of hazardous goods Gefahrguttransport *m*
transportation technology Transporttechnik *f*
transshipment of hazardous goods Gefahrgutumschlag *m*
trap Falle *f*, Abscheider *m*, Auffanggefäß *n*, Sammelgefäß *n*
trap/to abfangen, fangen, einfangen
treatment and disposal facility Behandlungs- und Beseitigungsanlage *f*
treatment centre Behandlungszentrum *n*
treatment method Behandlungsverfahren *n*
treatment of leach water Sickerwasserbehandlung *f*
treatment of rubble Bauschuttaufbereitung *f*
treatment of sludge Schlammbehandlung *f*
treatment of the waste site leaching (percolating) water Deponiesickerwasserbehandlung *f*
treatment of wastes Abfallbehandlung *f*
treatment plant Behandlungsanlage *f*
tree line Baumgrenze *f*
tree resin Baumharz *n*
tree species Baumart *f*
tree structure of errors for disposal sites Fehlerbaumstruktur *f* für Deponien

tree-structurized operator pathway baumartiger Bedienpfad *m*
tree surgeon Baumchirurg *m*
trench system Grabensystem *n*
trial Probe *f*, Versuch *m*
trial-and-error method Näherungsmethode *f*
trial operation Probebetrieb *m*
trial plantation Versuchsanpflanzung
trial shipment Probelieferung *f*
trickling filter Rieselfilter *n*, Tropfkörper *m*
tropical tropisch; Tropen-
tropical climate Tropenklima *n*
tropical conditions Tropenbedingungen *fpl*, tropische Bedingungen *fpl*
tropical forest Tropenwald *m*
tropical medicine Tropenmedizin *f*
tropical rain forest tropischer Regenwald *m*
troposphere Troposphäre *f*
trout culture Forellenzucht *f*
tube furnace Röhrenofen *m*
tubing Rohrleitung *f*; Rohrmaterial *n*; Schlauch *m*
tubing resistance thermometer Schlauch-Widerstandsthermometer *n*
tubular furnace Röhrenofen *m*
tumour Tumor *m*, Geschwulst *f*, Wucherung *f*
tungsten Wolfram *n*
tunnel dryer Durchlauftrockner *m*, Tunneltrockner *m*
turbid trübe, schmutzig
turbid water Schmutzwasser *n*
turbidimeter Trübungsmeßeinrichtung *f*
turbidity Trübung *f*, Schleierbildung *f*

turbine 142

turbidity measurement Trübungsmessung *f*
turbine mixer Schaufelradmischer *m*
turbulence Turbulenz *f*, Unruhe *f*, Durchwirbelung *f*, Wirbelbewegung *f*
typhus Fleckfieber *n*, Typhus *m*
tyre Reifen *m*, Autoreifen *m*
tyre capping Reifenrunderneuerung *f*

U

ubiquitous ubiquitär, allgegenwärtig
ubiquity Allgegenwart *f*, Allgegenwärtigkeit *f*, Überall-Vorhandensein *n*, Ubiquität *f*
ulcer Ulkus *n*, Geschwür *n*
ulceration Geschwürbildung *f*
ultimate pressure Enddruck *m*
ultimate stress Bruchbelastung *f*
ultracentrifuge Ultrazentrifuge *f*
ultrafiltered sample flow ultrafiltrierter Probenstrom *m*
ultrafiltration Ultrafiltration *f*
ultrafiltration tube Ultrafiltrationsrohr *n*
ultrapure ultrarein, extrem rein
ultrasensitive überempfindlich
ultrasensitivity Überempfindlichkeit *f*
ultrasonic generator Ultraschallgenerator *m*
ultrasonic sensor Ultraschallsensor *m*
ultrasonic wave treatment Ultraschallbehandlung *f*
ultrasound detector (sensor) Ultraschallaufnehmer *m*

ultrasound sonar measuring method Ultraschall-Echolot-Meßmethode *f*
ultraviolet light ultraviolettes Licht *n*
ultraviolet radiation Ultraviolettstrahlung *f*
ultraviolet spectrum Ultraviolettspektrum *n*
unaerated zone unbelüftete Zone *f*
unalloyed unlegiert, unvermischt
unauthorized rubble dump wilde Bauschuttkippe *f*
unavoidable risk unvermeidbares Risiko *n*
uncertainty relation Unschärferelation *f*
unclean unsauber
uncleanness Unreinheit *f*, Unsauberkeit *f*
uncontaminated nicht verunreinigt, unverschmutzt; nicht infiziert, unverseucht
uncooked food Rohkost *f*
uncultivable unbebaubar, unkultivierbar
underdevelopment Unterentwicklung *f*
underground erdverlegt, unterirdisch
underground cable Erdkabel *n*
underground cavity unterirdischer Hohlraum *m*
underground deposition untertägige Ablagerung *f*
underground mining Untertagebau *m*
underground pipeline erdverlegte Rohrleitung *f*
underground water Grundwasser *n*
undernourishment Unterernährung *f*

undernutrition Unterernährung *f*
underpopulated unterbevölkert
undetectable nicht nachweisbar
undiluted unverdünnt; unverfälscht
undissolved ungelöst, unaufgelöst
undrinkable nicht trinkbar
unfertile unfruchtbar
unfiltered ungefiltert
unfruitful unfruchtbar
unfruitfulness Unfruchtbarkeit *f*, Fruchtlosigkeit *f*
unhealthy ungesund, kränklich, gesundheitsschädlich
unimolecular monomolekular
unimproved unverbessert; nicht kultiviert, nicht melioriert
unit cell Elementarzelle *f*
unit factor Erbfaktor *m*
unit for oxygen entry Sauerstoffeintragsaggregat *n*
unlabelled unmarkiert, nichtmarkiert
unleaded bleifrei, unverbleit, ungebleit
unmanned sewage works unbemannte Kläranlage *f*
unmixed unvermischt
unmodified unverändert
unplasticized weichmacherfrei
unprocessed unbehandelt, unbearbeitet, roh
unreacted unumgesetzt, nicht in Reaktion getreten, unverbraucht
unreactive reaktionslos, nicht reaktionsfähig
unresolved unaufgelöst
unrisky unbedenklich
unrottable unverrottbar
unsafety Unsicherheit *f*
unsaturated ungesättigt
unsolvable unauflösbar, unlösbar

unstable isotope instabiles Isotop *n*, Radioisotop *n*
untested ungeprüft, ungetestet
untilled unbebaut, nicht bestellt
untreated sewage Rohabwasser *n*
unventilated unbelüftet, nicht belüftet
update cycle Aktualisierungszyklus *m*
upgrade/to verbessern, verfeinern; ausbauen
upgrading degree of the configuration Ausbaugrad *m* des Systems
upland Hochland *n*
upper layer Deckschicht *f*
upstream water Oberwasser *n*
uptake Aufnahme *f*, Auffassungsvermögen *n*
uptake of a substance Substanzaufnahme *f*
uraemia Urämie *f*, Harnvergiftung *f*
uranium Uran *n*
urban städtisch
urban planning Stadtplanung *f*
urban population städtische Population *f*, Stadtbevölkerung *f*
urea Harnstoff *m*
urea cycle Harnstoffzyklus *m*
urgency of redevelopment Sanierungsdringlichkeit *f*
used air Abluft *f*
used-air cleaning Ablufteinigung *f*
used battery Altbatterie *f*
used glass Altglas *n*
used oil Altöl *n*
used-oil regulation Altölverordnung *f*
used plant Altanlage *f*
used tyre Altreifen *m*

used-tyre utilization Altreifenverwertung *f*
used tyres in the oyster breeding Altreifen *mpl* in der Austernzucht
useful effect Nutzeffekt *m*
useful work Nutzarbeit *f*
user interface Anwenderschnittstelle *f*
user memory Anwenderspeicher *m*
user program Anwenderprogramm *n*
user-specific anwenderspezifisch
utilizable verwertbar
utilizable material verwertbares Material *n*, verwertbarer Stoff *m*
utilization Ausnutzung *f*, Nutzung *f*, Verwertung *f*
utilization licence Nutzungslizenz *f*
utilization method Verwertungsverfahren *n*
utilization of rainwater Regenwassernutzung *f*, Nutzung *f* von Niederschlagswasser
utilization of sewage Abwasserverwertung *f*
utilization of waste Abfallverwertung *f*
utilization potential Nutzungspotential *n*
utilization system Verwertungssystem *n*
utilization tariff Nutzungspauschale *f*
utilize/to ausnutzen, verwerten
UV disinfection plant UV-Entkeimungsanlage *f*
UV radiation UV Strahlung *f*
UV radiation unit UV Bestrahlungseinheit *f*

V

vacancy Fehlstelle *f*, Leerstelle *f*, Lücke *f*, unbesetzte Stelle *f*
vaccine Impfstoff *m*
vacuole Vakuole *f*
vacuum broom sweeper Vakuumkehrmaschinenbesen *m*
vacuum cabinet Vakuumschrank *m*
vacuum chamber Unterdruckkammer *f*
vacuum distillation Vakuumdestillation *f*
vacuum evaporation Vakuumverdampfung *f*
vacuum filter Vakuumfilter *n*
vacuum filtration Vakuumfiltration *f*
vacuum flotation Vakuumflotation *f*
valency Valenz *f*, Wertikeit *f*
validity check Funktionstest *m*
valuable material Wertstoff *m*
valuation of the exposure Expositionsbeurteilung *f*
valve Ventil *n*; Absperrvorrichtung *f*, Klappe *f*, Hahn *m*
valve station Schieberstation *f*
vanadium Vanadium *n*
vandalism Vandalismus *m*, Zerstörungswut *f*
vapor adsorption Dampfadsorption *f*
vaporizability Verdampfbarkeit *f*, Verdunstbarkeit *f*
vaporizable verdampfbar, verdunstbar, vergasbar
vaporization Verdampfung *f*, Verdunstung *f*
vaporize/to verdampfen, verdunsten, eindampfen, vergasen

vaporizer Verdampfer *m*
vaporous dampfförmig
vapour Brüden *m*, Brüdendampf *m*; Dampf *m*; Dunst *m*, Nebel *m*
  vapour hood Dampfhaube *f*
  vapour line Dampfleitung *f*
  vapour nozzle Dampfventil *n*
  vapour pressure Dampfdruck *m*
  vapour recompression Brüdenverdichtung *f*
  vapour sterilization Dampfsterilisation *f*
  vapour supply pipe Dampfleitung *f*
varnish remover Lackentferner *m*, Abbeizmittel *n*
vascular system Gefäßsystem *n*
vasoconstriction Gefäßverengung *f*
vat Faß *n*, Bottich *m*, Kübel *m*, Trog *m*
vegetable pflanzlich
  vegetable butter Pflanzenfett *n*
  vegetable earth Düngeerde *f*
  vegetable manure Gründünger *m*
  vegetable mould Pflanzenerde *f*
  vegetable oil Pflanzenöl *n*, pflanzliches Öl *n*
  vegetable poison Pflanzengift *n*
vegetables Gemüse *n*; Gemüsepflanze *f*; Futtermittel *n*
vegetate/to wachsen, vegetieren
vegetation Vegetation *f*, Pflanzenwachstum *m*, Pflanzenwelt *f*; Bewuchs *m*
  vegetation area Vegetationsgebiet *n*
  vegetation geography Vegetationsgeographie *f*
  vegetation period Vegetationsperiode *f*, Wachstumsperiode *f*, Vegetationszeit *f*
vegetative vegetativ

vent Abzugsöffnung *f*, Abzug *m*, Entlüftung *f*, Lüftung *f*
vent/to lüften, entlüften, belüften, ventilieren
ventilate/to belüften, durchlüften, entlüften, lüften, ventilieren
ventilation Belüftung *f*, Entlüftung *f*, Ventilation *f*
  ventilation aggregate Belüfteraggregat *n*
  ventilation automat Entlüftungsautomat *m*
  ventilation device Belüftungseinrichtung *f*
  ventilation fan Propeller *m*
  ventilation group Belüftergruppe *f*
  ventilation system Belüftungssystem *n*
venturi scrubber Venturiwäscher *m*
vertebrate Wirbeltier *n*
vertical pipe Fallrohr *n*
vessel for liquid manure Jauchebecken *n*
veterinary medicine Veterinärmedizin *f*, Tiermedizin *f*
veterinary Veterinär *m*, Tierarzt *m*
viability Lebensfähigkeit *f*
vibrator Vibrator *m*, Rüttelgerät *n*
vinyl acetate Vinylacetat *n*
vinyl chloride Vinylchlorid *n*
viral infection Virusinfektion *f*
virgin rein, unvermischt, roh, unbehandelt
  virgin gasoline Rohbenzin *n*
virosis Viruskrankheit *f*
virotoxic virotoxisch
virus Virus *n*
  virus culture Viruskultur *f*
  virus disease Viruskrankheit *f*
  virus vaccine Virusimpfstoff *m*

viscosity Viskosität *f*, innere Reibung *f*
viscous viskos, dickflüssig, zähflüssig
visible sichtbar
visible sign of alarm sichtbares Alarmzeichen *n*
vital lebensnotwendig, lebenswichtig, vital
vitality Vitalität *f*, Lebenskraft *f*
vitalization Vitalisierung *f*, Belebung *f*, Aktivierung *f*
vitamin Vitamin *n*
vitamin deficiency Vitaminmangel *m*
vitrification Verglasung *f*
vitrification of fly ash Flugascheverglasung *f*
vitrify/to verglasen
vivify/to beleben
volatile flüchtig, verdampfbar
volatile matter flüchtige Bestandteile *mpl*
volatility Flüchtigkeit *f*, Verdampfbarkeit *f*
volatilization Verflüchtigung *f*, Verdampfung *f*
volatilize/to verflüchtigen
volumetric volumetrisch, maßanalytisch
volumetric analysis Volumetrie *f*, Maßanalyse *f*
volumetric flask Meßkolben *m*, Meßflasche *f*
volumetry Volumetrie *f*, Titrimetrie *f*, Maßanalyse *f*
voluminous voluminös, umfangreich
vomit Erbrochenes *n*
vomitive Brechmittel *n*
vulnerability Verwundbarkeit *f*, Verletzlichkeit *f*, Ungeschütztheit *f*

W

walk-in ability Begehbarkeit *f*
walk-in collection canal begehbarer Sammelkanal *m*
wall Mauer *f*, Wall *m*
wall salpeter Mauersalpeter *m*
war Krieg *m*
war gas Kampfgas *n*
war material industry Rüstungsindustrie *f*
warm-blooded animal Warmblüter *m*
warm-blooded animal toxicology Warmblütertoxikologie *f*
warm water Warmwasser *n*
warming of waters Gewässererwärmung *f*
warning and report system Warnmeldesystem *n*
warning colour Warnfarbe *f*
warning device Warngerät *n*
warning light Warnlicht *n*
wash water Waschwasser *n*
wash bottle Waschflasche *f*
wash water container Waschwasserbehälter *m*
wash water vessel Waschwasserbehälter *m*
wash/to waschen
washability Waschbarkeit *f*, Abwaschbarkeit *f*
washable waschecht, waschbar, abwaschbar
washer Wäscher *m*, Waschvorrichtung *f*
washer cycle Wäscherkreislauf *m*
washing agent Waschmittel *n*
washing by attrition Attritionswäsche *f*

washing efficiency Waschkraft *f*, Waschwirkung *f*
washing liquid Waschflüssigkeit *f*
washing method Waschverfahren *n*
washing plant Waschanlage *f*
washing powder Waschpulver *n*
washing programme Spülprogramm *n*
washing suspension Waschsuspension *f*
washing tower Waschturm *m*
washing water Waschwasser *n*
washproof waschecht
wastage Verlust *m*, Verschleiß *m*, Abnutzung *f*, Schwund *m*
waste öde, verödet, wüst, unfruchtbar, unbebaut; nutzlos, überschüssig, ungenutzt, unbrauchbar; Abfall *m*, Abfallgut *n*; Abraum *m*; Ausscheidungsprodukt *n*, Ausschuß *m*, Müll *m*; Verschwendung *f*, Vergeudung *f*; Wüste *f*, Öde *f*, Einöde *f*
waste acid Abfallsäure *f*
waste air Abluft *f*
waste-air purification Abluftreinigung *f*
waste-air purification system Abluftreinigungssystem *n*
waste assimilation Abfallaufnahme *f*
waste avoidance Abfallvermeidung *f*
waste bag Abfallsack *m*
waste charges Abfallgebühren *fpl*
waste code Abfallschlüsselnummer *f*
waste collection vehicle Müllsammelfahrzeug *n*
waste compacter Abfallverdichter *m*, Müllverdichter *m*, Müllpresse *f*

waste compacter plant Abfallverdichtungsanlage *f*
waste composition Abfallzusammensetzung *f*
waste container Abfallbehälter *m*, Abfallcontainer *m*
waste cotton Putzbaumwolle *f*
waste crusher Abfallzerkleinerer *m*
waste disposal measure Abfallentsorgungsmaßnahme *f*
waste disposal plant Abfallbeseitigungsanlage *f*
waste disposal programme Abfallentsorgungsprogramm *n*
waste dump Halde *f*
waste economical abfallwirtschaftlich
waste economical planning abfallwirtschaftliche Planung *f*
waste economical supervisory board abfallwirtschaftliche Aufsichtsbehörde *f*
waste export Abfallexport *m*
waste from livestock breeding Tierhaltungsabfall *m*
waste from slaughtering Schlacht[ungs]abfall *m*
waste fuel Abfallbrennstoff *m*
waste gas Abgas *n*; Deponiegas *n*
waste gas cleaning plant Abluftreinigungsanlage *f*
waste gas combustion Abgasverbrennung *f*
waste-gas component Abgasbestandteil *m*
waste-gas disposal Abgasentsorgung *f*
waste gas emission Abgasemission *f*
waste gas from a disposal site Deponiegas *n*

waste gas mixture Abgasmischung *f*
waste gas of aircraft Flugzeugabgas *n*
waste gas particulate trap Abgaspartikelabscheider *m*
waste gas purification Abgasreinigung *f*
waste-gas purification system Abgasreinigungssystem *n*
waste heap Abfallhaufen *m*
waste heat Abwärme *f*, Abhitze *f*
waste-heat aggregate Abwärmeaggregat *n*, Abhitzeaggregat *n*
waste-heat boiler Abwärme[dampf]kessel *m*
waste-heat loss Abwärmeverlust *m*
waste-heat recovery plant Abwärmerückgewinnungsanlage *f*
waste-heat utilization plant Abwärmeverwertungsanlage *f*
waste incineration Müllverbrennung *f*, Abfallverbrennung *f*
waste incineration plant Müllverbrennungsanlage *f*
waste laboratory Abfallabor[atorium] *n*
waste land recovery Brachlandbegrünung *f*
waste law Abfallgesetz *n*
waste legislation Abfallrecht *n*
waste load Schmutzfracht *f*
waste management Abfallwirtschaft *f*
waste management concept Abfallwirtschaftskonzept *n*
waste management market Abfallwirtschaftsmarkt *m*
waste management planning Abfallwirtschaftsplanung *f*
waste market Abfallmarkt *m*
waste material Abfallstoff *m*, Abfallmaterial *n*
waste mixture Müllgemisch *n*
waste of animal origin Abfall *m* tierischen Ursprungs, tierischer Abfall *m*
waste of mineral origin Abfall *m* mineralischen Ursprungs, pflanzlicher Abfall
waste of plant origin Abfall *m* pflanzlichen Ursprungs, pflanzlicher Abfall *m*
waste oil Abfallöl *n*; Altöl *n*
waste oil collection Altölsammlung *f*
waste package Abfallgebinde *n*
waste paper Altpapier *n*
waste paper processing Altpapieraufbereitung *f*
waste pile Abfallhaufen *m*
waste political situation abfallpolitische Situation *f*
waste press Müllpresse *f*
waste product Abfallprodukt *n*, Abfallerzeugnis
waste pyrolysis Abfallpyrolyse *f*
waste recycling Abfallrecycling *n*
waste reutilization Abfallwiederverwertung *f*
waste site managers Deponieverwalter *m*
waste sludge Abfallschlamm *m*
waste solidification Abfallverfestigung *f*
waste sorter Abfallsortierer *m*
waste sorting plant Abfallsortieranlage *f*
waste to be especially inspected (observed, supervised) besonders zu überwachender Abfall *m*
waste tourism Mülltourismus *m*

waste treatment Abfallbehandlung f
waste treatment method Abfallbehandlungsverfahren n
waste type Abfallart f
waste utilization Abfallverwertung f
waste utilization plant Abfallverwertungsanlage f
waste utilization programme Abfallverwertungsprogramm n
waste volume Abfallvolumen n
waste water Abwasser n, Schmutzwasser n
waste water biology Abwasserbiologie f
waste water canalization Abwasserkanalsystem n, Abwasserkanalisation f
waste water charge Abwasserabgabe f
waste water chemistry Abwasserchemie f
waste water clarification Abwasserklärung f
waste water cleanup Schmutzwasserreinigung f
waste water composition Abwasserzusammensetzung f
waste water concentrate Abwasserkonzentrat n
waste water conveyance Abwasserförderung f
waste water discharge Abwasserabfluß m
waste water disinfection Abwasserdesinfektion f, Abwasserentkeimung f
waste water filtration plant Abwasserfiltrationsanlage f
waste water freight Abwasserfracht f

waste water purification Abwasserreinigung f
waste water system Abwasser[beseitigungs]system n
waste water technology Abwassertechnik f
waste wood Abfallholz n
waste/to verschwenden
waste water collection system Abwassersammlungssystem n
water Wasser n
water absorption Wasseraufnahme f
water alarm cable Wassermeldekabel n
water analysis Wasseranalyse f
water analytics Wasseranalytik f
water balance Wasserhaushalt m
water burst Wasserdurchbruch m
water channel Wasserkanal m
water chemistry Wasserchemie f
water column Wassersäule f
water conditionning Wasseraufbereitung f
water consumption Wasserverbrauch m
water cycle Wasserkreislauf m
water distribution plant Wasserverteilungsanlage f
water distribution system Wasserverteilungssystem n
water endangering substance wassergefährdender Stoff m
water evaporation Wasserverdunstung f
water famine Wasserknappheit f
water hardness Wasserhärte f
water investigation Wasseruntersuchung f
water level of the disposal site Deponiewasserstand m

water management Wasserwirtschaft *f*
water ozonization Wasserozonisierung *f*
water pollution Gewässerverschmutzung *f*, Wasserverschmutzung *f*
water preparation Wasseraufbereitung *f*
water processing Wasseraufbereitung *f*
water protection Gewässerschutz *m*
water protection area Wasserschutzgebiet *n*
water purification Wasserreinigung *f*
water quality Gewässergüte *f*, Wasserqualität *f*
water requirement Wasserbedarf *m*
water-resistant wasserbeständig
water resource Wasserressource *f*, Wasservorrat *m*
water reuse Wasserwiederverwendung *f*
water reutilization Wasserwiederverwendung *f*
water sampling Wasserprobenahme *f*
water saving Wassereinsparung *f*
water softening plant Wasserenthärtungsanlage *f*
water solubility Wasserlöslichkeit *f*
water-soluble wasserlöslich, in Wasser löslich
water storage capability Wasserspeicherfähigkeit *f*
water supply Wasserversorgung *f*
water-supply and distribution management Wasserwirtschaft *f*
water supply network Wasserversorgungsnetz *n*
water supply plant Wasserversorgungsanlage *f*
water supply tank Wasservorratstank *m*
waterglass-improved mineral sealing system wasserglasvergütetes mineralisches Dichtungssystem *n*
watering Berieselung *f*
watering basin Bewässerungsbekken *n*
waterproof wasserdicht
watertight wasserdicht, wasserundurchlässig
water/to bewässern; wässern; tränken; sprengen
waterway Wasserweg *m*
wave breaker Wellenbrecher *m*
wave energy plant Wellenenergieanlage *f*
wear Verschleiß *m*
wear resistance Verschleißfestigkeit *f*, Abnutzungsbeständigkeit *f*
wear test Verschleißprüfung *f*
wearability Abnutzbarkeit *f*
weather Wetter *n*
weahter conditions Wetterlage *f*
weather observation by radar Wetterbeobachtung *f* durch Radar
weather resistance Wetterbeständigkeit *f*
weather-resistant wetterbeständig, witterungsbeständig
weathering Verwitterung *f*, Auswitterung *f*
weatherproof wetterbeständig
weatherproof painting wetterfester Anstrich *m*
weed Unkraut *n*
weed control Unkrautbekämpfung *f*
weekly load (output) curve Wochenganglinie *f*

weighing accuracy Wägegenauigkeit *f*
weighing appliance Wägevorrichtung *f*
weighing scoop Wägeschiffchen *n*
weight constancy Gewichtskonstanz *f*
weight control Gewichtskontrolle *f*
weight decrease Gewichtsabnahme *f*
weight gain Gewichtszunahme *f*
welding electrode Schweißelektrode *f*
well Quelle *f*; Brunnen *m*, Heilquelle *f*, Mineralbrunnen *m*, Zisterne *f*
well water Quellwasser *n*
wet analysis Naßanalyse *f*
wet cleaning Naßbehandlung *f*
wet crushing Naßzerkleinerung *f*
wet deposition Naßablagerung *f*, Naßdeposition *f*
wet desulphurization of flue gas Rauchgasnaßentschwefelung *f*
wet dust collection Naßentstaubung *f*
wet electrofilter Naßelektrofilter *n*
wet grinding Naßmahlen *n*, Naßmahlung *f*
wet-mechanical naßmechanisch
wet oxidation Naßoxidation *f*
wet oxidation of sewage Naßoxidation *f* von Abwässern
wet removal of ash Naßentaschung *f*
wet screening Naßabsiebung *f*
wet scrubber Naßwäscher *m*
wet sludge incineration Naßschlammverbrennung *f*
wet waste Naßmüll *m*
wettability Benetzbarkeit *f*
wettable benetzbar

wetting Befeuchtung *f*, Benetzung *f*
whiten/to bleichen, aufhellen, weißmachen
whiteness Weiße *f*, Blässe *f*, Weißgrad *m*
whitening agent Aufheller *m*, Aufhellungsmittel *n*
whizzer Schleuder *f*, Zentrifuge *f*, Trockenzentrifuge *f*
wholesale manufacturing Massenherstellung *f*
wide-mouth flask Weithalskolben *m*
wildlife Wildbestand *m*
wildlife conservation Naturschutz *m*
wildlife park Naturpark *m*
wilt/to verwelken, welk werden; verwelken lassen
wind gauge Anemometer *n*, Windmesser *m*
windpower Windkraft *f*
winter Winter *m*
wiring technique Anschlußtechnik *f*
withdraw by pipette/to abpipettieren
wither/to welken, absterben, verdorren, ausdörren, austrocknen
wood Holz *n*; Wald *m*, Waldung *f*; Gehölz *n*
wood ash Holzasche *f*
wood culture Holz-Waldwirtschaft *f*
wood fibre Holzfaser *f*
wood preservative agent Holzschutzmittel *n*
wood processing Holzverarbeitung *f*
woodfree paper holzfreies Papier *n*
woodland Waldland *n*
workplace Arbeitsplatz *m*
works chemist Betriebschemiker *m*

works clarification plant Betriebs-
kläranlage *f*
world climate Weltklima *n*
world energy consumption Weltenergieverbrauch *m*
world's supply Weltvorrat *m*
worm conveyor Förderschnecke *f*,
Schneckenförderer *m*
worm-destroying wurmvertilgend
worm dryer Schneckentrockner *m*
worst-case situation Notfallsituation *f*, Grenzfallsituation *f*
wreck Wrack *n*; Schiffbruch *m*
wreck/to abwracken, zertrümmern;
zugrunde gehen

X

X-radiation Röntgenstrahlung *f*
X-ray Röntgenstrahl *m*
X-ray absorption spectrum Röntgenabsorptionsspektrum *n*
X-ray analysis Röntgenanalyse *f*
X-ray apparatus Röntgenapparatur *f*
X-ray diagnostics Röntgendiagno-
stik *f*
X-ray diagram Röntgendia-
gramm *n*
X-ray/to röntgen, durchleuchten
xenobiosis Xenobiose *f*
xenobiotic xenobiotisch, körper-
fremd
xenon Xenon *n*
xerography Xerographie *f*
xerosols Xerosole *mpl*

Y

yeast Hefe *f*; Bierhefe *f*; Backhefe *f*
yeast fermentation Hefegärung *f*
yeast fungus Hefepilz *m*
yeast powder Trockenhefe *f*
yeast production Hefeproduktion *f*
yeasty hefig, gärend
yellow fever Gelbfieber *n*
yermosols Yermosole *mpl*
yield Ausbeute *f*, Ertrag *m*; Nutzleistung *f*, Produktion *f*; Ernte *f*
yield/to abgeben, abwerfen, ergeben, einbringen, hervorbringen
yoghurt Joghurt *n*
yoghurt beaker Joghurtbecher *m*
yperite Senfgas *n*
ytterbium Ytterbium *n*
yttrium Yttrium *n*

Z

zeolite Zeolith *m*
zeolitic zeolithisch
zero adjustment Nulleinstellung *f*,
Nullpunkteinstellung *f*
zero adjustment control Nullpunktkontrolle *f*
zero drift Nullpunktverschiebung *f*
zero point Nullpunkt *m*
zero/to auf Null einstellen
zinc Zink *n*
zinc plating Verzinken *n*
zirconium Zirkonium *n*
zone Zone *f*, Bereich *m*
zone melting Zonenschmelzen *n*
zoochemistry Zoochemie *f*
zooecology Zooökologie *f*
zoological zoologisch

zoology Zoologie *f*
zooplankton Zooplankton *n*
zymogeneous gärungsfördernd

zymogenic gärungsfördernd, zymogen
zymotechnology Gärungstechnik *f*

Deutsch/Englisch
German/English

A

Abart *f* variant, variety, modification
abätzen corrode/to
Abbau *m* decomposition, degradation
Abbau *m* **von Chemikalien** degradation of chemicals
abbaubar degradable
abbaubare Verpackung *f* degradable packaging
Abbaueigenschaft *f* degradation property
abbauen decompose/to, degrade/to
Abbaukonstante *f* degradation constant
Abbaumittel *n* decomposing agent
Abbauprodukt *n* decomposition product
Abbaurate *f* rate of degradation
Abbautest *m* degradation test
Abbauverbesserer *m* degradation improver
abbeizen pickle/to, dress/to
Abbeizmittel *n* paint remover, varnish remover
Abblaserohr *n* blow-off-pipe
Abblaseventil *n* discharge valve
Abbrand *m* burn-up, cinder, roasted ore; calcination
abbrennen burn down/to, burn off/to, deflagrate/to
Abbrucharbeiten *fpl* demolition work
Abbruchreaktion *f* termination reaction
abbruchreif due for demolition
Abdampf *m* waste steam
Abdampfen *n* evaporation, volatilization
Abdampfgefäß *n* evaporating vessel
Abdampfheizung *f* waste steam heater
Abdampfrückstand *m* evaporation residue
Abdampfverwertung *f* waste steam utilization
Abdeckerei *f* knackery
Abdeckungssicherung *f* **auf Silofutter** silage covering
abdekantieren decant/to
abdestillieren distil/to
abdichten seal/to, tighten/to
Abdichtmasse *f* sealing compound
Abdichtung *f* sealing
Abdichtung *f* **einer Deponie** sealing of a landfill
Abdichtungssystem *n* sealing system
Abdunstung *f* evaporation
Abfackelanlage *f* burn-off equipment (device), torch equipment
Abfall *m* waste; discard, garbage, refuse; scrap
Abfallabor[atorium] *n* waste laboratory
abfallarmes Verfahren *n* method poor in waste
Abfallart *f* waste type
Abfallaufbereitungsverfahren *n* processing method for waste
Abfallbehandlung *f* waste treatment, treatment of wastes
Abfallbeseitigungsanlage *f* waste disposal plant
Abfallboxaufnahmevermögen *n* debris (litter) box capacity, dust bin capacity
Abfallbrennstoff *m* waste fuel
Abfallbrennverhalten *n* burning behaviour of waste
Abfallcontainer *m* waste container

Abfallentsorgungsanlage *f* waste disposal plant
Abfallentsorgungseinrichtung *f* waste disposal facility
Abfallentsorgungsmaßnahme *f* waste disposal measure
Abfallentsorgungsmethode *f* waste disposal method
Abfallerzeugnis *n* waste product
Abfallexport *m* waste export
Abfallflüssigkeit *f* waste liquid
Abfallgebinde *n* waste package
Abfallgesetz *n* waste law
Abfallgips *m* waste gypsum
Abfallhaufen *m* waste heap, waste pile
Abfallholz *n* waste wood
Abfallkasten *m* debris box
Abfallmarkt *m* waste market
Abfallmaterial *n* waste material
Abfallmenge *f* quantity of waste
Abfallnachweisverordnung *f* proof-decree for waste, waste proof-decree
Abfallnutzung *f* waste utilization
abfallpolitische Situation *f* waste-political situation
Abfallproblem *n* waste problem, litter problem
Abfallprodukt *n* waste product
Abfallpyrolyse *f* waste pyrolysis
Abfallrecht *n* waste legislation
Abfallreststoff *m* residual waste substance
Abfallschlamm *m* waste sludge
Abfallschlüsselnummer *f* waste code
Abfallsortierer *m* waste sorter
Abfallsortierung *f* waste classification
Abfallsortierungsanlage *f* waste sorting plant
Abfallspezialbehälter *m* **für den medizinischen Gebrauch** special waste container for medicine use
Abfallstatistik *f* waste statistics
Abfallsäure *f* waste acid, spent acid
Abfallverbrennung *f* waste incineration (combustion)
Abfallverfestigung *f* waste solidification
Abfallvermeidung *f* waste avoidance
Abfallverwertung *f* waste utilization, waste recycling, use of waste
Abfallverwertungsanlage *f* waste utilization plant, recycling plant
Abfallverwertungsprogramm *n* recycling programme, waste utilization programme
Abfallvolumen *n* waste volume
Abfallwiederverwertung *f* waste recycling
Abfallwirtschaft *f* waste management
abfallwirtschaftlich waste-economical
abfallwirtschaftliche Aufsichtsbehörde *f* waste-economical supervisory board
abfallwirtschaftliche Planung *f* waste-economical planning
abfallwirtschaftliche Praxis *f* waste-economical practice
abfallwirtschaftliche Regelung *f* waste-economical regulation
abfallwirtschaftliches Verfahren *n* waste-economical method
Abfallwirtschaftsmarkt *m* waste management market
Abfallwirtschaftsplanung *f* waste management planning

Abfallwirtschaftspolitik *f* waste management policy
Abfallzerkleinerer *m* waste crusher (shredder)
Abfallzusammensetzung *f* waste composition
Abfallöl *n* waste oil
abfangen seize/to, trap/to
abfetten degrease/to
abfiltrieren filter off/to
Abfiltrierung *f* filtration, filtering off
abfließen discharge/to, drain/to
Abfluß *m* discharge, drainage, outlet, sink; outflow, effluent
Abflußbedingung *f* drainage condition
Abflußkanal *m* drain, sluice
Abflußleitung *f* drain pipe
Abflußmenge *f* discharge quantity
Abflußrohr *n* drain pipe, sink
Abflußwasser *n* drain water
abführen draw off/to, drain off/to
Abführung *f* outlet, discharge
Abfüllbetrieb *m* packaging plant
abfüllen fill/to; bottle/to
Abfüllvorrichtung *f* dispenser
Abgas *n* exhaust gas, flue gas, stack gas, waste gas
Abgasanalyse *f* flue gas analysis, waste gas analysis
Abgasanlage *f* exhaust gas plant
Abgasbehandlung *f* exhaust gas treatment
Abgasemission *f* waste gas emission
Abgasentstaubung *f* dust removal from exhaust gas
Abgaskamin *m* flue gas duct
Abgasreinigungstechnik *f* waste gas purification technology
Abgasreinigungsverfahren *n* waste gas purification method

abgefräste Straßenoberfläche *f* scarified road surface
abgepackt prepacked
abgesetzter Schlamm *m* sedimented sludge
Abhebern *n* siphonage
Abhitzeaggregat *n* waste heat aggregate
Abhitzekessel *m* waste heat boiler
Abholzung *f* deforestation, clearing
abiogen abiogenic
Abiose *f* abiosis
abiotisch abiotic
 abiotischer Abbau *m* abiotic degradation
Abiozön *n* abiocoen
Abklappen *n* dumping
abklären clarify/to, elutriate/to, filter/to
Abklären *n* clarification, elutriation, filtering
Abkömmling *m* derivative
Abladegebiet *n* dumping area
ablagern age/to; deposit/to; store/to
Ablagerung *f* deposition; deposit, sediment
Ablagerungsmöglichkeit *f* possibility for deposition
Ablagerungsverbot *n* prohibition for the deposition
ablandig offshore
Ablaß *m* discharge
ablassen discharge/to
Ablaßhahn *m* delivery cock
Ablaßschraube *f* bleeder screw, drain screw
Ablaßstopfen *m* plug
Ablaßventil *n* delivery valve
Ablauf *m* delivery, drain, outlet, draining board
ablaufen flow off/to

ablaufen lassen drain/to
Ablaufqualität *f* outflow quality
Ablaufschacht *m* drain shaft
Ablaufsteuerung *f* outflow control, delivery control; sequence control system, sequencing control, sequence control operation
Ableitungsleitung *f* delivery pipe
Ableitungsröhre *f* delivery pipe
Ablöschmittel *n* quenching agent
Abluft *f* exhaust air, used air, spent air
Abluftbestandteil *m* waste gas component
Abluftfilter *n* exhaust gas filter
Abluftreinigung *f* waste air (gas) purification, exhaust air purification, spent (used) air cleaning
Abluftreinigungsanlage *f* waste air (gas) cleaning plant
Abluftreinigungsprogramm *n* programme for the purification of exhaust air (gas)
Abluftreinigungssystem *n* waste air (gas) purification system
Abluftreinigungsverfahren *n* waste air (gas) purification method
Abluftstrom *m* exhaust air flow, waste gas flow
Ablufttechnik *f* exhaust air technology
Abnahmeprüfung *f* acceptance test
Abnormalität *f* abnormality
abnutschen filter by suction/to
Abnutzbarkeit *f* wearability
Abnutzung *f* wear, wastage
Abnutzungsbeständigkeit *f* wear resistance
Abort *m* abortion
Abortgrube *f* cesspool
Abpackung *f* packaging, packing

abpipettieren withdraw by pipette/to
Abraum *m* refuse, waste
Abraumsalze *npl* saline deposits
abreiben scrub/to
Abrieb *m* abrasion
Abriebbeständigkeit *f* resistance to abrasion
Abriebeigenschaft *f* abrasion property
Abriebfestigkeit *f* abrasion resistance, abrasion wear
Absauganlage *f* suction facility (plant), exhausting plant
absaugen exhaust/to, withdraw/to
Absauggerät *n* extraction unit, exhausting apparatus
Absaughaube *f* suction hood
Absaugung *f* exhausting by suction
Abschabevorrichtung *f* scraping device
abschälen peel/to
Abschaltventil *n* shut-off valve
Abschaum *m* scum
abschäumen scum/to
abscheiden separate/to, excrete/to, precipitate/to
Abscheiden *n* separating, separation
Abscheider *m* separator, trap
Abscheiderohr *n* separating tube
Abscheidung *f* separation
abschleifen polish/to
abschlämmen clarify/to
Abschreckmittel *n* quenching agent
Absenkung *f* **des Grundwasserspiegels** lowering of the groundwater level
absetzbar deposable
absetzbare Stoffe *mpl* suspended matter

absetzbarer Schlamm *m* deposible sludge
Absetzbassin *n* settling basin
Absetzbecken *n* precipitation tank, sedimentation basin
Absetzbehälter *m* precipitation tank
absetzen deposit/to, sediment/to
Absetzen *n* depositing, settling
Absetzgefäß *n* sedimentation basin (vessel)
Absetzgeschwindigkeit *f* sedimentation velocity, settling velocity
Absetztank *m* sedimentation basin, settling tank
Absetzungsverhalten *n* sedimentation behaviour
Absinkgeschwindigkeit *f* settling velocity
absitzen settle/to
Absondern *n* separating
Absorbat *n* absorbate
Absorbens *n* absorbent
absorbieren absorb/to
absorbierende Schicht *f* absorbing layer
absorbierender Stoff *m* absorbent
absorbierter Stoff *m* absorbate
Absorption *f* absorption
Absorptionfähigkeit *f* absorbency, absorptivity
Absorptionsbereich *m* absorption region
Absorptionsfilter *n* absorption filter
Absorptionsflasche *f* absorption bottle
Absorptionsgleichgewicht *n* absorption equilibrium
Absorptionskreislauf *m* absorption cycle
Absorptionssystem *n* absorption system

Absorptionstechnik *f* absorption technology
Absorptiv *n* absorbate
Absperrschieber *m* slide valve
Absperrventil *n* block valve, stop valve
Absperrvorrichtung *f* valve
absterben mortify/to, necrose/to, wither/to
abstrahlen emit/to
Abstrahlung *f* reflection
abstreifen strip/to
abtöten kill/to
Abtragung *f* erosion
abtreiben strip/to
abtrennen separate/to
Abtrennung *f* separation
Abtrieb *m* drift
Abtrift *f* drift, deviation
Abundanz *f* abundance
Abwärme *f* waste heat
Abwärmeverlust *m* waste heat loss
Abwärmeverwertungsanlage *f* waste heat recovery (utilization) plant
abwaschbar washable
Abwaschbarkeit *f* washability
Abwasser *n* sewage [water], waste water, foul water, effluent
Abwasser *n* **aus der Glasherstellung** sewage from the glass processing
Abwasser-Belebtschlamm-Gemisch *n* mixture of sewage and activated sewage sludge
Abwasser *n* **von Papierfabriken** paper mill sewage
Abwasserabfluß *m* waste water discharge, sewage discharge
Abwasserabgabe *f* waste water charge

Abwasserausfluß *m* sewage water discharge
Abwasserbehandlung *f* waste water treatment, sewage [water] treatment
Abwasserbehandlungsanlage *f* sewage treatment plant
Abwasserbeseitigung *f* sewage disposal, effluent disposal
Abwasserbeseitigungssystem *n* sewage disposal (removal) system, canalization system, waste water system, sewer system
Abwasserbiologie *f* waste water biology, sewage biology
Abwasserchemie *f* waste water chemistry, sewage chemistry
Abwasserdesinfektion *f* sewage (waste water) disinfection
Abwassereinlauf *m* sewage inflow, sewage inlet
Abwasserentkeimung *f* sewage (waste water) disinfection
Abwasserentsorgung *f* sewage disposal
Abwasserentsorgungskonzept *n* complete concept of sewage disposal
Abwasserfiltration *f* waste water filtration
Abwasserfracht *f* sewage load
Abwasserförderung *f* sewage (waste water) conveyance
abwasserfrei waste water free
Abwasserinhaltsstoff *m* waste water component
abwasserintensiv sewage-intensive
Abwasserkanal *m* sewer
Abwasserkanalsystem *n* sewer system, waste water canalization
Abwasserklärung *f* sewage (waste water) clarification
Abwasserkontrolle *f* effluent control

Abwassernorm *f* sewage standard values
Abwasserproblem *n* sewage problem
Abwasserpumpstation *f* sewage pumping station
Abwasserqualität *f* sewage quality
Abwasserreinigung *f* sewage purification
Abwassersammlungssystem *n* sewage (waste water) collection system
Abwasserstrom *m* waste stream
Abwassertank *m* sewage tank
Abwassertauchpumpe *f* sewage immersion pump
Abwassertechnik *f* sewage technology
abwassertechnisch sewage-technological
Abwasserteich *m* sewage pond
Abwassertransport *m* sewage transportation
Abwassertransportnetz *n* sewer network
Abwasserverbrennung *f* sewage incineration
Abwasserzusammensetzung *f* sewage composition, waste water composition
Abwehrmechanismus *m* defence mechanism
Abwehrreaktion *f* defence reaction
Abwehrstoff *m* antibody
abwracken wreck/to
abzapfen tap/to
Abziehvorrichtung *f* take-up mechanism
Abzug *m* chimney, [laboratory] hood, vent
Abzugsgas *n* chimney gas

Abzugsöffnung *f* vent
Abzugsrohr *n* fume pipe
Abzugsschrank *m* laboratory hood
Abzweigleitung *f* branch pipe
acidogen acidogeneous
Acker *m* field, soil
Ackerbau *m* tillage
Ackerboden *m* arable soil
Ackerland *n* ploughed land, tillage
Ackerschädling *m* field pest
Acrylamid *n* acrylamide
Acrylfaser *f* acryl fibre
Actinochemie *f* actinochemistry
Adaptation *f* adaption
adaptiver Reglerbaustein *m* adaptive controller, closed-loop adaptive control block
Additiv *n* additive
additive Rauchgasfiltertechnik *f* additive flue gas filter technology
Additionsverbindung *f* addition compound
Ader *f* vein
Adhäsion *f* adhesion
ADI-Wert *m* acceptable daily intake, no-effect level, permitted level
adipös adipose
Adrenalin *n* adrenaline
Adsorbat *n* adsorbate
Adsorbens *n* adsorbent, sorbent
Adsorber *m* adsorber
Adsorberharz *n* adsorption resin
adsorbierbar adsorbable
Adsorbierbarkeit *f* adsorbability
adsorbieren adsorb/to
adsorbierter Stoff *m* adsorbate
Adsorption *f* adsorption
Adsorptionseffektivität *f* adsorption efficiency
Adsorptionseigenschaft *f* adsorption property
Adsorptionskapazität *f* adsorbing capacity
Adsorptionsleistung *f* efficiency of adsorption
Adsorptionsmittel *n* adsorbent
Adsorptionsschicht *f* adsorption layer
Adsorptionssystem *n* adsorption system
adsorptive Reinigung *f* adsorptive purification
Adstringens *n* adstringent
aerob aerobic
aerobe Abwasserreinigung *f* sewage oxidation
aerobe Bakterie *f* aerobic bacterium
aerobe Schlammstabilisierung *f* aerobic stabilization of sludge
aerober Abbau *m* aerobic degradation
Aerobier *m* aerobe
Aerobiose *f* aerobiosis
aerodynamisch aerodynamic
Aeromedizin *f* aeromedicine
Aerosol *n* aerosol
Aerosolabscheider *m* aerosol separator
Aerosolsmogkammer *f* aerosol smog chamber
Affinität *f* affinity
Aflatoxin *n* aflatoxin
Agarkeimzahl *f* agar plate count
Agens *n* agent
Agglomerat *n* agglomerate
Agglomeration *f* agglomeration
agglomerieren agglomerate/to
Agglomerierung *f* agglomeration
Aggregat *n* aggregate

Aggregation *f* aggregation
Aggregatzustand *m* state of aggregation; plant status
Agrarökologie *f* agricultural ecology
agrarökologisches Modell *n* agricultural-ecological model
agrarökonomische Forderung *f* agricultural-economical demand
Agrarökosystem *n* agricultural ecosystem
Agrarstrukturmodell *n* model of agricultural structure
Agrartechnik *f* agrotechnology
Agrobiologie *f* agrobiology
Agrochemikalie *f* agricultural chemical
Agronomie *f* agronomy
agronomisch agronomic
Ahornsirupkrankheit *f* maple syrup urine disease
Akarizid *n* acaricide
Akklimatisation *f* acclimatization
akklimatisieren acclimatize/to
Akklimatisierung *f* acclimation
Akkumulation *f* accumulation
Akkumulationseigenschaft *f* accumulation property
Akkumulationspotential *n* accumulation potential
Akkumulationsrate *f* accumulation rate
akkumulieren accumulate/to
akkumulierte Konzentration *f* accumulated concentration
Akkumulierung *f* accumulation
Aktenvernichter *m* paper shredder
Aktinium *n* actinium
Aktionsradius *m* radius of action
aktive Biomasse *f* active biomass
aktive Luftbewegung *f* active movement of air

Aktivierung *f* activation, vitalization
Aktivierungsanalyse *f* activation analysis
Aktivierungsenergie *f* activation energy
Aktivitätsfaktor *m* activity factor
Aktivitätskoeffizient *m* activity coefficient
Aktivkohle *f* activated carbon, active carbon
Aktivkohleadsorption *f* activated carbon adsorption
Aktivkohlepulver *n* activated carbon powder
Aktivkohletechnik *f* activated carbon technology
Aktivkoks *m* activated coke
Aktualisierungszyklus *m* update cycle
akute Toxizität *f* acute toxicity
Akuttoxizitättest *m* acute toxicity test
akzeptabler Grenzwert *m* acceptable limit
Akzeptanz *f* acceptance
Alarmauslöseschwelle *f* alarm level
Alarmeinrichtung *f* signal alarm
Alarmierungssystem *n* alarming system, alarm system
Alarmmonitor *m* alarm monitor
Alarmschalter *m* alarm switch
Alarmsystem *n* alarm system
Albumen *n* albumen
Aldehyd *m* aldehyde
Alge *f* alga
Algenbekämpfung *f* algae control (removal)
Algenbekämpfungsmittel *n* algicide
Algenernte *f* algae harvesting
Algenteppich *m* carpeting of algae

Algenwachstum *n* algae growth
Algizid *n* algicide
algizid algicide
aliphatischer Kohlenwasserstoff *m* aliphatic hydrocarbon
Alkali *n* alkali
alkalibeständig lye-proof
Alkalichloridelektrolyse *f* alcaline chloride electrolysis
alkalifrei non-alkali
alkalisieren alkalify/to
Alkalität *f* alkalinity
Alkaloid *n* alkaloid
Alkohol *m* alcohol
Alkoholintoxikation *f* alcohol intoxication
Alkyl *n* alkyl
Alkylhalogenid *n* alkyl halide
Alkylierung *f* alkylation
Allelopathie *f* allelopathy
Allergen *n* allergen
Allergie *f* allergy
Allergiestoff *m* allergen
allergisch allergic
  allergische Reaktion *f* allergic reaction
  allergisches Potential *n* allergic potential
Allergologie *f* allergology
allgemeine Schutzmaßnahme *f* general safety measure
Allokation *f* allocation
Allopathie *f* allopathy
allotrop allotropic
alluvial alluvial
alpharadioaktiv alpha-active
Alpharadioaktivität *f* alpha radioactivity
Alphastrahl *m* alpha ray
alphastrahlend alpha-emitting
Alphastrahler *m* alpha emitter
Alphateilchen *n* alpha particle
Alphazerfall *m* alpha decay
Altablagerung *f* old deposit[ion]
Altablagerungstyp *m* type of old depositions
Altanlage *f* old plant, old equipment, spent plant, used plant
Altasphaltaufbereitung *f* asphalt regeneration
Altauto *n* used car
Altbatterie *f* used battery, waste-accumulator, spent battery
Altdeponiesanierung *f* remedial of old disposal sites
Alteisen *n* scrap iron
altern age/to, mature/to
Altern *n* aging
Alterung *f* aging
Altglas *n* waste glass, used glass
Altglasrecycling *n* waste glass recycling
Altgummi *m* scrap rubber
Altkabel *n* used cable
Altlast *f* contaminated site (area), old site, polluted area
Altlastenbehandlung *f* **mit Mikroben** microbial old site treatment
Altlastensanierung *f* redevelopment (remedial) of old sites
Altlastensicherung *f* securing of old sites
Altlasterfassung *f* old site registration, registration of polluted area
Altmaterial *n* used material, waste material, spent material
Altmaterialsammler *m* waste material collector
Altmetall *n* scrap metal
Altpapier *n* waste paper

Altpapier-Stroh-Klärschlamm-Gemisch *n* mixture of waste paper straw and sewage sludge
Altpapieraufbereitungsanlage *f* waste paper processing plant
Altreifen *m* used tyre
Altreifenbeseitigung *f* used-tyre disposal
Altreifenkarkasse *f* carcass of used tyre
Altreifenpyrolyse *f* pyrolysis of used tyres
Altreifenverwertung *f* used-tyre utilization
Altseite *f* old page
Altstadtsanierung *f* old-town redevelopment (renewal), towncentre renewal
Altstoff *m* waste material, spent material
Altstoffbewertung *f* valuation of waste material
Altöl *n* used oil, waste oil, spent oil
Altölverordnung *f* regulation of used (spent) oil
Altölsammlung *f* waste (spent, used) oil collection
Altölverwertung *f* processing of spent (used) oil
Aluminium *n* aluminium
Aluminiumdosenhersteller *m* aluminium can producer
Aluminiumguß *m* cast aluminium
Aluminiumkeramik *f* alumina ceramic
Aluminiumsulfat *n* aluminium sulphate
Amalgam *n* amalgam
Amalgamabfall *m* spent amalgam
Amalgamfüllung *f* amalgam filling
Amalgamierung *f* amalgamation

Amalgamverfahren *n* amalgam process
Americium *n* americium
Amin *n* amine
Aminobenzol *n* aminobenzene
Aminoplast *m* aminoplastic
Aminosäure *f* amino acid
Aminosäuresequenz *f* amino acid sequence
Amitose *f* amitosis
Ammoniak *n* ammonia
Ammoniakausscheidung *f* ammonia excretion
Ammoniumanalysator *m* ammonium analyzer
Ammoniumeinleitung *f* ammonium inlet
ammoniumgesteuerte Nitrifikation *f* ammonium-controlled nitrification
Amöbe *f* amoeba
amorpher Zustand *m* amorphous state
Amphibien *fpl* amphibians
Ampholyt *m* ampholyte
amphoter amphoteric
Amylase *f* amylase
Anabiose *f* anabiosis
Anabolismus *m* anabolism
anaerob anaerobic
anaerob-thermophile/mesophile Stabilisation *f* anaerobic-thermophilic/mesophilic stabilization
anaerobe Bakterie *f* anaerobic bacterium
anaerober Festbettreaktor *m* anaerobic fixed-bed reactor
Anaerobier *m* anaerobe
Anaerobiose *f* anaerobiosis
anaerobisch anaerobic

Analogwertausgabe *f* analogue output
Analogwertverarbeitung *f* analogue-value processing
Analogwertzustand *m* state of analogue output
Analysator *m* analyzer
Analyse *f* analysis, assay
Analysendaten *pl* analysis data
Analyseneichfunktion *f* analysis calibration function
Analysenergebnis *n* analysis result
Analysengerät *n* analyzer
Analysenmethode *f* analysis method
analysenrein analytically pure, of reagent purity
Analysenverfahren *n* analysis method
analysieren analyze/to
Analyt *m* analyte
Analytik *f* analytics
Analytiker *m* analyst
analytisch analytic[al]
  analytische Chemie *f* analytical chemistry
  analytische Daten *pl* analytical data
  analytische Methode *f* analytical method
  analytische Qualitätssicherung *f* analytical quality securing
  analytische Umweltdaten *pl* environmental analytical data
  analytisches Ergebnis *n* analytical result
Anbaufläche *f* arable area
anbaufähig tillable
Anbauversuch *m* cultivation trial
Anemometer *n* anemometer, wind gauge
Anfälligkeit *f* susceptibility
Anfangskonzentration *f* initial concentration
Anforderungskatalog *m* catalogue of demands
angeborene Mißbildung *f* congenital malformation
angereichertes Uran *n* enriched uranium
angeschwemmter Boden *m* alluvial soil
angewandte Chemie *f* applied chemistry
angewandter Umweltschutz *m* applied environmental protection
Angiographie *f* angiography
Anhaftung *f* adhesion
anhäufen accumulate/to
Anhäufung *f* accumulation, congestion
Anion *n* anion
anionenaktiv anion-active
Anionenaustausch *m* anion exchange
anisotrop anisotropic
Anisotropie *f* anisotropy
Anlagenabschnitt *m* plant sector
Anlagenausrüstung *f* plant equipment
Anlagendokumentation *f* plant documentation, system documentation
Anlagenfließbild *n* plant mimic, plant flow diagram, plant flow sheet
Anlagenkomponente *f* system component, plant device
Anlagenmanagement *n* plant management
Anlagensicherheit *f* plant safety
Anlagenübersicht *f* plant survey, general plan of the system
Anlagenverbund *m* integrated plants

Anlagerung f **von Wasserstoff** addition of hydrogen
Anlagerungsverbindung f addition compound
anlagespezifische Daten pl plant-specific data
Anlieferungsbedingung f delivery condition
anlösen solvate/to
Annahmeerklärung f acceptance certificate
annehmbare Grenze f acceptable limit
annehmbare Konzentration f acceptable concentration
Anode f anode
Anodenentladung f anode discharge
Anodenschlamm m anode slime
anodisch anodic
Anomalie f abnormality
anorganisch inorganic
anorganische Chemie f inorganic chemistry
anorganische Säure f mineral acid
anormal abnormal
anoxisch anoxic
Anpassung f adaption, adjustment, assimilation; fit
Anpassungsaufwand m adaption expenditure, extra work for adaption
Anpaßstück n adapter, fitting
anpflanzen plant/to
Anpflanzung f plantation
Anregung f excitation, stimulation
Anregungsbedingung f excitation condition
Anregungsenergie f excitation energy
Anregungsmittel n excitant, stimulant

anreichern enrich/to, concentrate/to, enhance/to, accumulate/to
Anreicherung f enrichment, enhancement, accumulation, adsorption
Anreicherung f **durch die Nahrungskette** food chain accumulation
Ansatzbad n starting bath
Ansatzrohr n connecting tube, nozzle, lateral adapter (tube)
Ansatzstück n joint adapter
ansäuern acidify/to, make acidic/to, sour/to
Ansäuerung f acidification
Ansaugbelüfter m suction aerator
ansaugen aspirate/to, suck in/to
Ansaugen n aspiration, sucking, suction
anschlämmen paste/to
Anschlämmen n depositing of mud
Anschlußbedingung f connection condition
Anschlußgebühr f connection charge
Anschlußkostenbeitrag m capital contribution to connection cost
Anschlußleitung f connecting cable, connecting lead
Anschlußtechnik f connecting technique, termination system, wiring technique
Anschwemmfilter n precoat filter, settling filter
Anschwemmschicht f precoat layer
anstecken infect/to
ansteckend infectious, infective, contagious
ansteckende Krankheit f infection (infectious) disease
Ansteckung f infection, contagion

Ansteckungsgefahr *f* danger (risk) of infection
ansteckungsgefährlicher Abfall *m* infectious endangering waste
Ansteckungsherd *m* centre of infection
Ansteckungsquelle *f* source of infection
Anstrich *m* paint[ing], coating
Anstrichfarbe *f* paint, coating compound
antarktisches Ozonloch *n* antarctic ozone hole
Anthracen *n* anthracene
Anthrazit *m* anthracite
anthropogen anthropogenic
anthropogene Chemikalie *f* anthropogenic chemical
anthropogene Einwirkungen *f* anthropogenic impacts
anthropogene Substanz *f* anthropogenic substance
anthropogener Stoff *m* anthropogenic substance
Antiallergikum *n* antiallergic agent
antiasthmatisch antiasthmatic
antibakteriell antibacterial, bactericidal
Antibiotikum *n* antibiotic
antibiotisch antibiotic, microbicidal
Antidepressivum *n* antidepressant
Antidottherapie *f* antidote therapy
Antigen *n* antigen
Antiklopfmittel *n* antiknock
Antikoagulationsmittel *n* anticoagulant substance
Antimon *n* antimon
Antischaummittel *n* defoaming agent
antiseptisch antiseptic
Anzeigebereitschaft *f* **der Bevölkerung** readiness for announcement of the population
Anzeigepflicht *f* obligation of announcement
anzeigepflichtig certifiable
Aphizid *n* aphicide
Appetitlosigkeit *f* inappetence
Appretur *f* finish, dressing, size
Aquakultur *f* mariculture
aquatisch aquatic
aquatische Flora *f* aquatic flora
aquatisches Feldsystem *n* aquatic field system
aquatisches System *n* aquatic system
Äquivalenzstudie *f* equivalence study
arbeitsbedingte Exposition *f* occupational exposure
Arbeitsbereichsanalyse *f* analysis of working areas
arbeitsmedizinisch industrial medical, occupational medical
arbeitsmedizinische Vorsorgeuntersuchung *f* occupational medical check-up, preventive medical check-up
Arbeitsschutz *m* occupational safety
Arbeitssicherheit *f* occupational safety
Argon *n* argon
Aroma *n* aroma, flavour, fragrance
Aromaten *pl* aromatic hydrocarbons, aromatics
aromatisch aromatic
aromatischer Kohlenwasserstoff *m* aromatic hydrocarbon
Arretierungseinrichtung *f* arresting device
Arsen *n* arsenic

Arsenschlamm *m* arsenic sludge
Arsenstaub *m* arsenic dust
Artefakt *n* artefact
Artenschutz *m* protection of species
Artenvielfalt *f* variety of species
Arterhaltung *f* preservation of species
artfremd foreign
artspezifisch species-specific
Arznei *f* drug, medicine, medicament, pharmaceutical, remedy
Arzneimittelangebot *n* medicine offer
Arzneimittelanwendung *f* medicine application
Arzneimittelforschung *f* drug research, pharmacological research
Arzneimittelsicherheit *f* drug safety
Arzneiverordnungsbericht *m* report of drug prescription
Asbest *m* asbestos
Asbestentsorgung *f* abestos disposal
Asbestfaser *f* asbestos fibre
Asbestfaserstaub *m* asbestos fibre dust
asbesthaltig asbestos-containing
Asbestkonzentration *f* asbestos concentration
Asbestose *f* asbestosis
Asbestprobenahme *f* asbestos sampling
Asbestsanierung *f* asbestos renovation
Asbeststaub *m* asbestos dust
asbeststaubgefährdet asbestos dust exposed
Asbestverfestigung *f* asbestos solidification
Asbestzementabfall *m* asbestos-cement waste
Asche *f* ash

Aschegehalt *m* ash content
Aschegemisch *n* mixture of ash
Aschenhalde *f* ash heap
Ascheverwertung *f* ash utilization
Asphalt *m* asphalt, mineral pitch
Asphaltbetonmischgut *n* mixed asphalt concrete material, asphalt concrete mix
Asphaltmischanlage *f* asphalt mixing plant
Asphalttragschichtmischgut *n* asphalt supporting layer mix
Aspirator *m* aspirator
Assimilation *f* assimilation
assimilieren assimilate/to
Astatium *n* astatine
astfreies Holz *n* branchless wood
astreiches Holz *n* branchy wood
atembar respirative
atemfähig respirative
Atemgift *n* respiratory poison
Atemgrenzwert *m* breathing capacity
Atemluft *f* respiratory air, breathable air
Atemnot *f* shortness of breath
Atemrate *f* respiratory rate
Atemschutz *m* breathing protection
Atemschutzgerät *n* respirator protection apparatus
Ätiologie *f* etiology
Atmosphäre *f* atmosphere
atmosphärisch atmospheric
  atmosphärische Schicht *f* atmospheric layer
  atmosphärische Spezies *f* atmospheric species
  atmosphärischer Druck *m* atmospheric pressure
  atmosphärischer Eintrag *m* atmospheric input

atmosphärischer Niederschlag *m* atmospheric precipitation
Atmung *f* breathing, respiration
Atmungsmesser *m* spirometer
Atmungswiderstand *m* breathing resistance
Atom *n* atom
Atombombe *f* atomic bomb
Atombombenüberlebender *m* atomic bomb survivor
Atomkraft *f* nuclear power
Atomkraftwerk *n* nuclear power station
Atomkrieg *m* nuclear war
Atommüll *m* radioactive waste
atomwaffenfreie Zone *f* nuclear weapon-free area
Atomnummer *f* ordinal number
atoxisch non-toxic, atoxic
Atrophie *f* atrophy
Attrition *f* attrition
Attritionsstufe *f* step of attrition
Attritionswaschtrommel *f* attrition washing drum, sand washing drum
Attritionswäsche *f* washing by attrition
Ätzbad *n* etching bath
ätzbeständig caustic resisting
ätzen etch/to
Ätzen *n* etching
Ätzflüssigkeit *f* caustic liquid
Ätzkali *n* caustic alkali
Ätzmittel *n* caustic
Ätzschlamm *m* etching sludge
Ätzstoff *m* caustic
Ätzung *f* etching
Audiogramm *n* audiogram
Aue *f* lea
Auenboden *m* alluvial soil
Aufarbeitung *f* preparation, working-up

Aufarbeitungsmethode *f* method of processing, preparation method, treatment process
Aufarbeitungszentrum *n* processing centre
Aufbautechnik *f* structural technique, installation technique, design technique
aufbereiten dress/to, process/to, prepare/to, refine/to
Aufbereitung *f* dressing, processing, preparation, conditioning
Aufbereitungsanlage *f* dressing plant, processing plant, preparation plant
Aufbereitungsverfahren *n* processing method
Auffangbehälter *m* receiver, collecting vessel
auffangen receive/to, collect/to
Auffanggefäß *n* trap, vessel
Auffangkolben *m* collecting flask
Auffangtrichter *m* collecting funnel
Auffangwanne *f* collecting tank, collection tank
auffinden detect/to
Aufforstung *f* afforestation
auffrischen refresh/to, renew/to
Aufgabebunker *m* feeding bunker
Aufgabetrommel *f* feeding drum
Aufgabevorrichtung *f* feeder, charging device
Aufguß *m* tincture
Auflage *f* **der Gewerbeaufsichtsbehörde** condition of the industrial inspection authority
auflandig onshore
auflösbar dissolvable, soluble, [re]solvable
Auflösbarkeit *f* dissolubility, resolvability, solubility

Auflösemittel

Auflösemittel *n* dissolvent, dissolving liquid
auflösen dissolve/to, resolve/to
Auflösung *f* dissolution, resolution, solution
Aufnahme *f* **durch Inhalation** inhalation uptake
Aufnahmegeschwindigkeit *f* absorption rate, uptake rate
Aufnahmerate *f* uptake rate
aufnehmen absorb/to, admit/to, adsorb/to
Aufräumungsarbeiten *fpl* salvage work
aufsaugen absorb/to, suck up/to
Aufsaugen *n* absorption, soakage
aufschlämmen slurry/to, suspend/to
Aufschlämmung *f* sludge, slurry, slime, suspension
aufschließen cook/to, decompose/to, digest/to
Aufschluß *m* cooking, decomposition, digestion
Aufschüttung *f* **mit Flugasche** fly ash dump
Aufschwemmen *n* suspending
Aufschwemmung *f* suspension
Aufsichtsbehörde *f* supervisory authority, supervisory board
Aufsichtspflicht *f* responsibility
aufspalten cleave/to, decompose/to, split/to, crack/to
Aufspaltung *f* cleavage, fission, decomposition, splitting, cracking
Aufstauen *n* damming-up
Aufstauhöhe *f* dam level
auftauen thaw/to, defrost/to
Auftaumittel *n* unfreezing material
aufwandsarm of low expenditure, of low expense
Aufwertung *f* valorization

Aufzeichnungsgerät *n* recording instrument
Augendusche *f* eye douche
Augenreizung *f* eye irritation
Ausatmung *f* respiration
Ausbaggern *n* dredging
Ausbeute *f* yield, harvest
Ausbeutung *f* exploitation, utilization, winning
Ausblasen *n* **flüchtiger Stoffe** blowing-off volatile substances
Ausbreitung *f* diffusion, dispersion, distribution, propagation, spread
Ausbreitungspfad *m* spreading pathway
Ausbreitungspfad *m* **von Schadstoffen** dispersion of pollutants
ausbrennen anneal/to, burn [out]/to
Ausbruch *m* outbreak, outburst
ausdünsten transpire/to, evaporate/to
Ausdünstung *f* transpiration, evaporation
Ausfall *m* breakdown, loss, fall-out, deficiency, deficit
Ausfallquote *f* failure rate
ausfließen flow out (off)/to, run out/to, discharge/to
ausflocken flocculate/to
ausflockend flocculating
Ausflockung *f* flocculation, coagulation
Ausflockungsmittel *n* flocculating agent, flocculant
Ausfluß *m* discharge, outflow
Ausfluß *m* **aus einem See** bayou
Ausflußkontrolle *f* outlet control
Ausflußrohr *n* outlet pipe
Ausflußzeit *f* time of discharge
Ausfrierverfahren *n* freezing process, freezing method

ausfällen precipitate/to
Ausfällung *f* precipitation; precipitate
Ausgangskonzentration *f* initial concentration
Ausgangslösung *f* initial solution
Ausgangsmaterial *n* feed (original, parent) material
Ausgangsstoff *m* base material
ausgebaggertes Material *n* dredged (excavated) material
ausgefaulter Schlamm *m* digested sludge
ausgeglühte Kohle *f* cinder
ausgelaugt leached out
ausgemusterte Mülltonne *f* rejected refuse bin
ausgewählter Schadstoff *m* selected pollutant
Ausgleichsbecken *n* equalizing tank
ausglühen anneal/to
Ausgrabung *f* disinterment, excavation
Ausguß *m* sink, discharge connection, outlet
Ausgußbecken *n* sink basin
Ausgußrohr *n* delivery pipe, outlet pipe, drain pipe
Aushebern *n* siphonage
ausheizen anneal/to
Aushub *m* spoil
Aushungern *n* starvation
auskleiden line/to, coat/to
Auskleiden *n* lining, coating
  Auskleiden *n* **mit Gummi** rubber lining
  Auskleiden *n* **von Deponien** lining of landfills
  Auskleiden *n* **von Deponien mit Bitumen** bitumen landfill lining
  Auskleiden *n* **von Deponien mit Polymeren** landfill lining with polymers
  Auskleiden *n* **von Deponien mit Zement** cement landfill lining
Auskleidung *f* lining, coating, jacket
  Auskleidung *f* **von Deponien mit Asphalt** asphalt landfill lining
Auskleidungsmaterial *n* liner material
Auskofferung *f* excavation
Auslauf *m* delivery, outlet
auslaufen leak out/to, run out/to, drain off/to
Auslaufventil *n* discharge valve, tap
auslaugen leach/to, extract/to
Auslaugergebnis *n* leaching result
Auslaugtest *m* leaching test
auslaugungsresistente Schlacke *f* leaching-resistant slag
Auslaugverhalten *n* leaching behaviour
Auslaugversuch *m* leaching test
Auslaugwiderstand *m* leaching resistance
Auslaßhahn *m* outlet cock
Auslaß *m* discharge, outflow
Auslaßventil *n* bleed valve, discharge valve, outlet valve
Auslesen *n* sorting
auslöschen extinguish/to; quench/to
Auslösekonzentration *f* threshold concentration
auslösen induce/to; release/to
auslösender Wert *m* threshold limit value
Auslöseschwelle *f* threshold limit
  Auslöseschwelle *f* **für einen Alarm** alarm level
Auslösung *f* induction

Ausmerzung *f* elimination
Ausnahmebestimmung *f* saving clause
Ausnahmeempfehlung *f* exceptional recommendation
Ausnahmefall *m* exceptional case
Ausnahmegenehmigung *f* exceptional permission
Ausnahmeregelung *f* exception regulation
ausrinnen leak/to
ausrotten eradicate/to, exterminate/to, extirpate/to
Ausrottung *f* eradication, extermination, extirpation
Ausrüstungsstandard *m* standard of the equipment
ausscheiden excrete/to
Ausscheidung *f* **im Urin** excretion in urine, urine excretion
Ausscheidungsprodukt *n* excretion product, waste
ausschlämmen elutriate/to
Ausschlämmung *f* elutriation
Ausschmelzung *f* smelting, melting
Ausschuß *m* rejection, rejects; scrap; waste
ausschwefeln sulphur/to, bloom/to
Ausschwemmen *n* **von Schadstoffen** flushing of harmful materials
Außenanstrich *m* exterior paint
Außenbeständigkeit *f* exterior durability, outdoor durability
aussenden emit/to
Aussender *m* emitter
Aussendung *f* emission
Außenexposition *f* outdoor exposure
Außenfarbe *f* exterior paint
Außenluft *f* external air
Außentemperatur *f* outdoor temperature

Außerbetriebsetzung *f* shutdown
aussondern eliminate/to, separate/to
aussortieren sort out/to
ausspülen rinse [out]/to, scour/to, wash out/to, flush/to
Ausspülung *f* elutriation, wash-out
Ausstoß *m* output, production, discharge, ejection
ausstrahlen emit/to, irradiate/to, radiate/to
Ausstrahlung *f* emission, irradiation, radiation
Ausstrahlungsvermögen *n* emittance
Ausströmung *f* emission, outflow, discharge, efflux
Ausströmung *f* **von Gas** emission of gas
Ausströmungsgeschwindigkeit *f* rate of discharge
austauschbares Filter *n* interchangeable filter
austragen carry out/to
Austrittsdosis *f* exit dose
austrocknen dry up/to, exsiccate/to, season/to
Austrocknung *f* drying up, desiccation, seasoning
auswandern emigrate/to
auswaschen wash out/to, scrub/to, scour/to
Auswaschmethode *f* wash-out method
Auswaschung *f* wash-out, leaching
Auswitterung *f* weathering
autark autarkic[al], self-supporting, self-sufficient
autarkes System *n* stand-alone system, autarkic system
Autoabgas *n* car exhaust gas

Autoabgaskatalysator *m* waste gas catalyst of automobile
Autoinfektion *f* autoinfection
Autointoxikation *f* autointoxication
Autokatalyse *f* autocatalysis, self-catalysis
Automatikbetrieb *m* automatic operation
automatisch arbeitende Probeaufgabeeinrichtung *f* autosampler
Autoökologie *f* autoecology
Autopsie *f* autopsy
Autoradiogramm *n* autoradiogram
Autoschnellstraße *f* motorway
Autoschrott *m* car scrap
Autotrophie *f* autotrophy
Autowrack *n* car wreck
Autoxidation *f* autoxidation
Axialgebläse *n* axial blast
azeotrop azeotropic
azeotrope Destillation *f* azeotropic distillation
Azidität *f* acidity
azyklisch acyclic

B

Bach *m* creek, stream
Backenbrecher *m* jaw crusher
backende Kohle *f* bituminous coal
Backhefe *f* yeast
Backofen *m* oven
Bad *n* bath
Badbeschickung *f* preparation of the bath
Badestrand *m* beach
Badeverbotzone *f* bathing prohibition zone
Badreaktion *f* bath reaction
Bagger *m* dredger; excavator
Baggerbetrieb *m* dredging

Baggertorf *m* dredged peat
Bajonettanschluß *m* bayonet fitting
Bakterie *f* bacterium
bakteriell bacterial
bakterielle Verschmutzung *f* bacterial pollution
Bakterienflora *f* bacterial flora
Bakteriengift *n* bacterial toxin
Bakterienimmobilisierung *f* immobilization of bacteria; immobilization by bacteria
Bakterienkrieg *m* bacterial war
Bakterienkunde *f* bacteriology
bakterienresistent bacteria-resistant
Bakterienverseuchung *f* bacterial contamination
Bakteriologe *m* bacteriologist
Bakteriologie *f* bacteriology
bakteriologisch bacteriological
bakteriologische Kultur *f* bacteriological culture
bakterizid bactericidal
Bakterizid *n* bactericide
Ballastbehälter *m* ballast tank
Ballaststoff *m* bulkage, ballast material, roughage
ballaststoffreich rich of roughage
Ballenpresse *f* baling press, bale press
Ballenschneider *m* bale cutter
Ballungsgebiet *n* overcrowded area
Bandbeschickung *f* belt feeding, belt charging
Bandfilter *n* belt filter
Bandfilterpresse *f* belt filter press
Bandförderer *m* belt conveyor
Bandförderung *f* belt conveyance
Bandwurm *m* tapeworm
Bandwurmmittel *n* taeniacide
Bankett *n* banquette

Barium n barium
Barometer n barometer
Base f base
Basenaustauscher m base exchanger
Basisabdichtung f bottom sealing
basisch basic, alkaline
  basischer Abfall m basic waste
basisch machen basify/to
Basiskomponente f base component
Basizität f basicity, alkalinity
Batterie f battery
Batteriebox f battery box
Batterietest m battery test
Bauakustik f building acoustics
Baubiologie f building biology
Baufälligkeit f disrepair
Bauholz n lumber
Baukasten-Räummaschinensystem n module system for broaching machines
Baukastenprinzip n modular construction principle, modular principle
Baukastensystem n modular construction system, modular system
Baulärmschutz m construction noise control
Baum m tree
Baumart f tree species
Baumchirurg m tree surgeon
Baumgrenze f timber line, tree line
baumreich arboreous
Baumrinde f bark, rind
Baumschaden m damage of a tree
Baumstamm m trunk, stem
Baumzucht f arboriculture
bauphysikalisch building-physical
Bauschutt m [building] rubble
Bauschuttaufbereitung f treatment of rubble
Bauschuttdeponie f rubble refuse dump

Bauschuttkippe f rubble dump
Bauschuttrecycling n rubble recycling
Baustellenmüll m waste of building sites
Baustoffrecyclinganlage f rubble recycling plant
bazillär bacillary
Bazillenherd m bacilli focus
Bazillenspore f spore of a bacillus
Bazillus m bacillus
Bearbeitungsabfall m scrap of processing
Beatmung f respiration, ventilation
Beatmungsgerät n respirator
Beauftragter m für Gefahrgut authorized person for hazardous goods
bebauen develop/to, build on/to; cultivate/to
Bebauung f development; cultivation
Bebauungsdichte f density of buildings
Bechersieb n bucket sieve
Becken n pool, basin, tank
Befall m affection, infestation
befeuchten humidify/to, moisten/to, wet/to, dampen/to
Befeuchtung f humidification, moistening, wetting
Befeuerung f heating, firing
Beförderung f von Gefahrgut auf Straßen transportation of hazardous goods on roads
Beförderungsgenehmigung f transportation permission
Beförderungspapier n transportation document
Befüllungsplan m filling plan
Befund m finding, result, state

Begehbarkeit *f* accessibility, walk-in ability
Begleiterscheinung *f* accompanying phenomenon
Begleitgift *n* accompanying poison
Begleitschein *m* accompanying document, way bill
Begleitscheinverfahren *n* method of the accompanying document
Begleitstoff *m* impurity, accompanying substance
begrünen grass/to, sow down/to
Begrünung *f* greenbelt setting
Behälterreinigung *f* container cleaning
Behältersystem *n* container system
Behandlung *f* **mit Ozon** ozonization
Behandlungsanlage *f* processing plant, treatment plant, treatment facility
Behandlungsanlage *f* **für Sonderabfall** plant for the treatment of special waste
Behandlungshäufigkeit *f* frequency of treatment
Behandlungsverfahren *n* treatment method
Behandlungszentrum *n* treatment centre, processing centre
Behördenentscheidung *f* authority decision
Behördenpflicht *f* duty of the authorities
Beimengung *f* additive
Beizflüssigkeit *f* pickle, caustic liquor, bate
Beizmittel *n* caustic, mordant
bekämpfen control/to
Bekämpfungsmaßnahme *f* control measure

Belastungsgrenze *f* load[ing] limit, maximum load
Belastungskonzentration *f* impact concentration
Belastungsprobe *f* loading test, tolerance test
Belastungssituation *f* situation of impact (burden), situation of load
Belastungsveränderung *f* load change
Belastungsverhältnisse *npl* loading situation, overloading situation
Belastungsvermögen *n* carrying capacity
Belastungswiderstand *m* impact resistance
beleben activate/to, revive/to, stimulate/to
Belebtschlamm *m* activated sludge
Belebtschlamm *m* **aus Kläranlagen** activated sludge of clarification plants
Belebtschlammbecken *n* activated-sludge basin
Belebtschlammeindickung *f* concentration of activated sludge
Belebtschlammflocken *fpl* activated sludge flocs
Belebtschlammmikrobiologie *f* activated-sludge microbiology
Belebungsanlage *f* activated sludge plant
Belebungsbecken *n* aeration tank (basin), activated-sludge basin
Belebungsmittel *n* stimulant, activator
belüften aerate/to; ventilate/to
Belüfter *m* aerator
Belüfteraggregat *n* blower aggregate, aeration aggregate, aerator, ventilation aggregate

Belüftergruppe 178

Belüftergruppe *f* aerator group, blower group, ventilation group,
Belüfterleistung *f* blower performance, aerator performance
belüftete Oberfläche *f* ventilated surface, aerated surface
Belüftung *f* aeration; ventilation
Belüftungsanlage *f* ventilation system
Belüftungseinrichtung *f* aeration device, ventilation device
Belüftungshaube *f* ventilation hood
Belüftungsintensität *f* aeration intensity
Belüftungskreiselrührer *m* aeration impeller
Belüftungsphase *f* aeration period
Belüftungsstein *m* aeration stone
Belüftungsstutzen *m* air admission tube
Belüftungssystem *n* ventilation system, aeration system
Belüftungsturbine *f* ventilation turbine
benetzbar wettable
benetzen wet/to, moisten/to, sprinkle/to
Bentonit *m* bentonite
Benzin *n* petrol, fuel, *{Am}* gas[oline]
Benzinabscheider *m* petrol separator, petrol trap
Benzindampfentsorgung *f* petrol vapour disposal
Benzindampfrückgewinnungsanlage *f* petrol vapour recovery plant
Benzinersatz *m* petrol substitute
Benzinleck *n* fuel leak
Benzinleitung *f* petrol pipe
Benzinzusatz *m* petrol additive

Benzol *n* benzene
Benzolvergiftung *f* benzene poisoning
Beobachtungsballon *m* observation balloon
Beobachtungsfunktion *f* observation (monitoring) function, surveillance function
Beobachtungsstelle *f* observation point
Beobachtungsturm *m* observation tower
Beobachtungswarte *f* observation tower
Beobachtungszeitraum *m* period of observation
Bepflanzung *f* plantation
Beregnung *f* sprinkler irrigation
bergen rescue/to, save/to
Bergung *f* rescue, salvage
Bergungsbehälter *m* recovery tank, recovery vessel
Bergungsfahrzeug *n* salvage vessel
Bergungstrupp *m* rescue party
Bergwerk *n* mine, pit
Bergwerksarbeiterpneumokoniose *f* coalminer's pneumoconiosis
Bergwesen *n* mining
berieseln irrigate/to, sprinkle/to, spray/to
Berieselung *f* irrigation, sprinkling, spraying, watering
Berieselungsapparat *m* sprinkler, sprayer
Berieselungsfläche *f* irrigation surface
Berieselungskühler *m* spray cooler, irrigation cooler
Berieselungsturm *m* scrubbing tower, spray tower
Berkelium *n* berkelium

berufsbedingte Atemwegserkrankung *f* occupational respiratory disease
berufsbedingte Exposition *f* occupational exposure
Berufskrankheit *f* industrial disease, occupational disease
Berufsrisiko *n* occupational hazard, occupational risk
Beryllium *n* beryllium
Beschaffungsrecht *n* right for provision, procuring right
beschichtete Säule *f* coated column
Beschichtung *f* coat[ing]
Beschichtungsfarbe *f* coating colour
Beschichtungsharz *n* coating resin
Beschichtungsmasse *f* coating compound
Beschichtungsmaterial *n* coating material, coating compound
Beschichtungsverfahren *n* coating process
Beschickungsanlage *f* charging device, feeder
Beschickungsansatz *m* batch
Beschickungseinrichtung *f* loading device, charging device
Beschickungszone *f* feed section
beschmutzen pollute/to, soil/to, stain/to
Beseitigung *f* removal, disposal
Beseitigung *f* **radioaktiver Abfälle** radioactive waste disposal
Beseitigungsanlage *f* disposal plant, disposal facility
Beseitigungskonzept *n* concept of disposal
Beseitigungsmethode *f* disposal method
Beseitigungspflicht *f* duty of disposal

Besiedlung *f* colonization
besprengen sprinkle/to
bespritzen sprinkle/to
besprühen spray/to
Besprühen *n* spraying
Bestandsaufnahme *f* resource inventory
Bestandteil *m* component, constituent
Beständigkeit *f* consistency, constancy, durability, stability, resistance
Bestäubung *f* dusting; pollination
Bestimmungsfehler *m* determination error
Bestimmungsgrenze *f* determination limit
bestrahlen irradiate/to
bestrahltes Saatgut *n* irradiated seed
Bestrahlung *f* irradiation, radiation
Bestrahlungsdauer *f* irridiation time
Bestrahlungsdosis *f* exposure dose
Bestrahlungskonservierung *f* radiation preservation
Bestrahlungsmutation *f* radiation mutation
Bestrahlungsschaden *m* irradiation damage
beta-aktiv beta-active
Betastrahl *m* beta x-ray
Betäubungsmittelgesetz *n* anaesthetic law
Betazerfall *m* beta decay
Beton *m* concrete
Betonabschirmung *f* concrete shield
Betonierung *f* concreting
Betonkonstruktion *f* concrete construction
Betonschicht *f* layer of concrete

Betonumhüllung *f* concrete encasement
Betonunterlage *f* concrete base
Betriebsarmatur *f* industrial fitting
Betriebsart *f* operating mode
Betriebsführungsebene *f* level of plant management
Betriebsmeldung *f* process signal
Betriebsparameter *m* operational parameter
Betriebssicherheit *f* operational safety
Betriebswasser *n* process water
betroffenes Gebiet *n* response area
Bevölkerungsdichte *f* density of population
Bevölkerungsdosis *f* collective dose
Bevölkerungsexplosion *f* population explosion
Bevölkerungswachstum *n* population growth
Bevorratungsbehälter *m* stock vessel, stock container
bewaldet arboreous
bewässern water/to, irrigate/to
Bewässerung *f* watering, irrigation
  Bewässerung *f* **von Wäldern** irrigation of forests
Bewässerungsbecken *n* watering basin, irrigation basin
Bewässerungsgraben *m* irrigation canal
Bewässerungstechnik *f* irrigation technology
Bewertungsmaßstab *m* parameter of valuation
Bewitterung *f* weathering
Bewitterungsprüfung *f* weathering test
bewohnbar inhabitable
bewohnen inhabit/to

Bewohnen *n* habitation
Bewuchs *m* vegetation
bewuchsverhindernd antifouling
Bezugsüberwachung *f* supply monitoring
Bienenwachs *n* beeswax
Bierhefe *f* yeast
Bilgewasser *n* bilge water
Bimsbeton *m* pumice concrete
Bimsstein *m* pumice
binäres Gemisch *n* binary mixture
Bindemittel *n* binder, binding agent (material), cement
Bindungsbruch *m* bond breaking
Bindungsenergie *f* binding energy
Bindungskraft *f* binding force
Bindungswertigkeit *f* linkage valency
Binnengewässer *fpl* inland water
Binnenklima *n* continental climate
Binnenland *n* inland
Binnenmeer *n* inland sea
Bioabfall *m* biological waste
Bioakkumulation *f* bioaccumulation
bioakkumulativ bioaccumulative
bioaktiv biochemically active
Bioauslaugung *f* bioleaching
biobewucherte Fläche *f* bio-overgrown area
Biochemikalie *f* biochemical
biochemisch biochemical
  biochemischer Abbau *m* biochemical degradation
  biochemischer Sauerstoffbedarf *m* biochemical oxygen demand
biodynamisch biodynamic
bioenergetisch bioenergetic
Bioenergie *f* bioenergy
Biofilter *n* biofilter
Biofiltration *f* biofiltration
Biogas *n* fermentation gas, biogas

Biogasproduktion *f* biogas production
Biogasverwertung *f* biogas utilization
biogene Emission *f* biogenic emission
Biogenese *f* biogenesis
biogenetisch biogenetic
Biogeographie *f* biogeography
Biogeologie *f* biogeology
Biogeosphäre *f* biogeosphere
Bioindikator *m* bioindicator
Biokatalysator *m* biocatalyst
Biologie *f* biology
biologisch abbaubarer Kunststoff *m* biologically degradable plastic
biologisch schwer abbaubar biologically heavily degradable
biologische Abbaubarkeit *f* biodegradation
biologische Bodenaktivität *f* biological soil activity
biologische Behandlung *f* biological treatment
biologische Kläranlage *f* biological clarification plant
biologische Kontrolle *f* biological control
biologische Reinigung *f* biological purification
biologische Stabilität *f* biological stability
biologische Testanlage *f* biological pilot plant
biologische Trocknung *f* biological drying
biologische Überwachung *f* biological monitoring
biologischer Kreislauf *m* biological cycle
biologischer Parameter *m* biological parameter
biologischer Rasen *m* biological slime
biologischer Trägerstoff *f* biological carrier substance
biologisches Verfahren *n* biological method
Biomanipulation *f* biomanipulation
Biomasse *f* biomass
Biomassekonzentration *f* biomass concentration
Biomedizin *f* biomedicine
Biomüll *m* biowaste
Biomüllkompost *m* biowaste compost
Bioökologie *f* bioecology
Biopolymer *n* biopolymer
Bioreaktor *m* bioreactor
Bioreaktorsteuerung *f* bioreactor control
Bioschlamm *m* activated sludge, biosludge
Biosphäre *f* biosphere
Biosynthese *f* biosynthesis
biosynthetisch biosynthetic
Biosystem *n* biosystem
Biotechnologie *f* biotechnology
biotechnologische Aktivität *f* biotechnological activity
Biotest *m* biotest
biotisch biotic
biotischer Abbau *m* biotic degradation
biotisches Potential *n* biotical potential
Biotonne *f* bio-bin
Bioumwandlung *f* bioconversion
Bioverfügbarkeit *f* bioavailability
Biowäscher *m* bioactive scrubber
Biozid *n* biocide

Biozönose

Biozönose *f* biocoenosis
Bitumen *n* bitumen, asphalt
Bitumendecke *f* bituminous pavement
Bitumenrückgewinnung *f* bitumen recovery
Bitumenwiedergewinnung *f* bitumen recovery
Blähmittel *n* blowing agent
Blasenzähler *m* bubble counter
Blattanalyse *f* leaf analysis
Blattdüngung *f* leaf fertilization
Blattgrün *n* chlorophyll
Blattherbizid *n* leaf herbicide
Blattlaus *f* greenfly, aphid
Blattschädigung *f* leaf damage (impairment)
Blattwachse *fpl* leaf waxes
Blaualge *f* blue-green alga
Blauasbest *m* blue asbestos
Blaufäule *f* blue stain
Blausäure *f* hydrocyanic acid, hydrogen cyanide
Blausäurevergiftung *f* hydrocyanic acid poisoning
Blechabfall *m* plate scrap
Blechdose *f* can
Blei *n* lead
Bleiabschirmung *f* lead shield
Bleiakkumulator *m* lead accumulator
Bleiamalgam *n* amalgam of lead
bleibende Immunität *f* permanent immunity
bleichen bleach/to, whiten/to, blanch/to
bleichend decolorant
Bleichmittel *n* bleaching agent
Bleierz *n* lead ore
bleifrei unleaded
bleifreier Kraftstoff *m* unleaded fuel
Bleigehalt *m* lead content
Bleigummischürze *f* lead-rubber apron
bleihaltig plumbic
Bleilegierung *f* lead alloy
bleireiches Abwasser *n* lead-rich waste water
Bleischutz *m* lead protection
Bleistaub *m* lead dust
Bleivergiftung *f* lead poisoning
blinder Alarm *m* false alarm
Blindprobe *f* blank [test]
Blindwert *m* numerical value of the blank, blank
Blindwertmessung *f* blank measurement
Blinkfeuer *n* signal fire
Blitzeinschlag *m* flash of lightning
Blitzschutz *m* lightning protection
Blut *n* blood
Blutalkoholkonzentration *f* blood-alcohol level
Blutanalyse *f* blood analysis, haemanalysis
Blutaustausch *m* exchange transfusion
Blutbank *f* blood bank
Blutbild *n* blood picture
blutbildend haematogenic
Blütenstaub *m* pollen
blutgerinnend haemocoagulative
Boden *m* soil
Bodenablagerung *f* bottom sediment
Bodenanalyse *f* soil analysis
Bodenatmung *f* soil respiration
Bodenaushub *m* soil excavation
Bodenauswaschung *f* soil erosion
Bodenbakterie *f* soil bacterium
Bodenbedingungen *fpl* soil conditions
Bodenbehandlung *f* soil treatment

Bodenbelag *m* floor covering
Bodenbelastung *f* soil impact
Bodenbeschaffenheit *f* nature of the soil
Bodenchemie *f* soil chemistry
Bodendurchlüftung *f* soil aeration
Bodenentseuchung *f* soil decontamination
Bodenerosionskontrolle *f* soil erosion control
Bodenfauna *f* soil fauna
Bodenfilter soil filter
Bodenflora *f* soil flora
Bodengefüge *n* soil structure
bodengenetisch soil genetic
Bodenkarte *f* soil map
Bodenkunde *f* soil science, pedology
Bodenluft *f* soil atmosphere
Bodenluftabsaugung *f* soil atmosphere exhaustion
Bodenluftkonzentration *f* soil atmosphere concentration
Bodenmechanik *f* soil mechanics
Bodenmikroorganismus *m* soil microorganism
Bodenmorphologie *f* soil morphology
Bodenorganismus *m* soil organism
Bodenprobe *f* soil sample
Bodenreinhaltung *f* soil pollution control
Bodenreinigungsverfahren *n* soil purification method
Bodensanierung *f* soil remediation
Bodensäule *f* soil column
Bodenschadstoff *m* soil pollutant
Bodenschicht *f* soil layer
Bodenschutz *m* soil protection
bodenspezifisch soil-specific
Bodenvergiftung *f* soil poisoning
Bodenwaschanlage *f* soil washing plant
Bohrschlamm *m* drilling mud, drilling sludge
Bohrloch *n* borehole
Bohrturm *m* derrick
Bor *n* boron
Borke *f* bark, rind
bösartiger Tumor *m* malignant tumour
Böschungsabdichtung *f* slope sealing
Böschungsschutz *m* slope protection
Botanik *f* botany
Boxenkompostierung *f* soil composting in boxes
Brachland *n* fallow, fallow land
Brachlandbegrünung *f* waste land recovery, fallow recovery
Brackwasser *n* brackish water
brandfest fire-proof
Brandgefahr *f* fire risk
Brandrodung *f* fire clearing
Brandschaden *m* fire damage
Brandschutz *m* fire protection
Brandung *f* surf, breakers
Brauchwasser *n* process water, water for industrial use
Brauchwasseraufbereitung *f* process water treatment, processing of raw water
Brauchwasserversorgung *f* process water supply
Brauerei *f* brewery
Braunalge *f* brown alga
Braunkohle *f* brown coal, lignite
Braunkohlefeuerung *f* lignite firing
Braunkohleförderung *f* brown coal mining
Braunkohlelager *n* lignite bed

Braunkohlenanlage *f* brown coal plant
Braunkohlenkoks *m* browncoal coke
Braunkohletagebau *m* lignite open cut
Brecher *m* breaker, crusher, cracker, grinding mill
Brecheranlage *f* crushing plant
Brechmittel *n* emetic, vomitive
breiartiger Schlamm *m* semifluid slurry
Bremsbelag *m* brake lining
brennbar combustible, inflammable
Brennbarkeit *f* combustability, flammability, inflammability
Brennelement *n* fuel rod
brennen burn/to, fire/to
Brenner *m* burner, torch
Brennerdüse *f* burner nozzle
Brenngas *n* fuel gas
Brennkammertemperatur *f* temperature of the burning chamber
Brennmaterial *n* fuel, combustible material
Brennofen *m* calcining furnace, oven, stove
Brennstoff *m* fuel, combustible
Brennstoffausnutzung *f* fuel utilization
Brennstoffeuchte *f* moisture of the fuel
Brennstoffgewinnung *f* fuel recovering
Brennstoffverbrauch *m* fuel consumption
Brennstoffzelle *f* fuel cell
Brennversuch *m* burning test
Brennwert *m* fuel value, calorific value
Brennwertkonvertierung *f* conversion of the calorific value

Brennwertschwankung *f* fluctuation of the calorific value
Brikettieranlage *f* briquetting plant
Brom *n* bromine
Bronchialasthma *n* bronchial asthma
Bronchialkatarrh *m* bronchitis
Bronchitis *f* bronchitis
bronchitisch bronchitic
Brüden *m* vapour
Brüdenverdichtung *f* vapour recompression
Brühe *f* liquor, lye
Brunnen *m* well
Brustkrebs *m* breast cancer
Brut *f* hatch
Brutkasten *m* incubator
Brutofen *m* incubator
Brutstoff *m* breeder material
Bruttoformel *f* total molecular formula, empirical formula
Bryéreholz *n* brier wood
Bundesrecht *n* federal law
Bunker *m* bunker, bin
Bunkereinbaufilter *n* built-in bunker filter
Bürgerinitiative *f* citizens' action committee
Bußgeld *n* fine
Bypassventil *n* bypass valve

C

Cadmium *n* cadmium
Cadmium-Reduktionssäule *f* cadmium reduction column
cadmiumhaltig cadmiferous
Calcinieren *n* calcining, calcination
Calcitlösevermögen *n* calcite solubility

Calcium *n* calcium
Californium *n* californium
Calorimeter *n* calorimeter
Calorimetrie *f* calorimetry
calorimetrisch calorimetric
Cambium *n* cambium
Carbolineum *n* carbolineum
Carbonat *n* carbonate
Carbonathärte *f* carbonate hardness
carbonisieren coke/to
carcinogen carcinogenic, cancerogenic
Carzinom *n* carcinoma
Cäsium *n* caesium
Cellulose *f* cellulose
Celluloseaufschluß *m* celluluse decomposition (disintegration), cellulose hydrolysis
Celluloseionenaustauscher *m* cellulose ion exchanger
Cer *n* cer
Charge *f* charge
Chargenbetrieb *m* batch production
Chargenprozeß *m* batch process
Chargenrezept *n* batch recipe
chargieren charge/to, load/to
Chelat *n* chelate
Chelatbildner *m* chelating agent
Chemie *f* chemistry
Chemieabfall *m* chemical waste
Chemiefaser *f* chemical fibre
Chemikalie *f* chemical
chemikalienbeständig resistant to chemicals
Chemikalienemission *f* emission of chemicals
Chemikaliengesetz *n* chemical law
Chemikalienrückgewinnung *f* chemicals recovery
Chemikalientanker *m* chemicals tanker

chemisch chemical
chemisch reinigen dry-clean/to
chemische Adsorption *f* chemical adsorption
chemische Eigenschaft *f* chemical property
chemische Energie *f* chemical energy
chemische Epidemiologie *f* chemical epidemiology
chemische Reaktion *f* chemical reaction
chemische Verbindung *f* chemical compound
chemische Verfahrenstechnik *f* chemical engineering
chemischer Abbau *m* chemical degradation
chemischer Katastrophenschutz *m* chemical emergency protection
chemischer Pflanzenschutz *m* chemical plant protection
chemischer Sauerstoffbedarf *m* chemical oxygen demand
Chemisorption *f* chemisorption
Chemoresistenz *f* chemoresistance
chemosensitiv chemosensitive
Chemosensor *m* chemosensor
Chemotechniker *m* laboratory technician
Chemotherapie *f* chemotherapy
Chlor *n* chlorine
Chlorakne *f* chloracne
Chloralkalielektrolyse *f* chlorine-alkali electrolysis
Chlorbildung *f* chlorine generation
chlorieren chloridize/to, chlorinate/to
chlorierter Kohlenwasserstoff *m* chlorinated hydrocarbon

chloriertes Wasser *n* chlorinated water
Chlorierung *f* chlorination, chlorinating
Chlorierung *f* **von Abwasser** chlorination of waste water
Chlorierungsanlage *f* chlorinating plant
Chlorierungsgrad *m* chlorination degree
Chlorierungsmittel *n* chlorinating agent
Chlorkalkanlage *f* chloride of lime plant
Chlorkalkbleiche *f* chloride of lime bleaching
Chlorkohlenwasserstoff *m* chlorinated hydrocarbon
Chlorophyll *n* chlorophyll
Chloroplast *m* chloroplast
Chlorung *f* chlorination
Chrom *n* chromium
Chromabwasser *n* chromium sewage
chromathaltig chromine containing
Chromatogramm *n* chromatogram
Chromatographie *f* chromatography
chromatographische Säule *f* chromatographic column
Chromosom *n* chromosome
Chromstaub *m* chromium dust
chronisch schädigend chronically affecting
Citratcyclus *m* citrate cycle
Clearingstelle *f* clearing authority
Cobalt *n* cobalt
Coenzym *n* coenzym
Colikeimzahl *f* coli germ number
Computerschrott *m* computer scrap
Containersystem *n* container system
Cortison *n* cortison, cortisone
Crackanlage *f* cracking plant
Curium *n* Curium n
Cyanid *n* cyanide
cyanidfrei free from cyanide
Cyanidlauge *f* cyanide liquor
Cyankali[um] *n* potassium cyanide
Cyclamat *n* cyclamate
Cyclotron *n* cyclotron
Cytochemie *f* cytochemistry
Cytogenetik *f* cytogenetics
Cytolyse *f* cytolysis
Cytoplasma *n* cytoplasm
Cytoskop *n* cystoscope
Cytoskopie *f* cystoscopy
Cytostatikum *n* cytostatic agent

D

Damm *m* barrage, dam
dämmen insulate/to, dam/to, block/to
Dammschüttmaterial *n* bulk material for dams
Dampf *m* steam, vapour
Dampfabführung *f* steam discharge
Dampfableitung *f* steam discharge
Dampfbildung *f* steam generation
Dampfdichte *f* vapour density
Dampfdruck *m* vapour pressure
dämpfen steam/to, damp/to; suppress/to, attenuate/to
Dampfentwässerer *m* steaming trap
dampfflüchtig steam-volatile
dampfförmig vaporous
Dampfsterilisation *f* vapour sterilization
Dämpfung *f* attenuation, absorption, suppression, dampening
Dämpfungsbereich *m* range of attenuation

Daphnientest *m* daphnia test
Darmbakterien *fpl* intestinal flora
Darmflora *f* intestinal flora
Darmgrippe *f* intestinal influenza
Datenbank *f* data bank, data base
Datenbank-Spezialdienst *m* specialized database service
Datenbasis *f* data pool
Datenfluß *m* data flow
Datenkanal *m* data channel
Datennetz *n* data network
Datensenke *f* data drain
Datensicherheit *f* data security
Datenspeicher *m* data logger
Datenübertragung *f* data transfer
Datenverarbeitung *f* data processing
Dauerbeanspruchung *f* continuous load, permanent load
Dauerbetrieb *m* continuous operation
Dauereinsatz *m* continuous use
Dauerfestigkeit *f* fatigue limit
Dauergebrauch *m* permanent use
dauerhaft durable, heavy-duty, lasting, permanent
Dauerhaftigkeit *f* durability
Dauerregen *m* continuous rain
Deckanstrich *m* paint finish, finishing coat
Deckglas *n* slide cover
Deformation *f* deformation, distortion
Deformierbarkeit *f* deformability
Deformierung *f* deformation, distortion
Degeneration *f* degeneration
degenerieren degenerate/to
degradieren degrade/to
Dehalogenierung *f* dehalogenation
Dehydratisierung *f* dehydration
Deich *m* dike

Deichbruch *m* dike burst (failure)
Deichsiel *n* dike sluice
dekantieren decant/to
Dekantierung *f* decantation
Dekantierungszentrifuge *f* decantation centrifuge
Dekontamination *f* decontamination
Dekontaminationstechnik *f* decontamination technology
Delokalisierung *f* delocalization
demographisch demographical
Demökologie *f* ecology of population
Demonstrationsanlage *f* demonstration plant
Demontage *f* disassembly, dismounting
Demontagewerk *n* disassembling plant
Demulgator *m* demulsifier
Demulgierungsreaktion *f* deemulsification reaction
Denaturierung *f* denaturation, denaturization
Denitrifikation *f* denitrification
Denitrifikationsbecken *n* denitrification basin
Denitrifikationszone *f* denitrification zone
Denitrifizierung *f* denitrification
Depolarisierung *f* depolarization
Deponie *f* landfill, refuse dump, disposal site
Deponieabdichtung *f* sealing of the waste (disposal) site, sealing of a landfill (refuse dump)
Deponieaufbau *m* structure of disposal site, disposal site structure
Deponiebinder *m* refuse dump binder

Deponiedichtung *f* sealing of the disposal site, (landfill)
Deponieentsorgung *f* landfill disposal
Deponiegas *n* waste gas from a disposal site, site [digestion] gas
Deponiegasbehandlung *f* treatment of dump site (oligestion) gas
Deponiegaskraftwerk *n* gas power station for site digestion gas
Deponiegasqualität *f* site digestion gas quality
Deponiegassammlung *f* site digestion gas collection
Deponiegelände *n* area for landfill sites, waste disposal site
Deponiekapazität *f* [disposal] site capacity
Deponiekataster *m* disposal site register, site register
Deponiekomponente *f* site component
Deponiekörper *m* site body
deponieren deposit/to, dispose of/to, landfill/to
deponierter Verbrennungsrückstand *m* landfilled incinerator residue
Deponierung *f* disposal
Deponierbarkeit *f* dumping ability
Deponiesickerwasserbehandlung *f* treatment of the waste site leaching (percolating) water, treatment of the percolating water of disposal
Deponiesohle *f* bottom of the refuse dump (site)
Deponiestandort *m* site location
Deponietechnik *f* disposal technology
Deponieumgebung *f* environment of the disposal site
Deponieverhalten *n* disposal behaviour
Deponieverwalter *m* waste site manager
Deponievolumen *n* refuse dump (site) volume
Deponiewasserstand *m* water level of the disposal site
Depositionsparameter *m* deposition parameter
Derivat *n* derivative
dermatisch dermatic
Dermatitis *f* dermatitis
Dermatophyt *m* cutaneous fungus, dermatophyte
Dermatose *f* dermatosis
desaktivieren deactivate/to
desaktivierte Säule *f* deactivated column
desensibilisieren desensitize/to
Desensibilisierung *f* desensitization
Desinfektion *f* disinfection
Desinfektionsbecken *n* disinfection basin
Desinfektionsmittel *n* disinfectant, sanitizer
desinfizieren disinfect/to, sterilize/to
desinfiziert disinfected, sterilized
Desodorierung *f* deodorization
desorbieren desorb/to, strip/to
Desorption *f* desorption
destabilisieren destabilize/to
Destabilisierung *f* destabilization
Destillat *n* distillate
Destillation *f* distillation
Destillationsabwasser *n* distillation effluent
Destillationsapparat *m* rectifier
Destillationskolben *m* retort

Destillationsrückstand *m* distillation residue
destillieren distil/to
destilliertes Wasser *n* distilled water
Destruktion *f* destruction
destruktiv destructive
Desulfurierung *f* desulphuration
Detektor *m* detector
Detektorverschmutzung *f* detector fouling
Detergens *n* detergent
Detonation *f* detonation
detonieren detonate/to, explode/to, blow up/to
Dezentralisierung *f* decentralization
Diagnose *f* diagnosis
Diagnosesystem *n* **für Kernkraftwerke** diagnosis system for nuclear power plants
Diagnostik *f* diagnostics
diagnostizieren diagnose/to
Dialyse *f* dialysis
Dialyseeinheit *f* dialysis unit
Dialysierflüssigkeit *f* dialysis fluid
Diamant *m* diamond
Diaphragma *n* diaphragm
Diaphragmafilterpresse *f* diaphragm filter press
Diaphragmapumpe *f* diaphragm pump
Diaphragmaverfahren *n* diaphragm process
Diarrhöe *f* diarrhoea
Diastereomer *n* diastereomer
Diatomit *m* diatomite
Dichte *f* density; tightness
Dichtebestimmung *f* density determination
dichten tighten/to, seal/to
Dichtfett *n* joint grease
Dichtlack *m* sealing lacquer
Dichtmanschette *f* packing, sealing packing
Dichtstoff *m* sealing material
Dichtung *f* seal, sealing, joint
Dichtungsfett *n* joint grease
Dichtungsfuge *f* sealing joint
Dichtungsmasse *f* sealing compound
Dichtungsmittel *n* sealing agent, sealant
Dichtungspappe *f* packing cardboard
Dichtungssystem *n* sealing system
Dichtwand *f* sealing wall
Dichtwandsystem *n* sealing wall system
die freie Natur *f* the great outdoors
Dieselabgaspartikelabscheider *m* diesel waste gas particulate trap
Dieseldrehstromaggregat *n* three-phase diesel aggregate
Dieselkraftstoff *m* diesel [fuel]
Differentialblutbild *n* differential blood picture
Differentialdiagnose *f* differential diagnosis
differentialdiagnostisch differential diagnostic
diffundieren diffuse/to
Diffundieren *n* diffusing
Diffusion *f* diffusion
Diffusionsfähigkeit *f* diffusibility
Diffusionsgeschwindigkeit *f* diffusion velocity
Diffusionskoeffizient *m* diffusion coefficient
Diffusionspotential *n* diffusion potential
Diffusionsprozeß *m* diffusion process
digerieren digest/to
Digestorium *n* fume cupboard

Dioxin n dioxin
Dioxin-Abfallentsorgung f disposal of dioxin waste
Dioxinaufnahme f dioxine uptake
Dioxinemission f dioxin emission
Dioxinkatastrophe f dioxin disaster
Dioxinprobe f dioxin sample
Dioxinskandal f dioxin scandal
Dioxintoxizität f dioxin toxicity
Dioxinvergiftung f dioxin poisoning
Dipolmolekül n dipole molecule
Dipolmoment n dipole moment
Direktbestimmung f direct determination
Direktdampfeinleitung f direct vapour feeding
Direkteinleiter m direct introducer
Direkteinleitung f direct release (introduction)
Direktfällung f direct precipitation
Dispergens n dispersant
disperse Phase f disperse phase
disperse System f disperse system
Dispersion f dispersion
Dispersionsmittel n dispersant
Dispersionsphase f dispersed phase
Dispersionszerstäuber m dispersion mechanism (nozzle)
Dissimilation f dissimilation
Dissoziation dissociation
dissoziieren dissociate/to
Dolomit m dolomite
Donator m donor
Doppelbestimmung f repeated determination (test)
Doppeleinspritzung f duplicate injection
doppelwandiger Tank m double-walled tank
Dose f tin, can
Dosenkonserve f canned food

Dosenmilch f canned milk
Dosierbereich m dosage area
Dosiereinrichtung f dosage facility
Dosieren n dosing
Dosieranlage f dosage plant
Dosierpumpe f dosage pump
Dosierstation f dosage station
Dosiersystem n dosage system
Dosiertechnik f dosage technique
Dosierung f dosage, dosing
Dosierungsintervall n dosing interval
Dosierzählerbaugruppe f module for dosage counting, dosage-counting module
Dosis f dose
Dosisanzeigegerät n dose indicating apparatus
Dosis-Wirkung-Beziehung f dose-response relationship
Dosis-Wirkung-Kurve dose-effect-curve
Drahtglas n wire glass
Drahtwurm m wireworm
Dränage f drainage
dränieren drain/to
Dränrohr n drain pipe
Dränschicht f drain layer
Dränwasser n drainage water
Drehkolbengebläse n rotary-piston blast
Drehofen m rotary furnace
Drehrohrofen m rotary-tube furnace
Drehrohrofenanlage f rotary-tube furnace plant
Drehsprenger m rotary sprayer, rotary sprinkler
Drehtrockner m rotary dryer
Dreihalskolben m three-necked bottle

Dreikomponentensystem *n* ternary system
dreischichtige Wand *f* three-layered wall
Dreiwegekatalysator *m* three-way catalyst
Dreiwegeventil *n* three-way valve
Droge *f* drug
Drogenmißbrauch *m* drug abuse
Drossel *f* choke, throttle
Drosselklappe *f* choke, throttle valve
drosseln throttle/to, choke/to, baffle/to, restrain/to
Drosselventil *n* throttle valve
Druckabfall *m* pressure drop
Druckanstieg *m* pressure rise
Druckanzeiger *m* manometer, pressure gauge
Druckausgleich *m* pressure equalization
Druckbelüftungssystem *n* pressure aeration system
druckelektrisch piezoelectric
Druckentspannungsflotation *f* pressure-release flotation
Druckfilteranlage *f* pressure filter plant
Druckfiltration *f* pressure filtration
drucklose Rohrleitung *f* pressureless piping, pressureless pipeline
Druckluftfilter *n* compressed-air filter
Druckmeßumformer *m* pressure transducer
Drucksonde *f* pressure probe
Druckänderung *f* change of pressure
Duft *m* odour, smell
Duftstoff *m* odorous substance
duktil ductile
Düne *f* dune
Dünenpflanze *f* beach plant
Dünenschutz *m* dune protection
Dung *m* dung, manure
Düngeerde *f* humus, mould, vegetable earth, compost
Düngemittel *n* fertilizer, fertilizing agent
Düngemittelfabrik *f* fertilizer factory, manure works
Düngemittelgabe *f* application of fertilizer
Düngemittelherstellung *f* fertilizer production
Düngemittelkombination *f* fertilizer combination
Düngemittelphosphor *m* fertilizer phosphorus
Düngemittelverteilung *f* fertilizer distribution
düngen fertilize/to, manure/to
Düngen *n* manuring
Düngepulver *n* powdered manure
Dünger *m* fertilizer, manure
Düngesalz *n* manuring salt
Düngeschlamm *m* fertilizing sludge
Düngung *f* fertilization, manuring
Dünnschichtchromatographie *f* thin-layer chromatography
Dünnschichtelektrophorese *f* thin-layer electrophoresis
Dünnschichtfilm *m* thin emulsion film
Dünnschlamm *m* thin-sludge
Dünnsäure *f* spent acid, waste acid, dilute acid
Dünnsäureverklappung *f* barging of waste acid, ocean dumping of waste acid, barging of spent acid
dünnwandig thin-walled
Dunst *m* fume, mist, steam, vapour
Dunstabzug *m* hood

Dunstabzugsrohr *n* vent pipe
Dunstglocke *f* layer of smog, haze, pall of smog
Dunstschleier *m* haze
Durchblutungsstörung *f* vascularity disorder, disturbed circulation
Durchfall *m* diarrhoea
Durchfluß *m* flow, passage, percolation
Durchflußbegrenzer *m* flow limiter
Durchflußblende *f* flow aperture, flow orifice
Durchflußcytometrie *f* flow cytometry
Durchflußgeschwindigkeit *f* velocity of flow
Durchflußmenge *f* flow [rate]
Durchflußmessung *f* flow measurement
Durchforstung *f* thinning of woodland, thinning
Durchführungsverordnung *f* implementing order
Durchlaufarmatur *f* flow fitting
Durchlauftrockner *m* tunnel dryer
Durchlaß *m* opening, passage
durchlässig permeable, penetrable, porous
durchlässige Schicht *f* permeable layer
Durchlässigkeit *f* permeability, porosity, transparency
Durchlässigkeit *f* **des Bodens** permeability of the soil
durchleiten pass through/to, conduct/to
durchleuchten x-ray/to
Durchleuchtung *f* fluoroscopy, X-ray
durchlüften aerate/to, ventilate/to
durchmischen mix thoroughly/to, blend/to, interfuse/to
Durchmischungstiefe *f* depth of mixing
durchnässen soak/to, wet/to
durchrühren stir/to
Durchsatz *m* throughput, output
Durchsatzleistung *f* output capacity, maximum capacity
durchscheinend transparent
Durchschlagsfestigkeit *f* breakdown strength, breakdown resistance
Durchschnittsleistung *f* average performance
Durchschnittsmenge *f* average quantity
Durchschnittswert *m* average value
durchseihen strain/to, percolate/to, filter/to
durchsichtig transparent, clear
durchsickern trickle through/to, soak/to, infiltrate/to, leach/to, leak/to, percolate/to
Durchsickern *n* trickling, soakage, leaching
durchsieben sieve/to, screen/to
durchstrahlen irradiate/to, radiate/to
durchströmen flow through/to, pass through/to
durchtränken impregnate/to, saturate/to
Durchtränkung *f* impregnation, saturation
Durchwirbelung *f* turbulence
Dusche *f* douche, shower
Düse *f* jet, nozzle, tip
Düsenaggregat *n* jet engine
Düsenantrieb *m* jet propulsion
Düsenlärm *m* jet noise
Düsensystem *n* nozzle system

Düsentreibstoff *m* jet fuel
Düsentriebwerk *n* jet engine
Düsenöffnung *f* orifice
Dysfunktion *f* dysfunction
Dysplasie *f* dysplasia

E

Ebbe *f* ebb, low tide, falling tide
Ebene *f* level, plane
Ebenheit *f* evenness
Ebullioskopie *f* ebullioscopy
Echtzeitdatenerfassung *fpl* real-time data collection
Edelgas *n* inert gas, noble gas, rare gas
Edelholz *n* precious wood
Edelmetall *n* noble metal
Edelmetallrecycling *n* noble metal recycling
Effektivwert *m* effective value
Effizienz *f* efficiency
eichen calibrate/to, gauge/to, standardize/to
Eichfaktor *m* calibration factor
Eichkurve *f* calibration curve, calibration plot
Eichmarke *f* gauge mark
Eichmaß *n* gauge, standardization measure
Eichmischung *f* calibration mixture
Eichprotokoll *n* calibration record
Eichstrich *m* gauge mark
Eichung *f* calibration
Eigenfarbe *f* natural colour
Eiklar *n* albumen
Eimerkettenbagger *m* bucket chain dredger
einäschern incinerate/to
einatmen inhale/to

Einatmung *f* inhalation, inspiration
einbetten embed/to, encapsulate/to, immerse/to
Einbettung *f* embedment, encapralation
Einbettungsmittel *n* embedding medium
Einbrennfarbe *f* baking finish
Einbrennlack *m* ceramic varnish
Einbrennlackierung *f* baking varnishing
Einbrennofen *m* stove
Einbruch *m* invasion
eindampfen concentrate by evaporation/to, evaporate/to, vaporize/to
Eindampfen *n* concentrating, evaporating
Eindampfer *m* evaporator
Eindampfung *f* evaporation
Eindampfungsanlage *f* concentration plant, evaporation plant
Eindampfverfahren *n* concentrating process
eindeichen impolder/to
Eindeichung *f* diking
eindicken thicken/to, concentrate/to
Eindicker *m* thickener
Eindickerausrüstung *f* thickener equipment
Eindickung *f* thickening, concentration
Eindickung *f* **durch Flotation** thikkening by flotation
Eindickung *f* **durch Schwerkraft** thickening by gravity
Eindickungsanlage *f* thickening plant
Eindickungsgrad *m* degree of thikkening
Eindickungsvorrichtung *f* thickening device

Eindringtiefe *f* penetration depth
Eindringung *f* infiltration
eine Blindprobe machen run a blank/to
eine Probe nehmen sample/to
einebnen plane/to
einengen concentrate/to, evaporate/to
Einengung *f* concentrating; confinement
Einfachsystem *n* simple system, single system
Einfang *m* capture
Einfangsquerschnitt *m* capture cross section
Einfangswahrscheinlichkeit *f* capture probability
Eingangskontrolle *f* control of receipt
Eingangswert *m* input value
eingedämmtes Deponiegelände *n* dammed disposal site
eingeschlossener Analyt *m* occluded analyte
Eingeweide *n* intestines, bowels
eingraben dig in/to, bury/to
Eingrabung *f* burial
eingrenzen limit/to, localize/to, locate/to
Eingrenzung *f* limitation, localization
Eingrenzung *f* **von Altlasten** localization of polluted areas, localization of old-sites
Einhaltung *f* **von Gefahrgutvorschriften** observance of regulations for the transportation of hazardous goods
Einhaltung *f* **von Sicherheitsansprüchen** observance of safety requirements

einheimisch domestic
Einheitspackung *f* standard packing
Einheitssystem *n* standard system
Einheitsverfahren *n* standard method
einimpfen inoculate/to
Einjahrespflanze *f* annual plant, yearling
einkalken soak in lime water/to, lime/to
einkapseln encapsulate/to
Einkapselung *f* encapsulation
Einkapselung *f* **von radioaktivem Abfall in Beton** radioactive waste incasement in concrete
Einkapselung *f* **von Sonderabfall** encapsulation of special waste
Einkomponentenkleber *m* one-component adhesive
Einlaßöffnung *f* feed inlet
Einlaßrohr *n* inlet pipe
Einlaßsystem *n* inlet system
Einlaßventil *n* feed valve
Einlauf *m* inlet
Einlaufbauwerk *n* inflow building
Einlauföffnung *f* inlet port
Einlaufrohr *n* inlet pipe
Einlaufseite *f* intake
einleiten feed/to, introduce/to
Einleitung *f* feeding, introduction
Einöde *f* wilderness, waste
einpflanzen implant/to, plant/to
Einpflanzung *f* implantation
einregulierbar adjustable
einregulieren adjust/to, regulate/to
Einrichtung *f* **einer Deponie** setting-up of a landfill
einrosten rust/to
einsaugen suck in/to, imbibe/to, soak up/to

Einsaugen *n* sucking in, imbibition, soakage
Einschlämmung *f* sedimentation, deposition, settlement
Einschlußverbindung *f* inclusion compound
einschmelzen fuse/to, smelt/to, melt down/to
Einschmelzrohr *n* sealed tube
einschweißen weld/to
einsickern seep in/to, soak in/to
einspritzen inject/to
Einspritzort *m* injection port
Einspritzung *f* injection
einstampfen stamp/to, compress/to, ram down/to
Einstampfmaschine *f* pulping machine
Einstampfpapier *n* waste paper
einstäuben powder/to, dust/to
Einsteinium *n* einsteinium
Einstellung *f* **der Exposition** cessation of exposure
Einstieghilfe *f* entering facility
einströmen flow in/to, stream in/to
einstweilige Verfügung *f* interdict
Eintaucharmatur *f* immersion fitting
Eintauchen *n* immersion, dipping, plunging
Eintauchkammer *f* immersion chamber
Eintauchtiefe *f* immersion depth
Eintauchzeit *f* immersion time
Einteeren *n* tarring
Eintrag *m* entry
Eintrag *m* **von Schadstoffen** entry of pollutants
Eintragsaggregat *n* entry unit, feeding (charging) unit
Eintragsmenge *f* amount of input

Eintrittsöffnung *f* inlet hole
Eintrittsrohr *n* admission pipe
Einwaage *f* weighed sample; initial weight; test portion
Einwegartikel *m* disposable product
Einwegbehälter *m* one-way container (vessel), disposable container (vessel)
Einwegcontainer *m* one-way container, disposable container
Einwegfaß *n* disposable vat
Einwegfilter *n* disposable filter
Einwegflasche *f* non-returnable bottle, throw-away bottle, non-deposit bottle
Einweggeschirr *n* one-way dishes
Einweghahn *m* one-way stopcock
Einwegspritze *f* disposable syringe
Einwegstaubmaske *f* disposable dust mask
Einwegverfahren *n* disposable method, one-way method
einweichen soak/to, steep/to
Einwohnergleichwert *m* inhabitant equivalent (value)
Einzelanalyse *f* individual analysis
Einzelbestimmung *f* single determination
Einzeleinleiter *m* individual introducer
Einzeller *m* monocellular organism, protozoon
einzellig single-celled, unicellular
Eisenbakterie *f* iron bacterium
Eisenbeton *m* reinforced concrete, armored concrete
Eisenerz *n* iron ore
eisenfrei iron-free
Eisenguß *m* iron casting
eisenhaltig ferriferous, iron-containing

Eisenindustrie *f* iron industry
Eisensalz *n* iron salt
Eiskalorimeter *n* ice calorimeter
Eiszone *f* ice zone
Eizelle *f* egg cell, ovocyte, ovum
Ekzem *n* eczema
Elastomer *n* elastomer
Elektrizitätswerk *n* electric works, power station
Elektroabscheidung *f* electroprecipitation
Elektrochemie *f* electrochemistry
elektrochemisch electrochemical
Elektrode *f* electrode
Elektrodialyse *f* electrodialysis
Elektrofilter *n* electrofilter, electric separator, electrostatic precipitator
Elektrokardialyse *f* electrocardialysis
Elektrolyse *f* electrolysis
Elektrolysenschlamm *m* electrolytic slime
Elektrolyt *m* electrolyte
Elektrolythaushalt *m* electrolyte metabolism (balance)
elektrolytische Trennung *f* electrolytic separation
Elektrolytraum *m* electrolyt chamber (space)
Elektromagnet *m* electromagnet
elektromagnetisch electromagnetic
elektromagnetische Welle *f* electromagnetic wave
Elektron *n* electron
elektronegativ electronegative
Elektronegativität *f* electronegativity
Elektronenabsorption *f* electron absorption
Elektronenaffinität *f* electron affinity
Elektronenbeschuß *m* electron bombardment
Elektronenemission *f* electron emission
Elektronenreichweite *f* range of electrons
Elektroneurographie *f* electroneurography
Elektronikschrott *m* electronic scrap
elektronischer Meßzusatz *m* electronic measurement adapter
Elektropathologie *f* electropathology
Elektrophorese *f* electrophoresis
Elektrophysiologie *f* electrophysiology
elektropositiv electropositive
Elektroschmelzverfahren *n* electrosmelting process
elektrostatisch electrostatic
elektrostatische Ölreinigung *f* electrostatic oil purification
Elementaranalyse *f* elementary analysis
Elementarladung *f* elementary charge
Elementarzelle *f* elementary cell, unit cell, embryonal cell
Elfenbein *n* ivory
Elimination *f* elimination
eliminieren eliminate/to
Eloxieren *n* anodic treatment
Eloxieren *n* **von Kunststoffoberflächen** anodizing of plastic surfaces
Eluat *n* eluate
eluationsstabil elution-stable
Eluent *m* eluent
eluieren elute/to
Eluierung *f* elution
Eluierungsmittel *n* eluent

Elution *f* elution
Emaillelack *m* enamel varnish
Emaillierung *f* enamelling, enamel work
Embryo *m* embryo
embryonale Abnormalität *f* embryonic abnormality
Embryonalzelle *f* embryonic cell
Embryosterblichkeit *f* embryo mortality
embryotoxischer Effekt *m* embryotoxic effect
Empryotoxizität *f* embryo toxicity
emigrieren emigrate/to
Emission *f* emission
Emissionsdaten *f* emission data
Emissionsdichte *f* emission density
Emissionsgrenzwert *m* emission limit value
Emissionskataster *m* emission register
Emissionskonzentrationsmessung *f* measuring of emission concentration
Emissionsmenge *f* emission quantity
Emissionsmessung *f* emission measurement
Emissionsminderung *f* decrease of emission
emissionsneutral emission-neutral
Emissionsquelle *f* source of emission
Emissionsrechner *m* emission calculator
Emissionsursache *f* cause of emission
emittieren emit/to, eject/to
emittierter Stoff *m* emitted substance
empfindlich sensitive
Empfindlichkeit *f* sensitivity
Empfindlichkeitsbereich *m* range of sensitivity

Empfindlichkeitsgrad *m* degree of sensitivity
Empfängnis *f* conception
Empfängnisverhütung *f* contraception
Emphysem *n* emphysema
Emulgator *m* emulsifier
emulgieren emulsify/to
Emulsion *f* emulsion
emulsionsbelastetes Abwasser *n* emulsion-loaded sewage
Emulsionsbildung *f* emulsification
Emulsionsentmischung *f* demulsification
Emulsionsspaltung *f* emulsion separation (breaking), demulsification
Emulsionsstabilität *f* emulsion stability
Emulsionsverfahren *n* emulsion process
Enchondrom *n* enchondroma
endgültige Diagnose *f* final diagnosis
Endlagenschalter *m* limit switch
Endlagerung *f* final storage
Endlagerungsbedingung *f* repository condition
Endnutzung *f* final use
endogen endogenous
Endoplasma *n* endoplasm
Endothelzelle *f* endothelial cell
Endothelzellvermehrung *f* endotheliosis
endotherm endothermal, endothermic, heat consuming
Endprodukt *n* final product
Endschieber *m* final valve
Endstadium *n* final stage
Endtemperatur *f* final temperature
Energie *f* energy

Energie *f* aus Altöl energy from used oil
Energie aus Altreifen *f* energy from used tyres
Energieabgabe *f* energy output
energiearm energy-poor
energieaufwendig energy-consuming
energieautark energy-autarkic[al]
Energiebilanz *f* energy balance
Energieeinsparung *f* energy saving
Energiegewinnung *f* energy recovery
energieintensiv energy-intensive
Energielieferant *m* energy supplier
Energienutzung *f* energy utilization
energiereich high-energy
Energierückgewinnung *f* energy recovery
Energierückgewinnungssystem *n* energy recovery system
Energieverbrauch *m* energy consumption
Energieversorgung *f* energy supply
energiewirtschaftlich energy-economical
Enklave *f* enclave
Enkulturation *f* enculturation
entarten degenerate/to, deteriorate/to
entarteter Zustand *m* degenerate state
Entartung *f* degeneration, atrophy
Entartungsreaktion *f* reaction of degeneration
Enteisenung *f* de-ironing, de-ferrization, removal of iron
entfetten degrease/to, ungrease/to
Entfettungsbad *n* degreasing bath
entflammbar [in]flammable, ignitable

Entflammbarkeit *f* [in]flammability, ignitability
entflammen inflame/to, ignite/to
Entflammung *f* inflammation
Entflammungstemperatur *f* ignition temperature
entfärben decolour/to, bleach/to
Entfärbung *f* decolourizing, bleaching
Entfärbungsmittel *n* decolourizing agent, decolourant
entgasen degasify/to, outgas/to
Entgasung *f* degasification, outgassing, degassing
Entgasungsmaßnahme *f* degasification measure
entgiften detoxify/to, decontaminate/to
Entgiftung *f* decontamination, detoxification, detoxication
Entgiftungsanlage *f* detoxification plant
Entgiftungsresultat *n* detoxification result
Enthalpie *f* enthalpy
Enthärter *m* softener
Enthärtung *f* softening, hardness removal
entharzen remove resin/to
Entkalkung *f* decalcifation, deliming, unliming
Entkalkungsmittel *n* deliming agent
entkeimen sterilize/to, disinfect/to, degerminate/to
Entkeimung *f* sterilization, disinfection, degermination, pasteurization
Entkeimungsanlage *f* disinfection plant
Entkeimungsfilter *n* germ-proof filter
Entkeimungsmittel *n* disinfectant

entkieseln desilicate/to, desilicify/to
Entkieselung *f* desilicification
Entlastung *f* unloading, balancing
Entlaubungsmittel *n* defoliant, defoliation agent
entlüften deaerate/to, ventilate/to, deair/to, vent/to
Entlüftung *f* deaeration, ventilation
Entlüftungsanlage *f* deaerator, ventilation equipment
Entlüftungsautomat *m* ventilation automat
Entlüftungsleitung *f* exhaust duct
Entlüftungsrohr *n* vent pipe
Entmanganung *f* removal of manganese, demanganizing
Entmineralisierung *f* demineralization
Entölung *f* deoiling
Entomologie *f* entomology
Entparaffinierung *f* dewaxing
Entparaffinierungsanlage *f* dewaxer
Entphenolung *f* dephenolization
entrinden debark/to, peel/to
Entrindungstrommel *f* debarking drum
Entropie *f* entropy
Entrostung *f* rust removal
Entsalzung *f* demineralization
Entsäuerung *f* elimination of acid
entschlacken deslag/to
Entschlackung *f* deslagging
entschlammen remove sludge/to
entschwefeln desulphurize/to, remove sulphur/to
Entschwefelung *f* desulphurization, desulphurizing
Entschwefelungsanlage *f* desulphurization plant
Entschwefelungsprozeß *m* desulphurization process

Entschwefelungsverfahren *n* desulphurization process
entseuchen decontaminate/to, disinfect/to
Entseuchung *f* decontamination, disinfection
Entseuchungsmittel *n* decontaminant
Entseuchungstechnik *f* decontamination technology
entsorgbar disposable
entsorgen dispose of/to
Entsorgung *f* disposal
Entsorgung *f* **auf flachem Land** shallow-land disposal
Entsorgungsanlage *f* disposal plant
Entsorgungsfähigkeit *f* capability of disposal
entsorgungsfreundlich suitable (acceptable) for disposal
entsorgungsgerechtes Produkt *n* product compatible to disposal
Entsorgungskapazität *f* disposal capacity
Entsorgungskonzept *n* concept of disposal, disposal concept
Entsorgungslogistik *f* logistics of disposal
Entsorgungsmethode *f* disposal method
Entsorgungsmöglichkeit *f* possibility of disposal
Entsorgungsnachweis *m* disposal proof
Entsorgungsnetzautomatisierung *f* disposal network automation
Entsorgungsproblem *n* disposal problem
Entsorgungsprojekt *n* disposal project

Entsorgungssicherheit *f* disposal security
Entsorgungsstrategie *f* strategy of disposal
Entsorgungsstruktur *f* disposal structure
Entsorgungssystem *n* disposal system
Entsorgungstechnik *f* disposal technology
Entsorgungsverordnung *f* regulation for the disposal
Entsorgungsweg *m* disposal pathway
Entsorgungswirtschaft *f* disposal management
Entspannungsflotation *f* stress-relief flotation
entstauben dust off/to, free from dust/to
Entstaubung *f* dust removal
Entstaubungsanlage *f* dust removal plant, dust separator
Entstaubungsgrad *m* degree of dust removal
Entstaubungstechnik *f* dust removal technology
Entstickung *f* denitrification, removal of nitrogen oxides
Entstickungsanlage *f* plant for nitrogen removal
Entstickungsgrad *m* degree of nitrogen removal
Entstörbereitschaft *f* stand-by interference
entstrahlen decontaminate/to
Entwarnung *f* all-clear
entwässern dewater/to
entwässert anhydrous
Entwässerung *f* dewatering, drainage, draining, water removal
Entwässerungsanlage *f* dewatering plant
Entwässerungsrohr *n* drain pipe
Entwässerungssatzung *f* draining regulation
Entwässerungssystem *n* drainage system, dewatering system
Entwicklerflüssigkeit *f* developer
Entwicklungsland *n* developing country
Entwicklungsstrategie *f* development strategy
Entwöhnung *f* weaning
entzündbar ignitable, [in]flammable, inflammatory
  entzündbare Abgasmischung *f* ignitable waste gas mixture
Entzündbarkeit *f* ignitability, [in]flammability
entzünden ignite/to, inflame/to
entzündlich inflammatory, inflammable
Entzündung *f* ignition, inflammation
Entzündungspunkt *m* flash point
Entzündungstemperatur *f* ignition temperature
Enzym *n* enzyme, ferment
Enzymaktivierung *f* enzyme activation
enzymatisch enzymatic
  enzymatischer Abbau *m* enzymatic degradation
Enzymbestimmung *f* enzyme assay
Enzymimmunoassay *n* enzym immunoassay
Enzymologie *f* enzymology
Epidemie *f* epidemic
Epidemiologie *f* epidemiology
epidemiologisch epidemiological

epidemiologisch unbedenklich epidemiologically unrisky
epidemiologische Datenlage *f* epidemiological data situation
epidemiologische Erhebung *f* epidemiological research
epidemiologische Kurzzeitvorhersage *f* epidemiological short-term forecast
epidemiologische Longitudinalstudie *f* epidemiological longitudinal study
epidemiologische Prospektionsstudie *f* epidemiological prospective investigation (study)
Epikutantest *m* patch test
Epilimnion *n* epilimnion
Epistaxis *f* epistaxis
Erbeinheit *f* gene
Erbfaktor *m* gene, genetic factor, unit factor
erbgutverändernd mutagenic
Erbium *n* erbium
Erbkrankheit *f* hereditary disease
erblich hereditary, heritable
Erbmasse *f* genotype
Erbrechen *n* vomit
Erbrochenes *n* vomit
Erbschaden *m* hereditary defect
Erbschädigung *f* hereditary anomaly
Erdalkali *n* alkaline earth
Erdarbeiten *fpl* excavations, earth work
Erdatmosphäre *f* earth atmosphere
Erdbeben *n* earthquake
Erdbebenherd *m* earthquake focus
Erdbebenmesser *m* seismograph
Erdbebenschutz *m* earthquake protection
Erdbebenwelle *f* seismic wave
Erdboden *m* ground, soil

Erddruck *m* earth pressure
Erde *f* earth; ground, soil
Erdfunkstelle *f* satellite earth station
Erdgas *n* natural gas
Erdgasanlage *f* natural gas plant
Erdgasfeld *n* gas field
Erdgaslagerstätte *f* gas field
Erdgasnetz *n* natural gas supply system
Erdgeruch *n* earthy smell
Erdgeschichte *f* geology
erdgeschichtlich geological
erdig earthy
Erdkabel *n* underground cable
Erdkunde *f* geography
erdmagnetisches Feld *n* geomagnetic field
Erdoberfläche *f* earth's surface
Erdreich *n* ground
Erdreichsanierung *f* ground rehabilitation, ground remediation (remedial)
Erdrinde *f* earth's crust
Erdrutsch *m* landslide
Erdstrahlung *f* earth radiation
erdverlegt underground
erdverlegte Rohrleitung *f* underground pipeline
Erdöl *n* mineral oil, petroleum
Erdölchemie *f* petrochemistry, oil chemistry
Erdöldestillationsprodukt *n* distillation product of petroleum
Erdölindustrie *f* mineral oil industry
Erdöllagerstätte *f* oil deposit
Erdölraffinerie *f* oil refinery, petroleum refinery
Erdölverarbeitung *f* petroleum processing

ererbt congenital
Erfassungsgrenze *f* identification limit
Ergometer *n* ergometer
Ergometrie *f* ergometry
Ergonomie *f* ergonomics
ergonomisch ergonomical
Erhaltung *f* **von Biotopen** preservation of biotops
Erhaltungsdosis *f* maintainance dose
erhitzen heat/to
Erhitzen *n* heating
Erholungseffekt *m* recovery effect
Erkrankung *f* **der Atemwege** respiratory disease
Erlenmeyerkolben *m* conical flask
Ermittlungsausschuß *m* fact-finding committee
Ermittlungspflicht *f* fact-finding duty
Ernteausfall *m* crop failure
Ernteverfrühung *f* early cropping
ernährungsbedingt nutritional
Ernährungsforschung *f* alimentary research
Ernährungskette *f* food chain
Ernährungskrankheit *f* nutrition disease
Ernährungsphysiologie *f* nutritional physiology
Ernährungsstörung *f* nutritional disturbance
erodieren erode/to
Erosion *f* erosion
 Erosion *f* **durch Wasser** erosion by water
Erosionsbekämpfung *f* erosion control
Erosionsschutzeinrichtung *f* erosion protection facility
erosionsstabil erosion-stable
Erregbarkeit *f* excitability
Erreger *m* exciter
Erregung *f* excitation
Ersatzanspruch *m* claim for compensation
Ersatzpflicht *f* liability for compensation
Ersatzprodukt *n* substitution product
Ersatzstoff *m* substitute [material]
Ersatzwertstrategie *f* strategy of substitute values
erschöpfen exhaust/to
erschütterungsfest shock-proof
erschütterungsfrei shock-absorbent, vibrationless
Erstattungsanspruch *m* claim of refund
erstickend choking
Erstickung *f* suffocation, choking
 Erstickung *f* **von Fischen** fish suffocation
Erstickungsanfall *m* choke
Ertrag *m* yield, harvest, output
Eruptivgestein *n* igneous rock
erwachsen adult
Erweichen *n* softening
Erweichungsmittel *n* softener
Erweichungspunkt *m* point of softening
Erythrozyt *n* erythrocyte, red blood cell
Erythrozytose *f* erythrocytosis
Erz *n* ore
Erzabfall *m* ore tailing
Erzaufbereitung *f* ore benefication
Erzbergbau *m* ore mining
Erzlager *n* ore deposit
Erzlagerstätte *f* ore deposit
Erzschlamm *m* ore sludge, ore slime

Erzstaub *m* ore dust, fines
Etagenofen *m* multiple-hearth furnace, multiple-hearth roaster
Ethik-Kommission *f* ethics commission
Europium *n* europium
eutroph eutrophic
Eutrophierung *f* eutrophication
Evakuierung *f* evacuation
Evolution *f* evolution
Evolutionstheorie *f* theory of evolution
Exhumierung *f* disinterment
Exkrement *n* excrement
Exkret *n* excretion
exogene Substanz *f* exogenous substance
exogene Wirkung *f* exogeneous effect
exotherm exothermic, exothermal
expandieren expand/to
Expansion *f* expansion, dilatation
Experiment *n* experiment, test
Experimentalchemie *f* experimental chemistry
experimentelle Bedingungen *fpl* experimental conditions
experimenteller Aufbau *m* experimental set-up
experimenteller Wert *m* experimental value
Experimentiervorrichtung *f* experimental facility
Expertenanhörung *f* hearing of experts
explodieren explode/to, detonate/to, blow up/to
Explosion *f* explosion, detonation
Explosionsdruck *m* explosion pressure
explosionsdruckstoßfest explosion-pressure-resistant
explosionsfähig explosive
Explosionsgefahr *f* explosion hazard, explosion risk
explosionsgeschützt explosion-proof, explosion-protected
explosionsgeschützte Ummantelung *f* explosion-proof enclosure
Explosionsgrenze *f* explosion limit
Explosionsschutz *m* explosion protection
explosionssicher inexplosive
Explosionsursache *f* cause of explosion
Explosionswelle *f* explosion
explosiv explosive
Explosivstoff *m* explosive
exponiert exposed
Exponiertendatei *f* data file of exposed persons
Exportgenehmigung *f* **für Abfall** export licence for waste
Exposition *f* exposition, exposure
Expositionsanalyse *f* exposure analysis
Expositionsanlage *f* exposition facility
Expositionsbedingungen *fpl* exposure conditions
Expositionsdaten *pl* exposure data
Expositionseinrichtung *f* exposition device, exposition facility
Expositionsintensität *f* intensity of exposure
Expositionsparameter *m* exposure parameter
Expositionspotential *n* exposure potential
extrahierbar extractable
Extrahierbarkeit *f* extractability

extrahieren extract/to
Extrakt *m* extract
Extraktion *f* extraction
Extraktionsanalysator *m* extraction analyzer
Extraktionsanalyse *f* extraction analysis
Extraktionsmittel *n* extraction agent
Extrapolation *f* extrapolation
extrem rein ultrapure
Extremwert *m* extreme value
Exzenterschneckenpumpe *f* eccentric single-rotor screw pump, eccentric worm pump

F

Fabrikabwasser *n* industrial waste water
Fabrikofen *m* furnace
Fachmesse *f* **für Umweltschutz und Umwelttechnik** trade fair for environmental protection and environmental engineering
Fadenbakterium *n* trichobacterium
Fadenhygrometer *n* hair hygrometer
Fadenwurm *m* nematode, threadworm
fahrbares Absauggerät *n* mobile extraction unit, mobile exhausting unit
Fahrerschulung *f* **für Sondermülltransporte** driver training for hazardous waste transportations
fäkal faecal
Fäkalabwasser *n* faecal water
Fäkalannahmestation *f* faeces acceptance station
Fäkaldünger *m* poudrette
Fäkalien *fpl* faeces, faecal substances

Fäkalienbeseitigung *f* faeces disposal, excreta disposal
Fäkalienfaulanlage *f* faeces fouling plant
Fäkalkoli *f* faecal coliforms
Fäkalschlamm *m* faecal sludge
Faktortest *m* factorial test
fällen precipitate/to; cut/to
Fällmittelstation *f* precipitation station
Fallout *m* fallout
Falloutablagerung *f* fallout deposit
Falloutüberwachung *f* fallout monitoring
Fallrohr *n* vertical pipe
Fallstromverdampfer *m* falling-film evaporator
Fallstudie *f* case study
Fällung *f* flocculation, precipitation,
Fällungschemikalie *f* precipitating agent
Fällungsprodukt *n* precipitate
Familienpackung *f* economy size pack
Familienplanung *f* family planning
Farbabfall *m* **aus Spritzprozessen** waste paint from spraying processes
Farbbrühe *f* dye liquor
Farbechtheitsprüfung *f* colour fastness test
Farbenabbeizmittel *n* paint remover
Farbenindustrie *f* colour and dye industry
Färberei *f* dye house
Färbereimaschine *f* dye machine
farbgetrennte Sammlung *f* **von Altglas** colour-separated collecting of used glass
Farbnebel *m* paint spray
Farbnebelabsauganlage *f* paint spray extraction (exhausting) plant

Farbstoff *m* dye[stuff], pigment
Farbstoffabwasser *n* pigments containing sewage
faserartig fibrous
Faserasbest *m* fibrous asbestos
Faserkohle *f* fibrous coal
Faserstaub *m* fibre dust
Faßpumpe *f* vat pump
faul foul, rotten
Faulanlage *f* fouling plant
Faulbehälter *m* fouling vessel (tank), digestion vessel(tank), fouling basin
Faulbehälter *m* **in Erdaufschüttungen** fouling vessel in earth heapings
Faulbehälterausrüstung *f* fouling vessel equipment
Fäule *f* rot, rotteness, putrefaction
faulen rot/to, putrefy/to, spoil/to
Faulen *n* rotting, putrefying, decomposing
Faulgas *n* fouling gas, biogas, digester gas, sewer gas
Faulgasaufbereitung *f* biogas (fouling gas) processing
Faulgasleitung *f* digester gas pipe, fouling gas pipe, biogas pipe
Faulgasverwertung *f* biogas (fouling gas) utilization
Fäulnis *f* rot[tenness], putrefaction, decay, decomposition
Fäulnisbakterie *f* putrefaction bacterium
Fäulnisbeständigkeit *f* decay resistance, rotproofness,
fäulniserregend putrefactive
Fäulniserreger *m* putrefactive agent
Fäulnisgeruch *m* rotten odour
fäulnisverhindernd antiputrefactive, rot-preventing

Faulkammer *f* fouling chamber, fermentation chamber
Faulschlamm *m* digested sludge, foul sludge, sludge
Faulschlammbildung *f* sludge formation
Faulschlammentgasung *f* fouling sludge degassing
Faulschlammgas *n* sewer gas
Faulteich *m* anaerobic pond
Faulturm *m* fouling tower, digestion tower
Faulung *f* fouling
Fauna *f* fauna
Fehlbedienung *f* maloperation
Fehlerbaumstruktur *f* **für Deponien** tree structure of errors for disposal sites
Fehlernährung *f* malnutrition, subnutrition
Fehlfunktion *f* malfunction
Fehlgeburt *f* abortion
Feinabsiebung *f* fine-sieving
Feinaufbereitung *f* fine-preparation
Feinchemikalien *fpl* fine chemicals
feindisperses Fällungsprodukt *n* finely dispersed precipitation product
feiner Nebel *m* haze
Feinfilter *n* fine filter
Feinkorn *n* fine grain
feinkörnig fine-grained
feinkörniger Boden *m* fine-grained soil
feinkörniges Pulver *n* fine-grained powder
feinporig finely pored
feinpulverisiert finely pulverized
Feinrechen *m* fine-rake
Feinregulierventil *n* precision adjusting valve

Feinsiebrechen *m* fine sieve rake
Feinstaubmaske *f* fine-dust mask
feinverteilt finely dispersed, suspended
Feldbedingungen *fpl* field conditions
Felddiagnose *f* field diagnosis
Feldinstrumentierung *f* field instrumentation
Feldsystem *n* field system
Felduntersuchung *f* field investigation
Feldversuch *m* field experiment
Fels *m* rock
Felsblock *m* boulder
Felsriff *n* rockreef
Ferment *n* ferment, enzyme
Fermentation *f* fermentation
Fermenter *m* fermenter
fermentieren ferment/to
Fermentwirkung *f* enzymatic action
Fermium *n* fermium
Fernbedienung *f* remote control
Fernheiznetz *n* central heating network
Fernheizung *f* central heating system
Fernmeßsystem *n* telemetry system
Fernüberwachung *f* remote monitoring, telemonitoring
Fernwärme *f* district heating
Fernwärmeleitung *f* district heating pipe
Fernwirkinformation *f* telecontrol information
Fernwirksystem *n* telecontrol system
Fernwirkunterstation *f* telecontrol station
Fernwirkzentrale *f* telecontrol master station

ferromagnetisch ferromagnetic
Fertigbeton *m* precast concrete
Fertigkompost *m* ready-compost
Fertigprodukt *n* finished product
Festabfall *m* solid waste
Festbett *n* fixed bed, solid bed
Festbettreaktor *m* fixed-bed reactor
Festbettumlaufreaktor *m* fixed-bed circulating reactor
feste Bestandteile *mpl* solid matter
fester Abfallstoff *m* solid waste [material]
fester Zustand *m* solid state
Festkörper *m* solid
Festkörperchemie *f* solid-state chemistry
Festkörperdiffusion *f* solid diffusion
Festkraftstoff *m* solid fuel
Festphase *f* solid phase
Feststeller *m* arresting device
Feststoff *m* solid [matter]
Feststoffabscheidung *f* precipitation of solids, separating of solid substances
Feststoffabsiebung *f* screening of solids
Feststoffkatalysator *m* fixed catalyst
Feststoffmischung *f* mixture of solids
Feststofftrennung *f* separation of solids
Festwert *f* constant value
Festzustand *m* solid state
Fettabfälle *mpl* refuse fat
Fettablagerung *f* deposition of fat
Fettabscheider *m* fat separator
Fettabtrennung *f* separation of fat
fettähnlich lipoid
fettarm poor in fat

fettartig lipoid
Fettbildung *f* fat formation
Fettembolie *f* fat embolism
Fettemulsion *f* fat emulsion
fettfrei non-fatty
fettgespeichert fat-stored
Fettgewebe *n* adipose tissue
fetthaltig adipose, fat-containing, fatty
fettig adipose, fatty
Fettprodukt *n* fat product
Fettschlamm *m* fat sludge
Fettverbrennung *f* grease incineration
Feuchte *f* moisture, humidity, dampness
Feuchteanalysator *m* moisture analyzer
Feuchtemessung *f* humidity measurement
Feuchtigkeit *f* moisture, humidity, dampness
Feuchtigkeitsaufnahme *f* absorption of moisture
feuchtigkeitsbeständig moisture-resistant
Feuchtigkeitsbeständigkeit *f* water resistance
Feuchtigkeitsgehalt *m* moisture content
Feuchtigkeitskorrosion *f* aqueous corrosion
Feuchtigkeitsmesser *m* hygrometer
Feueralarm *m* fire alarm
feuerbeständig fire-proof
Feuerbeständigkeit *f* fire resistance, fireproofness
feuerfangend ignitible
feuerfest fire-proof, heat-proof, incombustible
feuerfeste Auskleidung *f* refractory lining
feuerfeste Farbe *f* fire-proof paint
Feuerführung *f* fire regulation, fire control
Feuergefahr *f* fire risk
feuergefährlich inflammable, combustible
Feuergefährlichkeit *f* inflammability
feuerhemmend flame-retardant
Feuerkitt *m* fire cement
Feuerlöscher *m* fire extinguisher
Feuerraumatmosphäre *f* fire chamber atmosphere
Feuerraumseitenwand *f* side wall of the fire chamber
Feuerungsanlage *f* firing plant
Feuerungsoptimierung *f* firing optimization
Feuerungstechnologie *f* combustion technology
Feuerverzinkung *f* hot galvanizing
Feuerverzinnung *f* hot-dipped tinning
Fiberglas *n* fibre glass
Fibroblastenkultur *f* fibroblast culture
fibrogene Wirkung *f* fibrogenic effect
Filter *n* filter, screen
Filterasche *f* filter ash
Filterelement *n* filter element
filterfähig filterable
Filtergehäuse *n* filter housing
Filtergewebe *n* filter tissue
Filterhalle *f* filtering hall (shed)
Filterkuchen *m* filter cake
Filtermedium *n* filter medium
filtern filter/to, filtrate/to
Filterpatrone *f* filter cartridge

Filterplatte *f* filter plate
Filterpresse *f* filter press
Filtersalz *n* filter salt
Filterschlauch *m* filter hose
Filterstaub *m* filter dust
Filterstaubrest *m* filter dust residue
Filtersystem *n* filter system
Filtertechnik *f* filter technology
Filtertuch *n* filter cloth
Filtrat *n* filtrate
Filtration *f* filtration
Filtrationsbedingungen *fpl* filtration conditions
Filtrationsphase *f* filtration phase
Filtrationsrückstand *m* filtration residue
Filtrationsverfahren *n* filtration method
Filtratvorlage *f* filtrate receiver
filtrierbar filtrable
Filtriereinheit *f* filtering unit
filtrieren filter/to, filtrate/to
Filtrieren *n* filtering
Filtriersand *m* filtering sand
Filtrierung *f* filtration
Firnis *m* varnish
Fischbauch-Wehrklappe *f* dam, weir, pound-lock, flap, gate
Fischgift *n* fish poison
Fischgiftigkeit *f* fish toxicity
Fischmehl *n* fish meal
Fischsterben *n* fish mortality, death of fish, fish kill
Fischtest *m* fish test
fischtoxisch fish-toxic
Fischzaun *m* fish corral, fish barrier
Fixierflüssigkeit *f* fixing bath
Flächenbrand *m* extensive fire
Flächenkontamination *f* area contamination
Flächennutzungskarte *f* land utilization map
Flächennutzungsplan *m* plan of land utilization
Flächenreaktivierung *f* area reactivation, area revitalization
Flächenschonung *f* land saving
Flammenbild *n* flame shape, flame picture
Flammendurchschlag *m* flashback
Flammenfront *f* flame front
Flammenionisationsdetektor *m* flame ionization detector
flammenlos flameless
Flammenphotometer *n* flame photometer
Flammenrückschlag *m* flashback
Flammenrückschlagsicherung *f* backfiring safeguard, flashback safeguard
Flammpunkt *m* fire point, point of ignition
Flammschutzmittel *n* fire retardant, flame retardant
Flammschutzmittelbeschichtung *f* fire retardant coating
flammwidrig uninflammable, flame-resistant
Flasche *f* **aus Kunststoff** plastic bottle
Flaschenverschluß *m* bottle cap
Flechte *f* lichen, moss
Fleckentfernungsmittel *n* spot remover, stain remover
fleischfressend carnivorous
Fleischkonserve *f* tinned meat, preserved meat
Fleischverarbeitung *f* meat-processing
Fleischwaren *fpl* meat products

flexible Anpassung *f* flexible adaption
flexibler Graphit *m* flexible graphite
Fliegengift *n* fly poison
Fliehkraftabscheider *m* centrifugal separator
Fliehkraftstaubabscheider *m* cyclon dust collector
Fließbett *n* fluidized bed
Fließbettkatalysator *m* fluidized-bed catalyst
Fließbettverfahren *n* fluidized-bed process
Fließeigenschaft *f* flow property
Fließgeschwindigkeit *f* velocity of flow, flow rate
Fließmittel *n* mobile solvent
Fließverhalten *n* flow behaviour
Flockung *f* flocculation
Flockungsbecken *n* flocculation basin
Flockungshilfsmittel *n* flocculation agent
Flockungsmittel *n* flocculation agent
Flockungsmittelaufbereitung *f* flocculation agent preparation
Flockungsreaktor *m* flocculator
Flora *f* flora
Flotation *f* flotation
Flotationsschlamm *m* flotation sludge
Flotationsverfahren *n* flotation method
Flotationszelle *f* flotation cell
Flotte *f* batch, dye liquor
Flöz *n* layer, seam, stratum, vein
flüchtig volatile
 flüchtige Bestandteile *mpl* volatile matter

 flüchtige Chemikalie *f* volatile chemical
 flüchtige Substanzen *fpl* volatile matter
 flüchtiges Spaltprodukt *n* volatile fission product
Flüchtigkeit *f* volatility
Flugasche *f* fly ash, light ash, airborne dust, quick ash
Flugaschedeponie *f* fly ash landfill
Flugaschenemission *f* fly ash emission
Flugaschequalität *f* fly ash quality
Flugascheverglasung *f* vitrification of fly ash
Fluglärm *m* aircraft noise
Flugsand *m* shifting sand
Flugstaub *m* air dust, flue dust, fluidized dust
Flugverkehr *m* air traffic
Flugzeugabgas *n* waste gas of aircraft
Flugzeugdesinfektion *f* aircraft disinfection
Flugzeugdüngung *f* airplane fertilizing
Flugzeugmotor *m* aircraft engine
Flugzeugsprühausrüstung *f* aircraft spray equipment
Fluor *n* fluorine
Fluoreszenz *f* fluorescence
fluoridieren fluorine/to
Fluoridierung *f* fluoridation
Fluorimetrie *f* fluorimetry
Fluorschaden *m* damage by fluorine
Fluorvergiftung *f* fluorosis
Flurbereinigung *f* consolidation of arable land
Flurbereinigungsgesetz *n* law of consolidation of arable land
Fluß *m* river; stream

Flußbecken *n* river basin
Flußbett *n* channel, river bed
Flüssig-Flüssig-Extraktion *f* liquid-liquid extraction
Flüssigbrennstoff *m* liquid fuel
Flüssigbrennstoffrakete *f* liquid-fuelled rocket
Flüssigchromatographie *f* liquid chromatography
flüssige Luft *f* liquid air
flüssige Phase *f* liquid phase
flüssiger Zustand *m* liquid state
Flüssiggas *n* liquefied gas
Flüssigkeitsabscheider *m* liquid separator
Flüssigkeitsaufnahme *f* liquid uptake, fluid intake
Flüssigkeitsfiltration *f* fluid filtration
Flüssigkeitshöhe *f* liquid level
Flüssigkeitskühler *m* liquid cooler
Flüssigkeitsmaß *m* liquid measure
Flüssigsauerstoff *m* liquid oxygen
Flüssigwerden *n* liquefaction
Flußmündung *f* estuary
Flußrichtung *f* direction of flow
Flußschlamm *m* river silt
Flußufer *n* river bank
Flußverunreinigung *f* stream pollution
Flußwasser *n* river water
Flutwelle *f* flood wave
Folgendiagnose *f* effect diagnosis
Folgereaktion *f* secondary reaction
Folgeschaden *m* secondary injury
Folie *f* film, foil, sheet
Folienabdichtung *f* foil sealing, foil covering
Folienabfall *m* diposed films
Förderbecher *m* elevator bucket
Förderdruck *m* delivery pressure
Fördereimer *m* elevator bucket
Fördereinrichtung *f* conveying facility
Förderplattform *f* conveyance platform
Förderschnecke *f* screw conveyor, worm conveyor
Forderungenkatalog *m* **zum Umweltschutz** catalogue of demands for environmental protection
Forellenzucht *f* trout culture
forensische Chemie *f* forensic chemistry
Formenlehre *f* morphology
Formerei *f* moulding
Formsand *m* foundry sand, moulding sand
Forschungsreaktor *m* research reactor
Forschungsstrategie *f* research strategy
Forst *m* forest
Forst-Ökosystem *n* forest ecosystem
Forstwirtschaft *f* forestry
Fortpflanzung *f* propagation, regeneration
fossil fossil
 fossil befeuertes Kraftwerk *n* fossil heated power plant
 fossile Energie *f* fossil energy
 fossiler Brennstoff *m* fossil fuel
Fossil *n* fossil
Fotochemikalie *f* photochemical
Fraktion *f* fraction
fraktionieren fractionate/to
fraktionierte Destillation *f* fractional distillation
Fraktionsschnitt *m* cut in distillation
Francium *n* francium
freies Radikal *n* free radical

Freilandfläche *f* open land area
Freilandversuch *n* open-air test
Freiluft *f* open-air
Freiluftbedingung *f* free air condition, open air condition
freischwebend air-borne
freisetzen liberate/to, release/to
Freisetzung *f* release
Freisetzungsverbot *n* für gentechnisch umgewandelte Mikroorganismen release prohibition for genetechnologically converted substances
Freizeitaktivität *f* leisure activity
Freizeitbeschäftigung *f* leisure activity
Freizeitforschung *f* leisure research
Freizeitgestaltung *f* recreational activity
Fremdkörper *m* foreign body
Fremdstoff *m* impurity
Fremdwasser *n* impurity water
Frequenz *f* frequency, periodicity
Frischen *n* refining
Frischerei *f* refinery
Frischkompost *m* fresh-compost
Frischluft *f* inlet air, fresh air
Frischschlamm *m* fresh sludge
Frischwasser *n* freshwater
Fritte *f* frit
Frost *m* frost
frostbeständig resistant to frost
Frostschutz *m* frost protection
Frostschutzmittel *n* antifreeze
fruchtbar fertile
Frucht *f* fruit
Fruchtbarkeit *f* fertility
Fruchtfolge *f* rotation of crops
Frühentdeckung *f* early detection
Früherkennung *f* early detection
Frühgeburt *f* premature birth
frühreif premature

Frühstadium *n* early stage
Fühler *m* sensor
Führungsdaten *pl* guiding data, prompting data
Füllanlage *f* filling plant
Füllanschluß *m* filling connection
Füllgut *n* filling material
Füllkörper *m* filling material, packing
Füllkörperhöhe *f* packed height
Füllkörperrohr *n* packed tube
Füllkörperturm *m* packed tower
Füllkörperwäscher *m* packed scrubber
Füllmasse *f* packing
Füllmaterial *n* filler
Füllstandanzeiger *m* level indicator
Füllstandskontrolle *f* level control
Füllstandsmessung *f* level measurement
Füllstandsüberwachung *f* level monitoring
Füllstoff *m* filler, filling material; chromatographic packing
Füllstoffkapazität *f* bulkage capacity
Füllstutzen *m* filler neck
Fungizid *n* fungicide
funktionale Güteziffer *f* functional quality index
funktionale Umweltdefinition *f* functional environmental definition
Funktionalität *f* functionality
Funktionsbaustein *m* function block
Funktionsbereitschaft *f* standby function, readyness for operation
Funktionssicherheit *f* functional safety
Funktionsstörung *f* disfunction, malfunction

Funktionstest *m* function test, validity check
Furan *n* furan
Furanemission *f* furan emission
Fuselöl *n* fusel oil
Fußbodenanstrich *m* floor finish
Fußbodenpflegemittel *n* floor polish
Futtermittel *n* feed, forage
Futterpflanze *f* forage plant
Futterzusatz *m* feed supplement

G

Gadolinium *n* gadolinium
Galaktose *f* galactose
Gallenausscheidung *f* excretion in bile
Gallertfilter *n* colloid filter
Gallertmasse *f* gelatinous substance
Gallertmoos *n* sea moss
Gallertsubstanz *f* gelatinous substance
Gallium *n* gallium
Galvanik *f* galvanics
Galvanikbetrieb *m* galvanizing company
Galvanikindustrie *f* galvanizing industry
Galvanikprozeß *m* electrodeposition
Galvanikschlamm *m* galvanic sludge
Galvaniküberzug *m* electrodeposit
galvanisch galvanic, voltaic
  galvanisch niederschlagen electrodeposit/to
  galvanisches Bad *n* galvanic bath
Galvanisieranlage *f* electroplating plant
Galvanisierung *f* electroplating, electrodeposition, galvanization

Gametoblast *m* gametoblast
Gametozyt *m* gametocyte
gammaaktiv gamma-active
Gammastrahler *m* gamma emitter
Gammastrahlung *f* gamma radiation
Gammazählrohr *n* gamma counter tube
Gammazerfall *m* gamma decay
Gangliniendarstellung *f* output curve representation, hydrograph curve
Ganzgesichtsmaske *f* full face mask
Ganzkörperbelastung *f* total body burden
Gäranlage *f* fermentation plant
Gärbottich *m* fermenter
Gärfutter *n* silage fodder
Gärtank *m* fermenter, fermenting tank
Gartenabfall *m* garden refuse
Gartenbau *m* gardening
Gärung *f* fermentation
Gärungserreger *m* ferment
gärungsfördernd zymogeneous, zymogenic
Gärungspilz *m* fermentation fungus
Gärungstechnik *f* zymotechnology
Gärungsvorgang *m* fermentation process
Gas *n* gas
  Gas-Dampf-Gemisch *n* gas-vapour mixture
Gasabfackelung *f* gas burning
Gasatmosphäre *f* gas atmosphere
Gaschromatographie *f* gas chromatography
gasdicht gas-tight
Gasdränage *f* gas drainage
Gasdruck *m* gas pressure
Gasentwicklung *f* gas evolution
Gasfackel *f* gas torch

Gasfilter *n* gas filter
gasförmig gaseous
Gasgemisch *n* gas mixture
Gashaube *f* gas hood
Gasmaske *f* gas mask
Gasmotor *m* gas engine, gas motor
Gasnetz *n* gas network
Gasphase *f* gas phase
Gasreinigung *f* gas purification
Gasspeicher *m* gas reservoir
Gasspeicherung *f* gas storage
Gasstrom *m* gas flow
Gastank *m* gas tank
Gasturbine *f* gas turbine
Gasverwertung *f* gas utilization
Gaswandler *m* gas transformer
Gaswarngerät *n* gas alarm detector
Gaswäscher *m* gas washer
Gaswerk *n* gasworks
Gaswerksgelände *n* gasworks area
GAU – größter anzunehmender Unfall maximum credible accident (MCA)
Gebläse *n* blast apparatus, blower, fan, ventilator
Gebläse-Abfackelanlage *f* blast flare-unit
Gebläseluft *f* blast air
Gebläsemaschine *f* fan engine
Gebläseofen *m* blast furnace
Gebläsestation *f* fan station
Gebläsewind *m* air blast
gebrannter Kalk *m* burnt lime, quicklime
Geburtsfehler *m* birth defect, connatal anomaly
gefährdete Arten *fpl* endangered species
Gefährdungsanalyse *f* risk analysis
Gefährdungsdosis *f* endangering dose, risk dose

Gefährdungspotential *n* endangering potential, risk potential
Gefahrenbereich *m* danger zone
Gefahrenherd *m* source of danger
Gefahrenidentifizierung *f* hazard identification
Gefahrenklasse *f* class of risk
Gefahrenpotential *n* hazard potential
Gefahrenquelle *f* source of danger
Gefahrensignal *n* danger signal
Gefahrenzulage *f* danger pay
Gefahrgut *n* hazardous goods, harmful goods
Gefahrgutbeauftragter *m* **ohne jegliches Weisungsrecht** authorized person for hazardous goods without any authority to issue directives
Gefahrguttransport *m* transportation of hazardous goods
Gefahrgutumschlag *m* transshipment of hazardous goods
Gefahrgutvorschrift *f* direction for transportation of hazardous goods
gefährlich dangerous, hazardous, risky, unsafe
gefährliche Güter *npl* hazardous goods
gefährlicher Arbeitsstoff *m* harmful working material
gefährlicher Stoff *m* harmful substance
Gefährlichkeit *m* danger, dangerousness
Gefährlichkeitsbewertung *f* danger assessment
gefahrlos safe
Gefahrpunkt *m* critical point
Gefahrstoff *m* harmful material, harmful substance, hazardous material, hazardous substance

Gefahrstoff *m* **am Arbeitsplatz** hazardous substance at the workplace
Gefahrstoff-Informationsdatenbank *f* information data-bank for harmful (hazardous) materials
Gefahrstoff-Informationssystem *n* information system of harmful (hazardous) substances
Gefahrstoffabfall *m* harmful waste, hazardous waste
gefahrstoffbezogen hazardous-substance-related
Gefahrstoffkataster *m* catalogue for hazardous substances
Gefahrstoffregistriersystem *n* registration system for hazardous substances
Gefahrstoffüberwachungsgerät *n* toxic hazard monitor
Gefäßversuch *m* pot experiment
Gefrier- und Auftaukreislauf *m* freeze-thaw cycling
gefriergetrocknet freeze-dried, lyophilized
Gefrierkonditionierung *f* freezing conditioning
Gefrierpunkt *m* point of freezing
Gefriertrockner *m* vacuum freeze dryer
Gefriertrocknung *f* freeze-drying, lyophilization
Gefriervorrichtung *f* congealer
gefügelos structureless
Gefügeuntersuchung *f* structural examination
Gefügeveränderung *f* structural change
Gegenanzeige *f* contraindication
Gegendruckventil *n* counter-pressure valve
Gegenfluß *m* counterflow
Gegengewicht *n* counter balance
Gegengift *n* antidote, antitoxin
Gegenmaßnahme *f* counteraction, countermeasure
Gegenmittel *n* antidote, remedy
Gegenreaktion *f* back reaction, reverse reaction
Gegenstrom *m* countercurrent, counterflow, reversed current
Gegenstromdestillation *f* countercurrent distillation
Gegenstromextraktion *f* counterflow extraction
Gegenstromprinzip *n* countercurrent principle
Gegenstromwaschung *f* countercurrent washing
gegorener Klärschlamm *m* fermented sewage sludge
Gehaltsbestimmung *f* analysis, assay
Gehölz *n* wood
Gehörschutz *m* hearing protection
Gehörverlust *m* hearing loss
Geiger-Müller-Zähler *m* Geiger-Müller counter
Geißeltierchen *n* flagellatum
gekörnt granular
Gel *n* gel
Gelände *n* terrain, field
Geländearbeit *f* field operation
Geländekunde *f* topography
Geländestufe *f* terrace
Gelatine *f* gelatine
Gelatinenährboden *m* gelatine culture medium
Gelbfieber *n* yellow fever
Gelbildung *f* gel formation, gelatinization, gelation

Gelchromatographie *f* gel chromatography
Gelelektrophorese *f* gel electrophoresis
Gelenkrohr *n* joint pipe, hinged pipe
Gelfiltration *f* gel filtration
Geliermittel *n* gelling agent
gelöschter Kalk *m* slaked lime
gemahlene Kohle *f* pulverized coal
Gemeinschaftskläranlage *f* communal clarification plant
Gen *n* gene
Genaktivität *f* gene activity
Genänderung *f* mutation
Genbank *f* gene pool
Genehmigung *f* permission, permit, licence, approval
Genehmigungsbestellung *f* collective order
Genehmigungspraxis *f* practice of approval
Genehmigungsrecht *n* right of approval, right of permission, approval right
Genehmigungsvoraussetzung *f* condition of approval
Genehmigungszuständigkeit *f* responsibility of approval
Generatorgas *n* generator gas
Genese *f* genesis
Genetik *f* genetics
Genetiker *m* geneticist
genetisch genetic
  genetischer Schaden *m* genetic damage
genießbar palatable
Genießbarkeit *f* palatability
Genmutation *f* gene mutation
Genotypus *m* genotype
Genregulation *f* gene regulation
Gensonde *f* gene probe

Gentechnik *f* gene technology
Gentechnologie *f* genetic engineering
gentoxischer Stoff *m* genotoxic substance
Gentoxizität *f* gene toxicity
Geobotanik *f* geobotanics
Geochemie *f* geochemistry
Geographie *f* geography
Geologe *m* geologist
Geologie *f* geology
geologisch geological
  geologische Formation *f* geological formation
Geomembran *f* geomembrane
Geometer *m* land surveyor
Geophysisk *f* geophysics
geordnete Müllkippe *f* ordered refuse dump, ordered landfill
Geosphäre *f* geosphere
Geotextilien *pl* geotextiles
geothermisches Kraftwerk *n* geothermal power plant
gepackte Säule *f* packed column
Geräusch *n* noise
geräuschabgebend noise-emitting
Geräuschbelästigung *f* noise nuisance
Geräuschemission *f* noise emission
Geräuschpegel *m* noise level
Geräuschpegelmesser *m* sound level meter
Gerben *n* tanning
Gerberei *f* tannery
Gerbereiwasser *n* tannery sewage
Gerbung *f* tanning
gerichtliche Chemie *f* forensic chemistry
gerichtliche Medizin *f* forensic medicine, legal medicine

Gerichtsmedizin f forensic medicine, legal medicine
geringwertig of low value, low-grade
geringwertiges Erz n lean ore
Gerinnbarkeit f coagulability
gerinnen coagulate/to, congeal/to
Gerinnungsfähigkeit f coagulability
Gerinnungsmittel n coagulant
Germanium n germanium
Germizid n germicide
Geröll n detritus, gravel, pebbles
Geruch m odour, smell
geruchlos free from odour, inodorous, odourless
Geruchsbekämpfung f deodorization
Geruchsbelästigung f odour nuisance
geruchsbindend deodorant, deodorizing
Geruchshygiene f deodorization hygiene
geruchsintensiv odour-intensive
Geruchssinn m smell
Geruchverbesserung f deodorization
Gesamtauswertung f complete evaluation
Gesamtdosis f total dose
Gesamtenergie f total energy
Gesamterzeugung f total production
Gesamtfracht f organischer Stoffe total load of organic substances
Gesamtkonzentration f total concentration
Gesamtsauerstoffbedarf m total oxygen demand
geschlossener Kreislauf m closed circuit [system]
Geschmackssinn m gustation

Geschmacksverbesserung f taste improvement
Geschmackszentrum n gustatory centre
geschütztes Gebiet n protected area
Geschwulst f tumour; swelling
Geschwulstbildung f neoplasm
Geschwür n ulcer
Geschwürbildung f ulceration
gesetzgeberische Maßnahme f legislative measure
gesetzliche Auflage f legal condition
gesicherte Diagnose f certified diagnosis
Gesichtsdusche f face douche
Gesichtsplastik f facioplastic
Gestank m bad smell, stench
Gestein n rock
Gesteinsmehl n rock flour, mineral powder, powdered stone
Gesteinsstaub m mineral dust
gestörtes Gleichgewicht n disequilibrium
gesund healthy, in good health
gesundes Wohnen n healthy living
Gesunderhaltung f von Gewässern maintenance of the health of waters
Gesundheit f health, healthiness
gesundheitlich sanitary
gesundheitlich unbedenklich - healthy unrisky
Gesundheitsamt n public health department
Gesundheitsgefahr f health risk
gesundheitsgefährdender Lärm m health-endangering noise
Gesundheitsgefährdung f health endangering
Gesundheitsministerium n Ministry of Health
Gesundheitspflege f sanitation

Gesundheitspolizei *f* sanitary police
Gesundheitsrisiko *n* health risk
gesundheitsschädlich harmful, injurious to health, insanitary, unhealthy
Gesundheitsstatus *m* health status, state of health
Gesundheitsverhaltensforschung *f* health behaviour research
Gesundheitswesen *n* health service
Gesundheitswirkung *f* health effect
Getreide *n* cereal, grain
Getreidebrand *m* smut
Getreideernte *f* cereal crop
Getreideschädling *m* cereal pest
Getreidesilo *m* grain silo
Getrenntsammlung *f* separate collection
Getrenntsammlung *f* **von Wertstoffen** separate collection of valuable materials
Getränkeindustrie *f* beverage industry
Getränkeverpackung *f* beverage packing
getränktes Holz *n* impregnated wood
Gewächs *n* neoplasm; plant
Gewässer *n* water
Gewässerbelastung *f* impact of waters, impact on aquatic systems
Gewässererwärmung *f* warming of waters
Gewässergüte *f* water quality, quality of waters
Gewässerkunde *f* hydrology
Gewässersanierung *f* inshore water renovation, remedial (rehabilitation) of the waters
Gewässerschutz *m* protection of waters, prevention of water pollution, river and lake protection
Gewässerschutzabkommen *n* agreement for water protection
Gewässerverschmutzung *f* water pollution
Gewebeanalysenprobe *f* tissue analysis sample
Gewebeauflösung *f* histodialysis
Gewebebildung *f* histogenesis
Gewebechemie *f* histochemistry
Gewebeeiweiß *n* tissue protein
Gewebefilter *n* tissue filter
Gewebeflüssigkeit *f* tissue fluid
Gewebekultur *f* tissue culture
Gewebelehre *f* histology
Gewerbeabfall *m* trade waste
Gewerbehygiene *f* industrial hygiene
Gewerbemedizin *f* industrial medicine
Gewerbemüll *m* trade waste, industrial waste
Gewerbetoxikologie *f* industrial toxicology
Gewichtsabnahme *f* decrease in weight
Gewichtsanalyse *f* analysis by weight, gravimetry
Gewichtsauslese *f* sorting by weight, classifying (separating) by weight
Gewichtserhöhung *f* weight increase
Gewichtskonstanz *f* weight constancy
Gewichtskontrolle *f* weight control
gewöhnlicher Abfall *m* ordinary waste
Gezeit *f* tide
Gießerei *f* foundry
Gießereisand *m* foundry sand
Gießform *f* mould

Gießmaschine *f* casting machine
Gießverfahren *n* casting process
Gift *n* poison, toxin
giftabtreibend antidotal
Giftarchivschlüssel *m* key of the poison archive
Giftaufnahme *f* poison uptake
Giftelimination *f* poison elimination
giftig poisonous, toxic
Giftinformationszentrum *n* poison information centre
Giftkonzentration *f* poison concentration
Giftstoff *m* poisonous matter, toxicant
Gips *m* gypsum
gipsartig gypseous
Gipserde *f* gypseous earth
Gipsmergel *m* gypseous marl
Gitternetzwerk *n* lattice network
Gittertherapie *f* grid therapy
Glanzmittel *n* polish
Glasfaserfilter *n* glass fibre filter
glasfaserverstärkter Beton *m* glass-fibre-reinforced concrete
Glasfaservlies *n* glass fibre fleece
Glasfiberfilter *n* glass fibre filter
Glasfläschchen *n* vial
Glasgefäß *n* glass vessel
Glasgranulat *n* glass granules
Glasindustrie *f* glass industry
Glaskapillare *f* glass capillary
Glasmatrix *f* glass matrix
Glasproduktion *f* manufacturing of glass
Glassammlungscontainer *m* glass collection container
Glatteisfrühwarnsystem *n* early-alarming system for ice
gleichförmige Strömung *f* uniform flow

gleichgerichtet equidirectional, unidirected
Gleichgewicht *n* **der Kräfte** equilibrium of forces
Gleichgewichtszustand *m* state of equilibrium
Gleitmittel *n* lubricant
Gleitsicherheit *f* sliding resistance
Glimmen *n* smouldering
globale Konzentration *f* global concentration
globales Umweltüberwachungssystem *n* global environmental monitoring system
Globalstrahlung *f* global radiation
Glockenkurve *f* bell-shaped curve
Glühdraht *m* annealed wire
glühen anneal/to; glow/to
Glühfaden *m* filament
Glühkatode *f* hot cathode
Glührückstand *m* annealing residue, ignition residue, residue of ignition
Glühtiegel *m* crucible
Gold *n* gold
Goldkatode *f* gold cathode
Gossenbesen *m* gutter broom
Grabensystem *n* ditch system, canal system, trench system
Graduierung *f* graduation, scale
Granalie *f* granule
Granit *m* granite
Granitfels *m* granite rock
Granulat *n* granules, granulated material, granular material, pellets
granulieren granulate/to, grain/to
Granuliermaschine *f* pelletizer
granulierte Aktivkohle *f* granulated activated carbon
Granulierung *f* **von Altreifen** granulation of used tyres

Granulierungsverfahren *n* granulation method
Granulozyt *m* granulocyte
Graphit *m* graphite
Graphitdichtung *f* graphite ferrule
Graphitelektrode *f* graphite electrode
graphitisierter Kohlenstoff *m* graphitized carbon
Graphitkohle *f* graphite carbon
Graphitkristall *m* graphite crystal
graphitmoderierter Reaktor *m* graphite-moderated reactor
Graphitofen *m* graphite furnace
Graphitpulver *n* graphite powder
Gras *n* grass
Grauguß *m* pig iron casting
Gravimetrie *f* gravimetry
gravimetrische Sortierung *f* gravimetric classifying
Gravitation *f* gravitation, gravity
Gravitationstrennung *f* gravitational separation
Greifvogelschutz *m* protection of prehensile birds
Grenzbedingung *f* limiting condition
Grenzbelastung *f* limiting load
Grenzfall *m* borderline case, limiting case
Grenzfallsituation *f* worst-case situation
grenzflächenaktiv surface-active
Grenzflächendiffusion *f* boundary diffusion
Grenzflächenerfassung *f* boundary layer determination
Grenzgebiet *n* border region
Grenzkonzentration *f* limiting concentration
Grenzlinie *f* borderline, limiting line, limit

Grenztemperatur *f* limiting temperature
Grenzwert *m* limit[ing] value, end value
Grenzwertregelung *f* regulation for limit values
grenzüberschreitender Abfalltransport *m* frontier-crossing waste transportation
grenzüberschreitender Verkehr *m* frontier-crossing traffic
Grenzwertverletzung *f* out-of-limit condition, limit violation
Grenzwertvorgabe *f* limit value selection
Grippe *f* influenza
Grobabtrennung *f* coarse separation
grober Sand *m* grit
Grobfiltration *f* coarse filtration
Grobkies *m* pebble
Grobkorn *n* coarse grain
Grobmüll *m* bulky refuse
Grobsiebung *f* coarse screening
Grobstoff *m* bulk material
Grobuntersuchung *f* gross examination
Großabnehmer *m* bulk purchaser
Großanlage *f* **zur Insektizidherstellung** insecticide plant
Größenfaktor *m* size factor
Größenordnung *f* magnitude, scale
Größenverteilung *f* size distribution
Großerzeuger *m* large producer
Großfeuerungsanlagenverordnung *f* regulation for large-scale incineration plants
Großhandel *m* wholesale trade
Großhersteller *m* large producer
Großkaufmann *m* merchant
Großklima *n* macroclimate

Großkompostieranlage *f* large-scale composting plant
Großlysimeter *n* large-scale lysimeter
großtechnische Herstellung *f* production on a large scale
großtechnische Kompostierung *f* large-scale composting
größter anzunehmender Unfall (GAU) maximum credible accident (MCA)
größtes anzunehmendes Risiko *n* greatest assumable (thinkable) risk
Großunternehmen *n* major enterprise
Großveranstaltung *f* big event
Großverbraucher *m* large consumer
Großversuch *m* large-scale test
Großwuchs *m* macrosomia
Grube *f* pit
Grubenarbeiter *m* miner
Grubenbau *m* mining
Grubenbrand *m* pit fire
Grubengas *n* mine gas, marsh gas
Grubenkohle *f* pit coal
Grubenlüftung *f* pit ventilation, mine ventilation
Grubenunglück *n* mine disaster
Grünabfall *m* green waste
Grünabfallkompostierung *f* green waste composting, composting of green waste
Grünalge *f* green alga
Grundbeschichtung *f* base coating
Grundbestandteil *m* basic component, base part
Grundbuchamt *n* land office
Grundchemikalie *f* basic chemical, key chemical
grundieren prime/to
Grundlagenforschung *f* basic research
Grundnährstoff *m* basic nutrient
Grundstoffgewerbe *n* trade of raw materials
Gründünger *m* green manure, vegetable manure
Grundwasser *n* groundwater, underground water
Grundwasserabsenkung *f* settlement of groundwater
grundwassergefährdend groundwater-endangering
grundwassergeschützt groundwater-protected
Grundwassermodell *n* groundwater model
Grundwasserpegel *m* groundwater level
Grundwasserqualität *f* quality of groundwater
Grundwasserschutz *m* groundwater protection
Grundwasserschutzgebiet *n* groundwater protection area
Grundwasserspiegel *m* groundwater level
Grundwasserstand *m* groundwater level
Grundwasserverunreinigung *f* groundwater pollution
Grüne Tonne *f* green bin
Grünfläche *f* lawn; park area
Grünglas *n* green glass
Grüngürtel *m* green belt
Grünspan *m* verdigris
Gülle *f* liquid manure, slurry
Gülleaufbereitung *f* processing of liquid manure

Gülleaufbereitung *f* **in großtechnischem Maßstab** large-scale processing of liquid manure
Güllebehandlung *f* treatment of liquid manure
Güllebeseitigung *f* disposal of liquid manure
Gülleverwertung *f* usage of liquid manure
Gummi *n* gum, rubber
Gummihandschuh *m* rubber glove
Gummiharz *n* gum resin
Gummikleber *m* rubber adhesive
Gummischlauch *m* rubber hose
Gummischürze *f* rubber apron
Gummistopfen *m* rubber stopper
Guß *m* cast
Gutachten *n* expertise, certificate, expert opinion
Gutachten *n* **zur Standortwahl** expert opinion on the site-selection
gutartiger Tumor *m* benign tumour
Güte *f* quality
Gütedaten *pl* efficiency data
Güterwagen *m* [goods] waggon, *{Am}* freight car
Gütetest *m* quality test
Gütezahl *f* index value
Güteziffer *f* quality index
Guttapercha *f* guttapercha

H

Haarhygrometer *n* hair hygrometer
Haarriß *m* capillary fissure
Habitat *n* habitat
Habitus *m* habit
Häckselgut *n* chaff
Hafnium *n* hafnium

haftbar liable, responsible, accountable
haften guarantee/to, be responsible/to, be liable/to; adhere/to, stick/to
Haftpflicht *f* liability, responsibility
Haftpflichtanspruch *m* legal liability claim
haftpflichtig liable
Haftpflichtversicherung *f* liability insurance
Haftreibung *f* adhesion friction, cohesive friction
Haftung *f* liability
Haftungsrecht *n* liability law
Haftungsrisiko *n* risk of liability
Haftvermögen *n* adhesion power
Haftwirkung *f* intensity of adhesion
Hagelschauer *m* hail shower
Hahnregulierung *f* stopcock control
Hahnventil *n* plug valve
Hakenwurm *m* hookworm
halbaktiv semiactive
halbautomatisch semiautomatic
halbdurchlässig semipermeable
halbfette Kohle *f* subbituminous coal
Halbleiterschalter *m* semiconductor switch
Halbmetall *n* semimetal
Halbmikroanalyse *f* semimicroanalysis
Halbperiode *f* half-period, half cycle, semiperiod
halbtechnisch semiindustrial
Halbwertszeit *f* period of half change, half change value, half[-life] period
Halde *f* waste dump, waste heap
Haldenschlacke *f* waste slag, dump slag

Halluzinogen *n* hallucinogen
Halogen *n* halogen
halogenfrei halide-free
Halogenid *n* halide
halogenidfrei halide-free
halogenierbar halogenatable
halogenieren halogenate/to
halogenierter Kohlenwasserstoff *m* halogenated hydrocarbon
halogeniertes organisches Lösemittel *n* halogenated organic solvent
Halogenierung *f* halogenation
Halogensäure *f* halogen acid
Halogenverbindung *f* halogen compound
Halogenwasserstoff *m* hydrogen halide
Halophyt *m* halophyte
haltbar machen preserve/to, sterilize/to, stabilize/to, conserve/to
Haltbarkeit *f* durability, shelf time, stability, storage quality
Haltbarkeitsgrenze *f* endurance limit
Haltbarkeitsverlängerung *f* shelf-life extension
Haltbarmachung *f* conservation, preservation
Hämatozyt *m* haematocyte
Hammerbrecher *m* hammer crusher
Hammermühle *f* hammer mill
Hammerwerk *n* hammermill
Hämoglobin *n* haemoglobin
Hämogramm *n* haemogram
Hämolyse *f* haemolysis
Hämotoxin *n* haemotoxin
Hämozytogenese *f* haemocytogenesis
Hämozytolyse *f* haemocytolysis
Handbedienung *f* hand control, manual operation

Handbetrieb *m* hand operation, manual operation
Handeingriffsmöglichkeit *f* possibility for manual intervention
handelsüblich commercial, customary
handelsübliche Verpackung *f* commercial packaging
handferngesteuert manually tele-controlled
Handfeuerlöscher *m* hand fire extinguisher, portable fire extinguisher
Handschutz *m* handguard
Handwerk *n* handicraft, craft
Hanffaser *f* hemp fibre
Hängekran *m* overhead crane
Harke *f* rake
Harn *m* urine
Harnbestandteil *m* urinary constituent
Harnblase *f* bladder
Harnsalz *n* urinary salt
Harnsäure *f* uric acid
Harnstoff *m* urea
Harnstoffharz *n* aminoplast resin
Harnstoffzyklus *m* urea cycle
Harnuntersuchung *f* urine analysis
Härteflüssigkeit *f* hardening liquid
Härtegrad *m* degree of hardness
Härtemeßgerät *n* hardness tester, hardness gauge
Härtemittel *n* hardener, hardening agent
Härter *m* hardener, curing agent, setting agent
Härtesalz *n* hardening salt
Hartfaserplatte *f* hardboard
Hartgesteinsaufbereitung *f* granite treatment
Hartgummi *m* hard rubber, ebonite, vulcanized rubber

Hartholz *n* hardwood
Hartschaum *m* hard foam
Harz *n* resin
Harzkleber *m* resin adhesive
Harzprodukt *n* resin product
Häufigkeitskurve *f* frequency curve
Häufigkeitverhältnis *n* abundance ratio
Häufigkeitsverteilung *f* frequency distribution, statistical distribution
Häufigkeitsverteilungskurve *f* abundance curve
Hauptinspektion *f* main inspection
Hausbrand *m* domestic fire
Hausbrandkohle *f* domestic coal
Haushaltskältegerät *n* household refrigerating device
Haushaltswasser *n* household water
Auskläranlage *f* domestic clarification plant
häuslicher Abfall *m* domestic waste, homeowner waste
häusliches Abwasser *n* domestic waste water
Hausmüll *m* household (domestic) waste, domestic refuse
Hausmüllanalyse *f* analysis of domestic waste
Hausmülldeponie *f* domestic waste dump
Hausmüllfraktion *f* fraction of domestic waste
Hausmüllkompost *m* domestic waste compost
Hausmüllkompostierung *f* domestic waste composting
Hausmüllverbrennungsanlage *f* incineration plant for domestic waste
Hautarzt *m* dermatologist
Hautatmung *f* cutaneous respiration
Hautentzündung *f* dermatitis

Havarie *f* average, accident
Havariesituation *f* accident situation
Heckenbepflanzung *f* hedge plantation
Hefe *f* yeast
Hefegärung *f* yeast fermentation
Hefepilz *m* yeast fungus
Hefeproduktion *f* yeast production
Heide *f* heath, heathland
Heilmittel *n* medicine, medicament, drug, remedy
Heilmittelchemie *f* pharmaceutical chemistry
Heilquelle *f* healing spring, mineral spring
Heilsalbe *f* salve, healing ointment
Heilverfahren *n* therapy
Heimwerken *n* do-it-yourself
Heißgaserzeuger *m* hotgas generator
Heißkleber *m* hot-setting adhesive
Heißluftgebläse *n* hot air blast
Heißlufttrocknung *f* hot air drying
Heißwasserbehandlung *f* von Ölsand hot-water processing of oil sand
Heißwassererzeugung *f* hot water production
Heizanlage *f* heating plant
Heizfläche *f* heating surface
Heizgasvorwärmer *m* economizer
Heizkessel *m* heating boiler
Heizkörper *m* heating body
Heizkörperverkleidung *f* radiator lining
Heizkraft *f* calorific power, heating power
Heizkreislauf *m* heating cycle
Heizleitung *f* heating pipe
Heizmaterial *n* fuel
Heizungsanlage *f* heating plant

Heizwert m calorific value
Heizwertschwankung f fluctuation of the calorific value
Helium n helium
Hemmstoff m inhibitor, growth retarder
Hemmung f inhibition, hindrance, restraint, retardation
Hemmungsstoff m inhibitor
Hemmwirkung f inhibitory effect
Hepatitis-Immunisierungsprogramm n hepatitis immunization programme
Hepatolyse f hepatolysis
herauslösen eluate/to, extract/to, dissolve/to
heraussickern trickle out/to, ooze out/to
herausspülen eluate/to, rinse out/to, wash out/to
Herbizid n herbicide, weed killer, weed killing agent
Herbizidrückstand m herbicide residue
hereinkommender Abfall m incoming waste
hermetisch hermetic
 hermetisch dicht hermetic tight
Herstellernachweis m list of producers
heterogen heterogenous
 heterogene Katalyse f heterogenous catalysis
Heterogenität f heterogeneity
heterotroph heterotrophic
Heterotrophie f heterotrophy
Heuschnupfen m allergic rhinitis, hay fever, pollen catarrh
Heuschrecke f grasshopper
Hilfsantriebssystem n servo drive system
Hilfsgröße f auxiliary quantity
Hilfsmittel n aid, auxiliary product
Hilfsquelle f resource
Hilfsreagens n auxiliary agent
Hilfsschütz m contactor relay, control relay
Hilfsstoff m aid, auxiliary product
Histochemie f histochemistry
Histologie f histology
Histolyse f histolysis
histopathologisch histopathological
Hitze f heat
hitzebeständig heat-resistant, heat-proof
hitzehärtbares Harz n thermosetting resin
Hitzemesser m pyrometer
Hitzewelle f heat wave
hochaktiv highly active
hocharomatisch highly aromatic
hochbeansprucht heavy-duty, high-duty
Hochbehälter m high-level tank
hochbelastet highly impacted, highly loaded
Hochdruck m high pressure
Hochdruckförderung f high-pressure conveyance
Hochdruckkessel m high-pressure boiler
Hochdruckleitung f high-pressure line
Hochdruckreiniger m high-pressure cleaner, high-pressure purifier
Hochdruckschlauch m high-pressure hose
Hochebene f plateau
hochempfindlicher Verstärker m high-sensitive amplifier
Hochglanzpapier n slick paper

Hochleistungs-Oberflächenbelüftung *f* high-performance surface ventilation
Hochleistungs-UV-Bestrahlungslampe *f* high-performance UV radiation lamp
Hochleistungsantrieb *m* heavy-duty drive
Hochleistungsfaulreaktor *m* high-performance fouling reactor
Hochleistungsfilter *n* high-performance filter
Hochleistungsreaktor *m* high-power pile
Hochmoor *n* high moor
Hochofen *m* blast furnace
Hochofenschlacke *f* blast furnace slag
Hochspannung *f* high voltage
Höchstabweichung *f* maximum deviation
Höchstabweichung *f* **vom Mittelwert** maximum mean deviation
Höchstbelastung *f* maximum load, peak load
Höchstdosis *f* maximum dose
höchste Aufnahmerate *f* maximum uptake rate
höchste Grenzkonzentration *f* maximum concentration limit
Höchstgehalt *m* maximum content
Höchstgeschwindigkeitsgrenze *f* maximum speed limit
Höchstkonzentration *f* maximum concentration
Höchstmengenverordnung *f* high-quantity regulation, maximum quantity regulation
höchstzulässige Dosis *f* maximum permissible dose

Hochtechnologiekeramik *f* high-technology ceramics
Hochtemperaturfackel *f* high-temperature torch
Hochtemperaturreaktor *m* high-temperature reactor
Hochtemperaturverbrennung *f* high-temperature incineration
Hochtemperaturverbrennungsanlage *f* high-temperature incineration plant
Hochvakuum *n* high vacuum
Hochvakuumtechnik *f* high-vacuum engineering
Hochwald *m* timber forest
Hochwasser *n* high tide
Hochwasserpumpwerk *n* high water pumping station
Hochwasserstand *m* high-water level
Hochwasservorhersage *f* high-water forecast
hohe Auflösung *f* high resolution
Hohe-See-Verbrennung *f* ocean incineration
Höhe *f* **der Konzentration** concentration level
Höhe *f* **des Grundrauschens der Basislinie** baseline noise level
Höhenförderer *m* lift conveyor
Höhenstandgrenzwert *m* level limiting value, limit value of the level
Höhenstrahlung *f* cosmic radiation
Höhlenbewohner *m* cave inhabitant, cavern inhabitant
Hohlkatode *f* hollow cathode
Hohlraum *m* cavity, hollow space
Holmium *n* holmium
Holz *n* wood, timber
Holzasche *f* wood ash
Holzbauweise *f* wood construction

holzbohrendes Insekt *n* wood-boring insect
Holzfaser *f* wood fibre
Holzfaserstoff *m* cellulose, lignin
holzfreies Papier *n* wood-free paper
Holzkohle *f* charcoal
Holzschutzmittel *n* wood preservative
Holzspanplatte *f* chipboard
Holzverarbeitung *f* wood processing
Holzveredelung *f* wood processing
homogen homogeneous
homogene Katalyse *f* homogeneous catalysis
homogenisieren homogenize/to
Homogenisiermaschine *f* homogenizer
Homogenisierung *f* homogenization
Homogenität *f* homogeneity
homöopathisches Arzneimittel *n* homoeopathic pharmaceutical [agent]
hörbarer Alarm *m* audible alarm
Hordentrockner *m* shelf dryer
Horizontalförderer *m* horizontal conveyor
Hormon *n* hormone
hormonarm hypohormonal
Hormongleichgewicht *n* hormone equilibrium (balance)
Hormonpräparat *n* hormone preparation
Hormonwirkung *f* hormone action (effect)
Hornabfall *m* horn shavings
Hornspäne *mpl* horn shavings
Hörschwelle *f* threshold of hearing
Hörvermögen *n* audition
Humangenetik *f* human genetics, population genetics
Humanisierung *f* **der Arbeitsplätze** humanization of working places
Humanmedizin *f* human medicine
Humanmilch *f* human milk
Humanökologie *f* human ecology
Humantoxikologie *f* human toxicology
Humantoxizität *f* human toxicity
Huminsäure *f* humic acid
Huminstoff *m* humic substance
Humus *m* humus, vegetable soil
humusartig humic
Humuszersetzung *f* humus degradation
Hüttenwerk *n* foundry
hybridisieren hybridize/to
Hydrant *m* hydrant
Hydratation *f* hydration
hydratisieren hydrate/to
Hydratisierung *f* hydration
Hydraulikflüssigkeit *f* hydraulic liquid
Hydrauliküberwachungssystem *n* hydraulic supervision (control) system, hydraulic monitoring system
hydraulisch hydraulic
Hydrierbarkeit *f* hydrogenability
Hydrierung *f* hydrogenation
Hydrobiologie *f* hydrobiology
Hydrodiaphragmapumpe *f* hydrodiaphragm pump
Hydrodynamik *f* hydrodynamics
hydrogeologisch hydrogeological
Hydrologie *f* hydrology
Hydrolyse *f* hydrolysis
Hydrolysereaktor *m* hydrolysis reactor
Hydrometallurgie *f* hydrometallurgy
hydrophil hydrophilic
hydrophob hydrophobic
hydrostatischer Antrieb *m* hydrostatic drive

Hydrothermalsynthese *f* hydrothermal synthesis
Hydroxid *n* hydroxide
Hydroxidschlamm *m* hydroxide sludge
Hydrozyklon *m* hydrocyclone
Hygiene *f* hygiene
Hygieneinspektion *f* hygiene inspection
Hygienevorschrift *f* hygiene regulation
hygienisch hygienic, sanitary
hygienische Deponierung *f* sanitary landfill
hygroskopisch hygroscopic
Hypertrophie *f* hypertrophy
Hypertonie *f* hypertension
Hypolimnion *n* hypolimnion
hypolimnisch hypolimnic

I

Identifikationssystem *n* identification system
Illuviation *f* illuviation
Imitation *f* imitation
imitieren imitate/to
Imker *m* bee-keeper
Immission *f* immission, intromission
Immissionsanalyse *f* immission analysis
immissionsbedingte Cadmiumdeposition *f* immission-conditioned cadmium deposition
Immissionsgrenzwert *m* immission limit value
Immissionskataster *n* immission register
Immissionsverteilung *f* immission distribution
Immissionspfad *m* immission pathway, intromission pathway
Immissionsschutz *m* immission protection
immobilisieren immobilize/to
Immobilisierung *f* immobilization
immun immune
Immunabwehrsystem *n* immune defence mechanism
Immunantikörper *m* immune antibody
Immunbiologie *f* immunobiology
immunbiologisch immunobiological
Immunchemie *f* immunochemistry
immuncytochemisch immunocytochemical
Immundiagnostik *f* immunodiagnostics
Immuneinheit *f* immunizing unit
Immungenetik *f* immunogenetics
immunisieren immunize/to
Immunkörper *m* immune body
immunologisch immunologic[al]
Immunserum *n* immune serum
Immunsystem *n* immune system
impfen inoculate/to, vaccinate/to
Impfstoff *m* inoculant, inoculum, vaccine
Impfschlamm *m* seeding sludge
Impfung *f* inoculation, vaccination
Implantation *f* implantation
implantieren implant/to
implodieren implode/to
Implosion *f* implosion
imprägnieren impregnate/to
Imprägnierflüssigkeit *f* impregnating fluid
Imprägnierlack *m* coating varnish
imprägnierter Beton *m* impregnated concrete

Imprägnierung *f* impregnation
Imprägnierungsmittel *n* impregnating agent
Impuls *m* impulse, pulse
inaktiv inactive, inert, neutral
inaktivieren inactivate/to
Inaktivierung *f* inactivation
Inaktivierungsgrad *m* degree of inactivation
Inaktivität *f* inactivity
Inanition *f* inanition
Inbetriebnahme *f* putting into operation, start-up [procedure]
Inbetriebnahmephase *f* start-up phase
Inbetriebsetzung *f* starting
indifferent indifferent, inert, neutral
Indikator *m* indicator, indicating instrument
Indikatorbakterien *fpl* indicator bacteria
Indikatorbereich *m* indicator range
Indikatororganismus *m* indicator organism
Indikatorpflanze *f* indicator plant
indirekte Zellkernteilung *f* mitosis
Indium *n* indium
indizieren index/to
Indizierung *f* indexing
Induktion *f* induction
Induktionsofen *m* induction furnace
industrialisieren industrialize/to
Industrialisierung *f* industrialization
Industrie *f* industry
Industrieabfall *m* industrial waste
Industrieabfalldeponie *f* industrial refuse dump
Industrieabfallstoff *m* industrial waste product
Industrieabgas *n* industrial waste gas
Industrieabwasser *n* industrial waste water, industrial sewage
Industrieabwasseranalyse *f* analysis of industrial sewage
Industrieemission *f* industrial emission
Industriegebiet *f* industrial area
Industriegesellschaft *f* industrial society
industrielle Aktivität *f* industrial activity
industrieller Feststoffabfall *m* industrial solid waste
Industrierückstände *mpl* industrial residues
Industrieschlamm *m* industrial sludge
Industriestandort *m* industrial location
induzieren induce/to
Induzierung *f* induction
ineffektiv ineffective
ineffizient inefficient
Ineffizienz *f* inefficiency
ineinander umwandeln interconvert/to
ineinander überführen interconvert/to
inert inert, inactive
inerte polymere Matrix *f* inert polymer matrix
inerter Rückstand *m* inert residue
inertes Abbauprodukt *n* inert degradation product
Inertgas *n* inert gas
inertisieren inertialize/to
Infektion *f* infection
Infektiongefahr *f* danger (risk) of infection

Infektionserreger *m* infective agent
Infektionskrankheit *f* infection disease, infectious disease
infektiös infectious
infektiöser Abfall *m* infectious waste
Infiltration *f* infiltration
infiltrieren infiltrate/to
infizieren infect/to
Informationsindustrie *f* information industry
Informationsnetzwerk *n* **für Umweltdaten** information network for environmental data
informationstechnische Vernetzung *f* information technical networking
Infrarot *n* infrared
Infrarotfilter *n* infrared filter
Infrarotspektroskopie *f* infrared spectroscopy
Infrarotspektrum *n* infrared spectrum
Infrarotstrahlung *f* infrared radiation
Infrastruktur *f* infrastructure
Infusion *f* infusion
Infusionsflasche *f* infusion bottle
Infusionslösung *f* infusion solution
Ingangsetzung *f* start-up, starting
Ingenieurgeologie *f* engineering geology
Ingestion *f* ingestion
Inhalation *f* inhalation
Inhalationsapparat *m* inhalator, inhaler
Inhalationsaufnahme *f* inhalation uptake
Inhaltsstoff *m* constituent
inhibieren inhibit/to

inhibierende Wirkung *f* inhibitory effect
Inhibition *f* inhibition
Inhibitor *m* inhibitor
inhomogen inhomogeneous
Inhomogenität *f* inhomogeneity
Injektion *f* injection
Injektionsgel *n* injection gel
Injektionsnadel *f* injection needle
Injektionsspritze *f* injection syringe
Injektionsvolumen *n* injection volume
Injektor *m* injector
Injektorblock *m* injector block
Injektortemperatur *f* injector temperature
injizieren inject/to
Inklusion *f* inclusion
inkompatibel incompatible
Inkompatibilität *f* incompatibility
inkonstant inconstant
Inkrementaldosis *f* incremental dose
Inkretion *f* incretion
Inkubationszeit *f* incubation time, latency period
Inkubator *m* incubator
Innenraumluft *f* indoor air
Innenraumluftfeuchte *f* indoor-air humidity
Innenraumluftverunreinigung *f* indoor-air pollution
innere Beanspruchung *f* internal strain
innere Reibung *f* viscosity
Innovation *f* innovation
Innovationsindustrie *f* innovation industry
Innovationsverhalten *n* innovative behaviour
Innovationsvorsprung *m* lead of innovation

in-situ Analyse *f* in-situ analysis
in-situ Experiment *n* in-situ experiment
Insekt *n* insect
Insektenfresser *m* insectivore
Insektenplage *f* insect pest
Insektenstich *m* insect bite, prick
insektenvernichtend insecticidal
Insektenvernichtungsmittel *f* insecticide
Insektizid *n* insecticide
insektizide Aktivität *f* insecticidal activity
Insektizidrückstand *m* insecticide residue
Inspektion *f* inspection
instabiles Isotop *n* unstable isotope
Installation *f* installation
installieren install/to, equip/to
Installierung *f* installation
Instandhaltungsarbeiten *fpl* repair work, maintainance routine work
Instandhaltungsmaßnahme *f* measure of repair
instandsetzen repair/to, restore/to
Instandsetzung *f* reconditioning, repair
instrumenteller Fehler *m* instrumental error
Insuffizienz *f* insufficiency
integrierte Vermeidungstechniken *fpl* integrated techniques of avoidance
intelligenter Abstandssensor *m* intelligent distance sensor
Intensität *f* intensity
Intensitätsabnahme *f* intensity decrease
Intensitätsbereich *m* intensity region
Intensivpflege *f* intensive care

Intensivrotte *f* intensive rot
Interaktion *f* interaction
Interessenkonflikt *m* **zwischen Überwachungs- und Durchführungsaufgaben** clash of interest between tasks of surveillance and execution
Interferenz *f* interference
intermediär intermediary, intermediate
intermittierend intermittent
internationaler Umweltausschuß *m* international environmental commission
interzellulär intercellular
Interzeption *f* interception
Intoxikation *f* intoxication
intrakutan intracutaneous
intramuskulär intramuscular
intravenös intravenous
intraventrikulär intraventricular
intrazellulär intracellular
Intromission *f* intromission
Invariable *f* invariable
Invasion *f* invasion
Inversionswetterlage *f* inversion weather [situation]
Inzidenz *n* incidence
Inzucht *f* inbreeding
Ion *n* ion
Ionenaustausch *m* ion exchange
Ionenaustauscher *m* ion exchanger
Ionenaustauschharz *n* ion-exchange resin
Ionenbindung *f* polar bond, ionic bond
Ionenselektivität *f* ion selectivity
Ionisationsstrom *m* ionization current
ionisierende Strahlung *f* ionizing radiation
Ionosphäre *f* ionosphere

Iridium *n* iridium
irreparabel irreparable
irreversibel irreversible
 irreversible Reaktion *f* irreversible reaction
Isocyanat *n* isocyanate
isoelektrisch isoelectric
Isolation *f* insulation, isolation
Isolationsmittel *n* insulator
Isolationsschicht *f* insulating layer
Isolationsstoff *m* insulator
Isolationsvermögen *n* insulating power
Isolationszustand *m* state of insulation
Isolator *m* insulator
isolieren insulate/to, isolate/to
isolierend insulating
Isoliermasse *f* insulating compound
Isoliermaterial *n* insulation material
Isolierschicht *f* insulating layer
Isolierschlauch *m* insulating tubing
Isoliervermögen *n* insulating power
Isomer *n* isomer
isomorph isomorph
isotherm isothermal
isotonisch isotonic
Isotop *n* isotope
Isotopeneffekt *m* isotopic effect
Isotopengemisch *n* isotopic mixture
Isotopentrennung *f* isotope separation
Istmenge *f* actual quantity
Istwert *m* actual value
Itai-Itai-Krankheit *f* itai-itai disease

J

Jagdwesen *n* hunting
Jahreserzeugung *f* yearly capacity, yearly output
Jahresganglinie *f* annual output curve, annual load curve
Jahreszeit *f* season
Jauche *f* liquid manure
Jauchebecken *n* manure vessel, vessel for liquid manure
Jod *n* iodine
Joghurt *m* yoghurt
Joghurtbecher *m* yoghurt beaker
justierbar adjustable
justieren adjust/to
Justierung *f* adjustment
Jutefaser *f* jute fibre
Jutegewebe *n* **zur Erosionsvermeidung** jute fabric for the avoidance of erosion

K

Kabelabfall *m* cable waste
Kabelüberzug *f* cable coating
Kabelummantelung *f* cable covering
Kadaver *m* cadaver, carcass
Kaffeerösterei *f* coffee roaster
Kahlschlag *m* clear-cutting, clear-felling
Kalandrieren *n* calendering
Kalbfleisch *n* veal
kalibrieren calibrate/to, gauge/to, graduate/to, standardize/to
Kalibrierfunktion *f* analysis calibration function
Kalibrierimpuls *m* calibration pulse
Kalibrierkurve *f* calibration curve

Kalibriermaß *n* standardization measure
Kalibrierstandard *m* calibration standard
kalibriertes Prüfgerät *n* calibrated test meter
Kalibrierung *f* calibration
Kalium *n* potassium
Kalk *m* chalk, lime, limestone
Kalkablagerung *f* calcification
Kalkanstrich *m* lime paint
Kalkboden *m* lime soil
Kalken *n* liming
  Kalken *n* **der Wälder** liming of forests
Kalkerde *f* lime
kalkhaltig calciferous
Kalkmangel *m* hypocalcaemia
Kalkmilch *f* lime milk
Kalksandstein *m* sand-lime-brick
Kalkschlamm *m* lime mud, lime sludge
Kalkstein *m* limestone
Kalksteinmehl *n* lime powder
Kalksteinsuspension *f* limestone suspension
Kalksteinwäscher *m* lime washer, limestone washer
Kalkung *f* **der Wälder** liming of forests
Kaloriengehalt *m* caloric content
Kalorienbedarf *m* caloric requirement
kalorienreiche Kost *f* high-caloric diet
Kalorimeter *n* calorimeter
kalorimetrisch calorimetric
Kälteanlage *f* refrigeration plant
Kältebeständigkeit *f* cold resistance
kälteerzeugend cryogenic
Kälteerzeugung *f* refrigerating

Kältemaschine *f* refrigerator
Kältemischung *f* mixture for freezing
Kältemittel *n* refrigerant
Kaltentseuchung *f* cold sterilization
kalthärtbare Reaktionsprodukte *npl* cold-curable reaction products
kaltlöslich cold-soluble
Kaltverformung *f* cold shaping
Kaltvermahlung *f* cold milling
Kaltverstreckbarkeit *f* cold drawability
Kaltvulkanisation *f* cold cure
Kalzinieranlage *f* calcination plant
Kalzinierofen *m* calcining furnace
Kamin *m* chimney, stack
Kammerfilterpresse *f* chamber filter press
Kammerofen *m* chamber furnace
Kammersäure *f* chamber acid
Kampfgas *n* poison gas, war gas
Kanal *m* canal, drain, sewer, channel
Kanal-Analyse-System *n* sewer analysis system
Kanalisation *f* sewerage
Kanalisationssystem *n* sewer system, drain and sewer system
Kanalkontrolle *f* canal control
Kanalnetz *n* pipeline network; sewer network
Kanalnetzüberwachung *f* sewer monitoring
Kanalreinigung *f* sewer purification, drain purification
Kanalsanierung *f* sewer remediation, redevelopment of drains (sewers)
Kanalsystem *n* sewer system, drain system
Kanteneffekt *m* edge effect

Kanüle *f* cannula
Kanzerogen *n* cancerogenic
Kapazität *f* capacity
Kapazitätsausgleich *m* capacity balance
Kapazitätseinbuße *f* loss of capacity
kapazitive Sonde *f* capacitive probe
kapillaraktiv capillary active
Kapillardepression *f* capillary depression
Kapillare *f* capillary
Kapillarflüssigkeit *f* capillary fluid
Kapillarkraft *f* capillary force
Kapillarsäule *f* **mit geringem Innendurchmesser** narrow-bore capillary column
Kapselverschluß *m* clip-on-cap
Karbidlampe *f* carbide lamp
Karies *f* caries
karieserzeugend cariogenic
kariös carious
Karst *m* karst
Kartonverpackung *f* cardboard packing
Kartusche *f* cartridge
karzinoembryonales Antigen *n* carcinoembryonic antigen
Karzinogen *n* carcinogenic
Karzinogenese *f* carcinogenesis
Karzinom *n* carcinoma, cancer
Karzinosarkom *n* carcinosarcoma
Karzinostase *f* carcinostasis
Kaskade *f* cascade
Kaskade *f* **mit Flächenbelüftung** cascade with surface ventilation
Kaskade *f* **von Reaktoren** cascade of reactors
Kaskadenkonfiguration *f* cascade configuration
Kaskadenmühle *f* cascade mill
Kaskadentest *m* cascade test

Kaskadenwirkung *f* cascade effect
Katalysator *m* catalyst
Katalysatorentechnologie *f* catalyst technology
Katalysatorlebensdauer *f* catalyst lifetime
Katalysatormasse *f* catalyst mass
Katalysatorträger *m* catalyst support
Katalysatortyp *m* catalyst type
Katalysatorvergiftung *f* catalyst poisoning, poisoning of the catalyst
Katalysatorvolumen *n* catalyst volume
Katalyse *f* catalysis
katalysieren catalyze/to
katalytisch catalytic
katalytische Abluftreinigung *f* catalytic waste air purification
katalytische Beschleunigung *f* catalytic acceleration
katalytische Reduktion *f* catalytic reduction
Katarrh *m* catarrh, rhinitis
Kataster *n* land register
Katastrophe *f* debacle, catastrophe
Katastrophengesetz *n* law of emergency
Katastrophenhilfe *f* disaster relief
Katastrophenschutz *m* emergency protection
Kation *n* cation
Kationenaustauscherharz *n* cation exchange resin
kationisch cationic
Katode *f* cathode
Kautschuk *m* caoutchouc
Kaverne *f* cavity
Kehrmaschine *f* [road] sweeper
Keim *m* germ, spore; sprout, bud
keimen germinate/to; sprout/to

keimfähig germinative
keimfrei sterile, germ-free, germless
keimfrei machen sanitize/to, sterilize/to
Keimfreiheit *f* sterility
Keimling *m* seedling, germ
keimtötend germicidal, sterilizing
keimtötendes Mittel *n* germicide
Keimzahl *f* bacterial count
Keimzelle *f* germ cell, spore
Kellerschwamm *m* wet rod
Keramik *f* ceramics
keramischer Regenerator *m* ceramic regenerator
Kerbtier *n* insect
Kernbaustein *m* nuclide
Kernbohrung *f* core drilling
Kernbrennstoff *m* nuclear fuel, reactor fuel
Kernenergie *f* nuclear energy
Kernexplosion *f* nuclear explosion
Kernforschung *f* nuclear research
Kernfusion *f* nuclear fusion
Kernkettenreaktion *f* nuclear chain reaction
kernmagnetische Resonanz *f* nuclear magnetic resonance
Kerosin *n* cerosine, kerosene
Kesselanlage *f* boiler equipment
Kesselfeuerung *f* boiler furnace
Kesselinjektionsverfahren *n* boiler injection method
Kesselschlamm *m* silt
Kesselstein *m* incrustation, boiler scale
Kesselwasser *n* boiler water
Kettenabbruch *m* chain breaking, termination of chains
Kettenförderer *m* chain conveyor
Kettenisomerie *f* chain isomerism
Kettenreaktion *f* chain reaction
Kettenstruktur *f* chain structure
Kettenverzweigung *f* chain branching
Kettenwachstum *n* chain growth
Kies *m* gravel, grit, pebble
Kiesaufbereitungsanlage *f* gravel treatment plant
Kiesbett *n* gravel bed
Kiesboden *m* gravel soil
Kiesel *m* pebble
Kieselgel *n* silica gel
Kieselgur *f* diatomaceous earth
Kieselstein *m* pebble
Kieselsäureprodukt *n* silicic acid product
Kiesfilter *n* gravel filter, pebble filter
Kiesgrube *f* gravel pit
Kiessieb *n* gravel screen
kindersicherer Sammelbehälter *m* **für Sonderabfall** child-proof collecting vessel for special waste
Kinetik *f* kinetics
kinetisch kinetic
kinetische Energie *f* kinetic energy
Kläranlage *f* clarification (purification) plant, sewage water treatment plant, sewage works, water treatment plant
Kläranlageneinlauf *m* clarifier inlet
Klärbassin *n* clarification basin
Klärbecken *n* filtering basin, sedimentation basin
Klärbehälter *m* precipitation tank
klären clarify/to, purge/to, settle/to
Klären *n* settling
Klärgas *n* sewer gas
Klärgasanlage *f* sewer gas plant
Klärgasnutzung *f* sewer gas utilization

Klärprozeß *m* clarification process
Klärschlamm *m* sewage sludge
Klärschlammaufbereitung *f* sewage sludge processing
Klärschlammbeseitigung *f* sewage sludge disposal
Klärschlammgranulat *n* sewage sludge granules
Klärschlammnachbehandlung *f* aftertreatment of sewage sludge
Klärschlammprobe *f* sewage sludge sample
Klärschlammpulver *n* sewage sludge powder
Klärschlammstabilisierung *f* sewage sludge stabilization, stabilization of sewage sludge
Klärschlammtrockengranulat *n* granules of dry sewage sludge
Klärschlammtrocknung *f* sewage sludge drying
Klärschlammverbrennung *f* sewage sludge incineration
Klärschlammverbrennungsanlage *f* sewage sludge incineration plant
Klärschlammverfestigung *f* **durch Tonmehldosierung** sewage sludge solidification by clay powder dosage
Klärschlammvolumen *n* sewage sludge volume
Klarspülmittel *n* clear rinsing agent
Klärtank *m* clarifier tank, precipitation tank, settling tank
Klärteich *m* settling tank
Klärung *f* clarification, settlement
Klärungsprozeß *m* clarification process
Klärwerkbetrieb *m* clarification plant operation
Klärwerksabschnitt *m* section of the clarification plant
Klärwerksbecken *n* basin of the clarification plant
Klärwerkskanal *m* drain of the clarification plant
Klärwerkskomponente *f* component of clarification works
Klärwirkung *f* clarification effect
Klassieren *n* grading
Klassierung *f* grading
Klassifikation *f* classification
Klebekitt *m* cement
Klebekraft *f* adhesion power
Klebeschicht *f* adhesion layer
Klebestreifen *m* [self-]adhesive tape, sealing tape
Klebstoff *m* adhesive, glue
Klebstoffindustrie *f* adhesive industry
Kleinklima *n* microclimate
Kleinmenge *f* small amount, small quantity
kleinste therapeutische Dosis *f* curative dose
kleinste wirksame Dosis *f* minimum effective dose
kleinster Wert *m* minimum [value]
Kleinversuch *m* small-scale experiment
Kleinwuchs *m* microsomia
Klima *n* climate
Klimaänderung *f* change of climate
Klimaanlage *f* air conditioning
Klimafaktor *m* climatic factor
Klimagefährdung *f* climate endangering
Klimakammer *f* climatic chamber
Klimakatastrophe *f* climatic disaster
Klimakontrolle *f* climatic control
Klimakunde *f* climatology

Klimaprüfschrank *m* climatic chamber
Klimaraum *m* conditioning room
Klimaregel *f* climatic rule
Klimasystem *n* climate system
klimatisch climatic
Klimatisierung *f* air-conditioning
Klimatologe *m* climatologist
Klimawechsel *m* change of climate, climatic change
klimawirksam climatic active
Klimazone *f* climatic zone
klinisch-toxikologische Analytik *f* clinical toxicological analytics
klinisch-toxikologischer Befund *m* clinical toxicological finding
klinische Symptomatik *f* clinical symptomatics
klinische Toxizität *f* clinical toxicity
klinische Untersuchung *f* clinical examination
klinischer Abfall *m* clinical waste
Kloake *f* sewer, drain, sink
Kloakenschlamm *m* faeces sludge
klopffrei knock-free, non-knocking
Knochenbildung *f* osteogenesis
Knochengerüst *n* skeleton
Knochenleimfabrik *f* bone glue factory
Knochenmark *n* bone marrow
Knochenmarkzelle *f* bone marrow cell
Knochenverarbeitung *f* bone processing
Knöllchenbakterien *fpl* nodule bacteria
Knospe *f* bud
Knötchen *n* nodule
knotig nodular

Koagulation *f* coagulation
Koagulationsmittel *n* coagulant
Koaleszenzabscheider *m* coalescence separator
Koaleszenzabscheidersystem *n* coalescence separator system
Koaleszenzabscheidung *f* coalescence separation
Kobalt *n* cobalt
kochen boil/to, cook/to, simmer/to
Kochen *n* boiling
Kochsalzinfusion *f* saline infusion
Kohäsion *f* cohesion, coherence
Kohle *f* coal
Kohleaufbereitung *f* coal dressing
Kohlebatterie *f* coal battery
Kohleentgasung *f* coal carbonization
Kohlefilter *n* charcoal filter
Kohlehydratbevorratung *f* carbohydrate supplies
Kohlehydratstoffwechsel *m* carbohydrate metabolism
Kohlekraftwerk *n* coal power plant (station)
Kohlendioxid *n* carbon dioxide
kohlensäurehaltiges Wasser *n* carbonic water
kohlenstoffarm low-carbon
Kohlenstoffkreislauf *m* carbon metabolism
Kohlenwasserstoff *m* hydrocarbon
Kohlenwasserstoffmischung *f* mixture of hydrocarbons
Kohlenwasserstoffmonitor *m* hydrocarbon monitor
Kohlenwasserstoffrückgewinnungsanlage *f* hydrocarbon recovery plant
Kohleveredelung *f* coal upgrading

Kohlungsgrad *m*   degree of carbonization
Koinzidenz *f*   coincidence
Kokerei *f*   coke-oven plant, coking plant, coal carbonizing plant
Koks *m*   coke
Koksbett *n*   bed of coke
Kokswäscher *m*   coke washer
Kolbenmotor *m*   piston engine
Kolbenpumpe *f*   piston pump
Kolibazillus *m*   coli-bacillus
Kolititer *n*   coli-titer
Kolloid *n*   colloid
kolloidales Teilchen *n*   colloidal particle
Kolloidchemie *f*   colloid chemistry
kolloidchemisch   colloid-chemical
Kolloiddiffusion *f*   colloid diffusion
kolloide Substanz *f*   colloidal substance
Kolloidzustand *m*   colloidal state
Kolonisierung *f*   colonization, settlement
Kolonnensumpf *m*   tower sump
Kolorimeter *n*   colorimeter
Kombi-Kraftwerk *n*   combination power plant
Kombinations-Kompaktkläranlage *f*   combination compact clarifier
kombiniert anaerob-aerobe Prozeßsteuerung *f*   combined aneorobe-aerobe process control
kommunale Aufsichtsbehörde *f*   communal authority, municipal authority
kommunale Kläranlage *f*   communal clarification plant
kommunale Wasserversorgung *f*   community water supply
kommunaler Humus *m*   municipal humus

kommunaler Klärschlamm *m*   communal sewage sludge
Kommunalfahrzeug *n*   communal vehicle, municipal vehicle
Kommunalhaushalt *m*   communal budget
kommunalpolitische Entscheidung *f*   communal political decision
Kommunalrecht *n*   municipal law
Kommunikation *f*   communication
Kommunikationssoftware *f*   communication software
Kommunikationssystem *n*   communication system
Kompaktentwässerer *m*   compact dewatering facility
Kompaktfilter *n*   compact filter
Kompaktschacht *m*   compact shaft
Kompartimentkonzentration *f* **von Wasser**   compartment concentration of water
Kompartimentmodell *n*   compartment model
Kompatibilitätsbedingung *f*   compatibility condition
Kompensationsschreiber *m*   compensation recorder
kompensieren   compensate/to
kompensierend   compensative
Komplettservice *m*   complete service
Komplexbildner *m*   complexing agent
Komplexbildung *f*   complex formation
Komplexchemie *f*   complex chemistry
Komplexität *f*   complexity
Komplexdünger *m*   complex fertilizer
Komplexitätsgrad *m*   degree of complexity

Kompost *m* compost
Kompost *m* aus Altpapier compost of waste paper
Kompost *m* aus Babywindeln compost of baby's nappies
Kompostanwendung *f* compost application
Kompostart *f* type of compost
Kompostaufbereitungsanlage *f* composting plant
Kompostbelüftung *f* compost aeration
Kompostenergieanlage *f* compost energy plant
Komposthalde *f* compost heap
kompostierbar compostable
kompostieren compost/to
kompostierter Siedlungsabfall *m* composted estate waste
Kompostierung *f* composting
Kompostierungsanlage *f* composting plant
Kompostierungsprojekt *n* composting project
Kompostierungsverfahren *n* composting method
Kompostmaterial *n* compost material
Kompostmiete *f* compost pile
Kompostprodukt *n* compost product
Kompostproduzent *m* compost producer
Kompostqualität *f* compost quality
Kompostrohmaterial *n* compost raw material
Kondensat *n* condensate
Kondensatabscheider *m* condensate separator
Kondensation *f* condensation
Kondensationsapparat *m* condenser
Kondensationskern *m* condensation core
Kondensator *m* condenser
Kondensatsammelgefäß *n* condensate accumulator vessel
Kondenser *m* condenser
kondensieren condense/to
konditionieren condition/to
Konditionierung *f* conditioning
Konditionierungsanlage *f* conditioning plant
Konditionierungszuschlag *m* conditioning agent (additive)
Konfiguration *f* configuration
Konfigurationsformel *f* space formula
konglomerieren conglomerate/to
Konservendose *f* tin, can
konservierbar conservable
Konservierung *f* conservation, preservation
Konservierungsmittel *n* preservative
Konsistenz *f* consistency, texture
Konstituent *m* constituent
Konsum *m* consumption
Konsument *m* consumer
Kontaktgift *n* contact poison, paralyzer
Kontaktinsektizid *n* contact insecticide
Kontaktoberfläche *f* contact surface
Kontaktfiltration *f* contact filtration
Kontaktzeit *f* von Mikroorganismen mit dem behandelten Abwasser contact time of the microorganisms with the processed waste water
Kontamination *f* contamination, pollution
Kontaminationsproblem *n* contamination problem
kontaminieren contaminate/to

kontaminierte Fläche *f* contaminated site
kontaminierter Rückstand *m* contaminated residue
Kontaminierung *f* contamination, pollution
Kontaminierungsgrad *m* degree of contamination
Kontrastuntersuchung *f* contrast examination
Kontrollampe *f* signal lamp
Kontrollbereich *m* controlled area
Kontrolle *f* control, monitoring, supervision, checking, inspection
Kontrolleinheit *f* control unit
Kontrolleinrichtung *f* monitoring equipment
Kontrollmessung *f* control measurement
Kontrollsystem *n* control system
Kontrollverfahren *n* control method
Konvektion *f* convection
Konvektionsheizung *f* convection heating
konventionelles Rückgewinnungsverfahren *n* conventional recovery process (method)
Konverter *m* converter
Konverterrauch *m* converterfume
konvertieren convert/to
Konvertierungsprodukt *n* conversion product
Konzentration *f* concentration
Konzentrationsbereich *m* concentration range
Konzentrationsprofil *n* concentration profile
konzentrieren concentrate/to
konzentrierter Schlamm *m* concentrated slurry
Konzentrierung *f* concentration

Koordinationsstörung *f* ataxia
Koordinationsverbindung *f* complex
Kopfprodukt *n* top product
Korallenbank *f* coral reef
Koralleninsel *f* coral island
Korallenriff *n* coral reef
Korkholz *n* cork wood
Kornbrand *m* smut
Körnchen *n* granule
Korndurchmesser *m* grain diameter
körnen grain/to, granulate/to
Kornfäule *f* smut
Korngröße *f* grain size
Korngrößenanalyse *f* grain size analysis
Korngrößenbereich *m* particle size range
körpereigene Substanz *f* endogenous substance
Körperflüssigkeit *f* body fluid, body liquid
Körpergewicht *n* body weight
körperliche Untersuchung *f* physical examination
Korrekturfaktor *m* correction factor; titrimetric factor
Korrelation *f* correlation
Korrelationsanalyse *f* correlation analysis
Korrelationskoeffizient *m* correlation coefficient
Korrosion *f* corrosion; rusting
korrosionsbeständig corrosion-resistant, stainless
korrosionsbeständiger Baustoff *m* corrosion-resistant building material
Korrosionschutzmittel *n* anticorrosion additive

korrosionsfester 240

korrosionsfester Edelstahl *m* high-quality stainless steel
korrosionsfrei non-corrosive, stainless
Korrosionsschutz *m* rust protection
korrosionsverhindernd anticorrosive
korrosive Flüssigkeit *f* corrosive liquid
kosmetisches Produkt *n* cosmetic product
kosmisch cosmic
kosmische Strahlung *f* cosmic radiation
Kosmochemie *f* space chemistry, cosmic chemistry
Kosmologie *f* cosmology
Kosmos *m* cosmos
Kosten-Ertrags-Gesichtspunkt *m* **in der Abfallwirtschaft** cost-profit aspect in the waste management
Kosten-Nutzen-Vergleich *m* cost-benefit analysis
Kot *m* dung, faeces, excrements, mud
Kotausscheidung *f* faecal excretion
Krabbe *f* crab, shrimp, prawn
Krach *m* noise
Kraft-Wärme-Kopplung *f* power-heat combination
Kraftstoff *m* fuel; petrol
Kraftstoffabscheider *m* fuel trap
Kraftstoffersparnis *f* fuel saving
Kraftwerk *n* power plant
Kraftwerkreststoff *m* residual material from power plants
Krankenhausabfall *m* clinical waste, hospital waste
krankenhausspezifischer Abfall *m* clinic-specific waste, hospital-specific waste

Krankheitsausbruch *m* outbreak of a disease
krankheitserregend pathogenic
Krankheitserreger *m* pathogen
Krankheitserscheinung *f* sign of a disease, symptom
Krankheitsherd *m* focus of disease
krankheitsübertragend infectious
krebsauslösend carcinogenic
Krebsbekämpfung *f* combat against-cancer
krebsbildend cancerogenic, carcinogenic
Krebsbildung *f* canceration
Krebsentstehung *f* carcinogenesis
Krebserkennung *f* cancer diagnosis
krebserzeugend carcinogenic
krebserzeugendes Potential *n* carcinogenic potential
Krebsforscher *m* cancerologist
Krebsforschung *f* cancer research
krebshemmend carcinostatic
krebshemmende Substanz *f* carcinostatic [agent]
Krebstiere *npl* crustaceans
krebszellenzerstörend carcinolytic
Kreideschicht *f* chalk bed
Kreisbecken *n* circular tank
Kreiselbelüfter *m* centrifugal aerator
Kreiselförderer *m* circuit conveyor
Kreiselmischer *m* impeller
Kreiselpumpe *f* centrifugal pump
Kreiselrad *n* impeller
Kreiselrührer *m* impeller
Kreislauf *m* cycle, circuit, circulation, revolution, rotation
Kreislauf *m* **von Natur und Industrie** cycle of nature and industry
Kreislaufsteuerung *f* circuit control

Kriechen *n* **dünner Flüssigkeitsschichten** film creep[ing]
Kristall *m* crystal
Kristallgitter *n* space lattice
kristalliner Zustand *m* crystalline state
Kristallisation *f* crystallization
Kristallisationsverdampfer *m* crystallization evaporator
Kristallkeim *m* crystal germ
Kristallwachstum *n* crystal growth
  kritische Dosis *f* threshold dose
  kritische Lage *f* critical situation
  kritische Masse *f* critical mass
  kritische Situation *f* critical situation
  kritische Temperatur *f* critical temperature
Krokidolithfaser *f* crocidolite fibre
Kryoanlage *f* cryotechnology plant
Kryochemie *f* cryochemistry
Kryostat *m* cryostate
Kryotechnik *f* cryotechnology
Kügelchen *n* pellet
Kugelhahn *m* ball stop-cock
Kugelmühle *f* ball mill
Kühlaggregat *n* refrigerating aggregate
Kühleigenschaft *f* cooling property
Kühlfalle *f* coolant trap, freezing trap
Kühlflüssigkeit *f* cooling fluid, refrigerant liquid
Kühlkreislauf *m* cooling circuit
Kühllagerung *f* cold storage
Kühlleistung *f* coolant performance
Kühlmantel *m* cooling jacket
Kühlmaschine *f* refrigerating aggregate, refrigerating engine
Kühlmittel *n* coolant, refrigerant, refrigerating agent

Kühlraum *m* refrigerating chamber
Kühlschlange *f* coil condenser
Kühlschmiermittel *n* cooling lubricant
Kühlschmierstoff *m* cooling lubricant
Kühlschrank *m* refrigerator
Kühlung *f* cooling, refrigerating
Kühlwagen *m* refrigerator truck (van)
Kühlwasser *n* cooling water, coolant water
kultivieren cultivate/to
Kultivierung *f* cultivation
Kultur *f* culture
Kulturboden *m* culture medium
kulturell cultural
Kulturlandschaft *f* land cultivated by man
Kulturpflanze *f* cultivated plant
Kulturröhrchen *n* culture tube
Kumulation *f* cumulation
kumulativ cumulative
kumulierend cumulative
Kunstdarm *m* artificial gut
Kunstdünger *m* fertilizer, artificial manure
Kunstdüngerherstellung *f* fertilizer production
Kunsteis *n* artificial ice
künstlich bewässern irrigate/to
künstliche Beatmung *f* artificial respiration
künstliche Belüftung *f* artificial ventilation
künstliches Ökosystem *n* artificial ecosystem
Kunststoff *m* plastic [material]
Kunststoffabfall *m* plastic waste
Kunststoffflasche *f* plastic bottle

Kunststoffmischabfall *m* mixed plastic waste
Kunststoffschlamm *m* plastic sludge
Kunststoffverwertungsanlage *f* plastic utilization plant
Kupfer *n* copper
Kurort *m* health resort
Kurzanalyse *f* proximate analysis
Kurztest *m* accelerated test, short test
Kurzversuch *m* accelerated test, short test
Kurzzeitanalysenautomat *m* short-term analysis automat
Kurzzeitanwendung *f* short-term application
Kurzzeitstabilität *f* short-term stability
Kurzzeitversuch *m* short-term test
Küste *f* coast, shore
Küstenerosion *f* coastal erosion
Küstenfischerei *f* inshore fishing
Küstengebiet *n* coastal area
küstennah inshore
Küstenrückgang *f* shore recession
Küstenschutz *f* share protection
Küstenverschmutzung *f* coast pollution
Küstenzone *f* coastal area
kutan cutaneous

L

Labor *n* laboratory, lab
Laborabzug *m* laboratory hood, fume hood, fume cupboard
Laborant *m* laboratory assistant, laboratory technician
Laborausstattung *f* laboratory equipment
Laborbefund *m* laboratory finding
Labordaten *pl* laboratory data
Laboreinrichtung *f* laboratory equipment
Laboruntersuchung *f* laboratory test, laboratory investigation
Lachszucht *f* salmon culture
Lack *m* lacquer, paint, varnish
Lackabbeizer *m* paint remover, varnish remover
Lackabfall *m* lacquer waste
Lackentferner *m* paint remover, varnish remover
Lackfarbe *f* lacquer, varnish paint, varnish colour
Lackfarbstoff *m* lacquer pigment
Lackhilfsmittel *n* lacquer auxiliary material
Lagerbeständigkeit *f* shelf time, storage life
Lagerfähigkeit *f* shelf time, storage life
Lagergefäß *n* storage vessel
Lagerhaus *n* warehouse
Lagerkapazität *f* storage capacity
Lagerungsbedingungen *fpl* storage requirements
lähmendes Gift *n* sedative poison
lahmlegen paralyze/to
Lähmung *f* palsy, paralysis, paralyzation
Laich *m* spawn
Lamelle *f* lamella, lamina
lamellenartig lamellar
Lamellenklärer *m* lamellar clarifier
laminar laminar, steady
 laminare Strömung *f* laminar flow
Laminarströmung *f* laminar flow
laminieren laminate/to
Landbau *m* agriculture
Landbevölkerung *f* rural population

Landesinnere *n* inland
Landesrecht *n* state law
ländlich rural
ländliche Entwicklung *f* rural development
ländliche Gegend *f* rural area
Landschaft *f* landscape, scenery
Landschaftsbiologie *f* landscape biology
Landschaftsdiagnose *f* landscape diagnosis
Landschaftsgestaltung *f* landscape architecture
Landschaftsökologie *f* landscape ecology
Landschaftspflege *f* landscape preservation
Landschaftsschutz *m* landscape protection, protection of the countryside
Landschaftszerstörung *f* landscape spoilation
Landwirtschaft *f* agriculture
landwirtschaftlich agricultural, rural
landwirtschaftliche Schlammverwertung *f* agricultural sludge utilization
landwirtschaftliche Tierhaltung *f* agricultural stock-breeding
landwirtschaftlicher Sprühwagen *m* agricultural spray vehicle
Landwirtschaftsgesetz *n* agricultural law
Landwirtschaftshydrologie *f* agricultural hydrology
langfaserig long-fibred
langkettig long-chain
langlebig long-lived
Langsamsandfilteranlage *f* low-velocity sandfilter plant
Langweggasküvette *f* long-path gas cuvette
langwirkend long-acting
Langzeitanwendung *f* long-term application
Langzeitbeanspruchung *f* sustained loading
Langzeitbehandlung *f* long-term therapy
Langzeitbeständigkeit *f* long-term resistance
Langzeiteffekt *m* long-term effect
Langzeitexperiment *n* long-term experiment
Langzeitmessung *f* long-term measurement
Langzeitreaktion *f* long-term reaction
Langzeitstabilität *f* long-term stability
Langzeitüberwachung *f* long-term monitoring
Lanthan *n* lanthanum
Lärm *m* loudness, noise
Lärmbelästigung *f* noise disturbance
Lärmbekämpfung *f* noise control
Lärmemission *f* noise emission
lärmempfindlich sensitive to noise
Lärmschutz *m* noise protection
Lärmschutzwand *f* noise protecting (prevention) wall
Lärmschutzwand *f* **aus Abfall** - noise protection wall from waste
Lärmschutzwand *f* **aus Kunststoffmüll** noise protection wall from plastic waste
Lärmverbot *n* noise ban
Larve *f* larva, maggot, grub
Larvenvernichtungsmittel *n* larvicide

Laseranwendung f laser application
Laserstrahl m laser beam
Lasertechnik f laser method
Laserüberwachungssystem n laser surveillance system
latenter Schaden m latent injury
Latenz f latency
Latenzperiode f latency period
Latenzzeit f latency period
Latex m latex
Latexschaum m foam rubber
Laub n leaf, foliage
Lauberde f leaf-mould
Laubwald m deciduous forest
Laufbandtrockner m screen belt dryer
Laufmittel n mobile solvent
Lauge f alkali, base, liquor, lye
laugen leach/to
Laugenbehälter m leaching vat
laugenbeständig lye-proof, alkaliproof
Laugenfaß n leaching vat
Laugenfestigkeit f alkali resistance
Laut m noise, sound
läutern percolate/to, purge/to
Läuterung f refining
Lautstärke f volume, loudness, sound intensity
Lautstärkemesser m volume indicator
Lautstärkemeßgerät n sound level meter
Lautstärkeregler m volume regulator
Lawine f avalanche
Lawrencium n lawrencium
Laxans n laxative
Leben n life, lifetime
lebende Hecke f quickset hedge

lebende Lärmschutzwand f quickset noise protection wall
Lebensabschnitt m period of life
Lebensalter n age
Lebensbedingungen fpl living conditions, life conditions
lebensbedrohlich life-threatening
Lebensdauer f life [span]; service life
Lebenserwartung f life expectancy
Lebensfähigkeit f viability
Lebensgefahr f danger of life, peril of life
Lebensgemeinschaft f biocenosis
Lebensmittel n food[stuff]
Lebensmittelabfall m food refusal
Lebensmittelchemie f food chemistry
Lebensmittelchemiker m food chemist
Lebensmittelfarbstoff m food colour
Lebensmittelgeschmacksstoff m food flavouring
Lebensmittelverarbeitung f food processing
Lebensmittelzusatz m food additive
lebensnotwendig vital
Lebensqualität f quality of life
Lebensraum m biotope, life space
Lebensstandard m standard of living
Lebensunfähigkeit f abiosis
Lebensweise f mode of life
lebenswichtig vital, essential
lebenswichtige Güter npl essential goods
Lebenszeit f lifetime
Lebenszeitrisiko n lifetime risk
Leberfunktionsstörung f liver dysfunction

Leberfunktionstest *m* liver function test
Lebergewebszelle *f* liver parenchymal cell
Leberschrumpfung *f* liver cirrhosis
Leberstoffwechsel *m* liver metabolism
leblos inanimate, lifeless
Leblosigkeit *f* inanimation
Leck *n* leak
Leckage *f* leakage
Leckageüberwachungssystem *n* leakage monitoring (control) system
Leckanzeigegerät *n* leakage indicator
leckes Schiff *n* leaky ship
Lecköluberwachung *f* leaking oil control (monitoring)
Lecköluberwachungssystem *n* leaking oil control system
lecksicher leaksafe
Lecksuche *f* leak detection
Lecksucher *m* leak detector
Lederabfall *m* leather waste, leather scrap
Leerversuch *m* blank test, blank run, blank experiment
Legierung *f* alloy, composition metal
Legierungsbestandteil *m* alloying component
Leguminose *f* leguminous plant, legume
Lehm *m* loam, clay
lehmartig loamy
Lehmboden *m* loamy soil
lehmig loamy
Lehmmergel *m* loamy marl
Lehmschicht *f* loam coat (layer)
Lehmstein *m* loam brick
Lehmziegel *m* loam brick
Leiche *f* corpse, body

Leichenbeschauer *m* coroner
Leichengift *n* ptomaine
Leichenöffnung *f* autopsy
Leichtbau *m* lightweight construction
Leichtbaustoff *m* light building material
Leichtbenzin *n* light distillate fuel
Leichtbeton *m* lightweight concrete
leichtflüchtiger halogenierter Kohlenwasserstoff *m* low-boiling halogenated hydrocarbon
Leichtflüssigkeit *f* light liquid
Leichtflüssigkeitsabscheider *m* separator for light-liquids
Leichtlegierung *f* light alloy
Leichtmetall *n* light metal
Leichtwaschmittel *n* light duty detergent
Leim *m* glue, size
Leinöl *n* linseed oil
Leinölfarbe *f* linseed oil paint
Leistungsabfall *m* power drop
Leistungsaufnahme *f* power input
Leistungsbedarf *m* power demand (requirement)
leistungsstarkes System *n* powerful system
Leiterplatte *f* printed circuit board
Leiterplattenherstellung *f* printed circuit board manufacturing
Leitfähigkeit *f* conductivity, conductance
Leitungswasser *n* tap water
Leseband *n* sorting band
letal lethal
letale Toxizität *f* lethal toxicity
Letaldosis *f* lethal dose
Lethargie *f* lethargy
lethargisch lethargic

Leuchtbakterien-Meßgerät *n* measuring device of luminous bacteria
Leuchtkraft *f* illuminating power
Leuchtöl *n* naphta, kerosine
Leukämie *f* leukaemia, leucosis
leukämisch leukaemic
Leukom *n* leucoma
Leukose *f* leucosis, leukaemia
Leukozyt *m* leucocyte
Leukozytenzerfall *m* leucocytosis
Leukozytose *f* leucocytosis
Lichtabsorption *f* light absorption
Lichtatmung *f* light respiration
Lichtausschluß *m* absence of light
Lichtausstrahlung *f* radiation of light
Lichtbedingungen *fpl* light conditions
lichtbeständig light-proof, photostable
Lichtbeständigkeit *f* light resistance, resistance to light
lichtdurchlässig light-permeable, transparent
lichtecht fade-proof
lichtgeschützt screened from the light
lichtkatalysiert light-catalyzed
Lienitis *f* lienitis
Lignin *n* lignin
Lignit *m* lignite, peat coal
lineare Regression *f* linear regression
lineare Skalierung *f* linear scaling
Linienerosion *f* line erosion
Lipämie *f* lipaemia
Lipoid *n* lipoid
lipophil lipophilic
Lipozyt *m* lipocyte, fat cell
Liquidation *f* liquidation
Lithium *n* lithium

Lithosphäre *f* lithosphere
litorale Ablagerung *f* littoral deposit
litoraler Transport *f* littoral transportation
Lizenz *f* licence, permit
Lizenzabkommen *n* licence contract
Lizenzerteilung *f* licensing
Lizenzgebühr *f* licence fee
Lochziegel *m* perforated brick
Lokalanästhetikum *n* local anaesthetic
lokale Konzentration *f* local concentration
lokale Störung *f* local disturbance
Lokalelement *n* local cell
Lokalisierung *f* localization
Lokalwirkung *f* local action
Lösbarkeit *f* dissolubility, solubility
löschen blow out/to, quench/to
Löschkalk *m* slaked lime
Löschkohle *f* quenched charcoal
Lösemittel *n* solvent, dissolvent
Lösemitteldampf *m* solvent vapour
Lösemittelemission *f* solvent emission
lösemittelfreie Farbe *f* solvent-free paint
Lösemittelgemisch *n* solvent mixture
lösemittelhaltige Abluft *f* solvent-containing exhaust air
lösemittelhaltige Suspension *f* solvent-containing suspension
Lösemittelrückgewinnung *f* **durch Adsorption** solvent recovery by adsorption
lösen solve/to
löslich soluble, dissolvable
lösliche Verbindung *f* soluble compound

Löslichkeit *f* solubility, dissolubility
Löslichkeitsprodukt *n* solubility product
loslösen separate/to
Löß *m* loess
Lösung *f* dissolution, solution
Lösungsenthalpie *f* solution enthalphy
Lösungsmittel *n* solvent
Lösungsvermittler *m* solubilizer
Lösungswärme *f* heat of solution
lückenlose Überwachung *f* complete monitoring
Luft *f* air
luftdichter Verschluß *m* airtight seal
luftdurchlässig breathable
Lufteinperlung *f* entry of air bubbles
Lufteintrag *m* air entry
lüften ventilate/to
luftentzündlich pyrophoric
Luftfeuchte *f* air humidity
Luftfilter *n* air filter
luftleer machen evacuate/to
luftleerer Raum *m* vacuum
Luftprobe *f* air sample
Luftprobenahme *f* air sampling
Luftqualität *f* air quality
Luftqualitätsdaten pl air quality data
Luftreinhaltemaßnahme *f* measure for the prevention of air pollution
Luftreinhaltung *f* prevention of air pollution
Luftreinhaltungssystem *n* air purification system
Luftschadstoff *m* air pollutant
Luftüberwachung *f* air monitoring
lufttrocken air-dry
luftundurchlässig air-proof

Lüftung *f* aeration, ventilation, airing
Lüftungsmaßnahme *f* aeration measure
Lüftungstechnik *f* aeration technique
Luftvergiftung *f* air poisoning
Luftverschmutzungsvorhersage *f* air pollution forecast
Luftverunreinigungskontrolle *f* air pollution control
Luftzerlegungsanlage *f* air separation plant
Luftzirkulation *f* air circulation
lumineszent luminescent
Lumpensammler *m* rag man
Lunge *f* lung
Lungenabszeß *m* lung abscess
Lungenaufnahme *f* lung uptake
Lungenemphysem *n* pulmonary emphysema
Lungenentzündung *f* pneumonia
Lungenfunktionsprüfung *f* pulmonary function test
lungengängiger Staub *m* respirable dust
Lungenspülung *f* pulmonary lavage
Lungenvolumen *n* lung capacity
Lutetium *f* lutetium
Lymphadenitis *n* lymphadenitis
Lymphdrüse *f* lymphatic gland
Lymphgewebe *n* lymphatic tissue
Lymphknoten *m* lymph node, lymphatic nodule
Lymphknotenentzündung *f* lymphadenitis
Lymphozyt *m* lymphocyte
Lymphozytose *f* lymphocytosis
Lysimeter *n* lysimeter

M

Made *f* maggot
Madenwurm *m* threadworm, pinworm
Magen *m* stomach, maw
 Magen-Darm-Kolik *f* intestinal colic
 Magen-Darm-Trakt *m* gastrointestinal tract
Magengeschwür *n* gastric ulcer
Magenschmerz *m* gastralgia
Magerkohle *f* dry burning coal, free-ash coal, non-baking (non-bituminous) coal
Magermilch *f* low-fat milk
Magma *n* magma
Magnesium *n* magnesium
Magnet *m* magnet
Magnetabscheider *m* magnetic separator
magnetisch magnetic
magnetisch-induktive Durchflußmessung *f* magnetic-inductive flow measurement
magnetisches Förderband *n* magnetic conveyor belt
Magnetisierung *f* magnetization
Magnetismus *m* magnetism
Magnetrührer *m* magnetic stirrer
Magnetschalter *m* magnetic switch, magnetic valve
Magnetscheider *m* magnetic separator
Magnettrommel *f* magnetic drum
Mahagoni *n* mahogany
Mahlanlage *f* milling plant
Mais *m* maize, *{Am}* corn
Maische *f* mash
Makroanalyse *f* macroanalysis
Makroklima *n* macroclimate
Makrokosmos *m* macrocosm
makromolekular macromolecular
Makromolekül *n* macromolecule
makroskopisch macroscopic
Makrostruktur *f* macrostructure
Malaria *f* malaria
Mammogramm *n* mammogram
Mammographie *f* mammography
Mangan *n* manganese
Mangrove *f* mangrove
Mangrovenbaum *m* mangrove
Mannigfaltigkeit *f* variety, diversity
Manometer *n* manometer, pressure gauge
Manometerablesung *f* pressure gauge reading
Manometerskala *f* manometer scale
manuelles Sortieren *n* manual sorting
marines Ökosystem *n* marine ecosystem
Mark *n* marrow; bone marrow; pulp
markiertes Atom *n* labelled atom
Markierungsfarbe *f* signal paint
Markierungsstift *m* **mit Diamantspitze** diamond-tipped scoring pen
Markkrebs *m* medullary cancer
Marsch *f* marshland
Maschensieb *n* mesh screen, mesh sieve
maschinell by machine, mechanical
Maschinenantrieb *m* machine drive
maschinenlesbar machine-readable
Maschinenöl *n* lubricating oil
Masernimpfstoff *m* measles vaccine
Maßabweichung *f* tolerance
Maßanalyse *f* titrimetric analysis, titrimetry, volumetric analysis, volumetry
maßanalytisch volumetric
Maßeinheit *f* unit

Massenanziehungsgesetz *n* law of mass attraction
Massenaufnahme *f* mass uptake
massenselektiver Detektor *m* mass selective detector
Massenspektrometrie *f* mass spectrometry
massenspektrometrisch mass-spectrometric
Massensterben *n* mass mortality
mäßig selektiv moderate selective
mäßig toxisch moderate toxic
Massivholz *n* solid timber
Maßnahme *f* **im Störfall** measure in case of accident
Maßnahmenkatalog *m* catalogue of measures
Materialbeanspruchung *f* material stress
Materialbedarf *m* material requirement
Materialeinsparung *f* saving in material
Materialfehler *m* material defect
Materialprüfung *f* material testing
Materialrückgewinnung *f* material recovery
Materialverbrauch *m* material consumption
Materialvorrat *m* stock
Materie *f* matter, substance
materiell material, substantial
Matrix *f* matrix
maximal zulässig maximum permissible
Mauersalpeter *m* wall saltpetre
Maximum *n* maximum, peak
mechanisch-biologische Kläranlage *f* mechanical-biological clarification plant

mechanische Eigenschaft *f* mechanical property
Medikamentenspende *f* medicament donation
Medikamentenverbrauch *m* drug consumption
Medikation *f* medication
Medium *n* medium
medizinisch-toxikologisches Gefährdungspotential *n* medical toxicological endangering potential
medizinische Umweltkontrolle *f* medical environmental control
medizinische Vorsorge *f* medical prevention
Meeresalge *f* seaweed
Meeresarm *m* estuary
Meeresboden *m* sea floor, sea bed
Meeresbodenerosion *f* sea floor erosion
Meeresklima *n* marine climate
Meereskunde *f* oceanography
Meeresküste *f* sea coast
Meerespflanzen *fpl* seaweed
Meeresspiegel *m* sea level
Meeresverunreinigung *f* sea pollution, ocean pollution, marine pollution
Meerwasser *n* sea water
Mehlstaub *m* mill dust
Mehltau *m* mildew
Mehrfachbestimmung *f* replicate determination (test)
Mehrfachnutzung *f* multiple use
Mehrfachverwendung *f* multiple use
Mehrkammercontainer *m* multi-chamber container
Mehrkanalanalysator *m* multi-channel analyzer

Mehrnährstoffdünger *m* multinutrient fertilizer
Mehrstoffanalyse *f* multi-component analysis
Mehrstoffgemisch *n* multi-component mixture
Mehrstoffsystem *n* multi-component system
mehrstufige Vorreinigung *f* multi-stage prepurification
mehrstufiges Verfahren *n* multi-stage process
Mehrwegbehälter *m* returnable container
Mehrwegflasche *f* returnable bottle
Mehrwegschieber *m* multi-way valve
Mehrwegverpackung *f* returnable packing
mehrzellig multicellular
Meilerverkokung *f* pile coking
Meiose *f* meiosis
Melanom *n* melanoma
Meldepflicht *f* duty of notification
Melioration *f* [a]melioration, betterment
meliorieren [a]meliorate/to
Membran *f* membrane, diaphragm
Membranauskleidungsmaterial *n* membrane lining material
Membrandurchlässigkeit *f* membrane permeability
Membranfilter *n* diaphragm filter
Membrangleichgewicht *n* membrane equilibrium
Membranhalterung *f* membrane support, membrane holder
Membrankammerfilterpresse *f* membrane chamber filter press
Membranpumpe *f* membrane pump

Membrantechnologie *f* membrane technology
Membranventil *n* diaphragm valve
Membranverfahren *n* membrane process
Mendelevium *f* mendelevium
Mengenbegrenzung *f* limitation of quantity
Mengenbestimmung *f* quantity determination
Mengenregelung *f* regulation of quantity
mengenmäßig quantitative
Mengenverhältnis *n* proportion, mass ratio
Menschenkraft *f* man power
Menschheit *f* mankind, humanity
menschliche Aktivität *f* man's activity
menschliches Versagen *n* human failure, human error
Mergel *m* marl
Mergelgrube *f* marl pit
Meristem *n* meristem
Merkaptan *m* mercaptane
mesomer mesomeric
Mesomerie *f* mesomerism; resonance
Mesomerieenergie *f* mesomeric energy
mesophil mesophile, mesophilic
Meßanordnung *f* measuring arrangement
Meßapparatur *f* measuring apparatus
meßbar measurable
Meßbarkeit *f* measurability
Meßbereich *m* measuring range, range of measurement
Meßdaten *pl* measured data, measuring data

Meßeinrichtung *f* measuring device
Meßelektrode *f* measuring electrode, working electrode
Meßelektronik *f* measuring electronics
Meßfühler *m* measuring sensor
Meßgas *n* measuring gas
Meßgut *n* measuring material (good)
Messing *n* brass
Meßküvette *f* measuring cuvette
Meßmedium *n* measuring medium
Meßstellenblatt *n* chart of measuring points (locations)
Meßstellenschema *n* scheme of measuring points (locations)
Meßstrategie *f* measuring strategy
meßtechnische Überwachung *f* **von Gefahrstoffen** monitoring of harmful materials by means of measurement techniques
Meßumformer *m* transducer, measuring transducer
Meßumformerausführung *f* transducer design
Meßverfahren *n* measuring method
Meßwert *m* measurand
Meßwertarithmetik *f* measured value arithmetics
Meßwertaufnahme *f* measurand pick-up
Meßwertaufnehmer *m* pick-up, measuring sensor
Meßzusatz *m* measurement adapter
metabolisch metabolic
metabolisch umsetzbare Verbindung *f* metabolic convertible compound
Metabolismus *m* metabolism
Metabolit *m* metabolite
Metalimnion *n* metalimnion
Metallabfall *m* scrap metal, waste metal
Metallauflage *f* metal plating
Metalldampf *m* metal vapour
Metalldetektor *m* metal detector
metallorganisch organometallic
Metallschlamm *m* metal sludge
Metallsuchgerät *n* metal detector
Metalltrennung *f* metal separation
metallurgische Anlage *f* metallurgical plant
Metamorphismus *m* metamorphism
Metamorphose *f* metamorphosis, transition
metastabil metastable
Metastase *f* metastasis
Metatoxizität *f* metatoxicity
Meteorologie *f* meteorology
meteorologisch meteorological
meteorologischer Transport *m* meteorological transport
Metereographie *f* metereography
Methanbakterie *f* methane bacterium
Methangas *n* methane gas
Methanisierung *f* methanization
methodischer Fehler *m* methodic error
Methylmethacrylat *n* methyl methacrylate
Mietenkompostierung *f* silo composting
Migräne *f* migraine
Migration *f* migration
Mikroanalyse *f* microanalysis
Mikro-Ultrafiltration *f* micro-ultrafiltration
Mikrobe *f* microbe
Mikrobenaktivität *f* microbial activity
Mikrobenkultur *f* microbe culture

mikrobentötend microbicidal
Mikrobenzüchtung *f* microbe culture
Mikrobiologie *f* microbiology
mikrobiologisch microbiological
mikrobiologische Reaktion *f* microbiological reaction
mikrobische Aktivität *f* microbial activity
mikrobischer Schlamm *m* microbial slime
Mikrochemie *f* microchemistry
mikrochemisch microchemical
Mikroelektronik *f* microelectronics
Mikroelementaranalyse *f* elementary microanalysis
Mikrofiltrationsmembran *f* microfiltration membrane
Mikroflora *f* microflora
Mikroklima *n* microclimate
Mikrokolloid *n* microcolloid
mikrokristallin microcrystalline
Mikroökosystem *n* micro-ecosystem
Mikroorganismus *m* microorganism
mikroporös microporous
mikroprozessorgesteuert microprocessor-controlled
mikroskopisch microscopic
Mikrosom *n* microsome
Mikrowellentechnik *f* microwave technology
Milbe *f* mite
Milchabsonderung *f* lactation
Milchpulver *n* powdered milk
Milchsäurebakterie *f* lactobacillus
Milchtrocknungsanlage *f* milk drying plant
Milchverarbeitung *f* milk processing
Milieu *n* milieu
Milzbrand *m* anthrax

Milzentzündung *f* lienitis, splenitis
Mindergewicht *n* underweight
minderwertiger Brennstoff *m* low-grade fuel
Mindestanforderung *f* minimum requirement, minimum demand
Mindestdruck *m* minimum pressure
Mindestgewicht *n* minimum weight
Mine *f* mine, lead, refill
Mineral *n* mineral
Mineralanalyse *f* mineral analysis
Mineralbodengemisch *n* mineral soil mixture
Mineralbrunnen *m* well
Mineraldünger *m* inorganic fertilizer
Mineralfaser *f* mineral fibre
Mineralfaserindustrie *f* mineral fibre industry
mineralischer Füllstoff *m* mineral filler
mineralisieren mineralize/to
Mineralisierung *f* mineralization
Mineralwasserabfüllstation *f* mineral water filling station
Mineralöl *n* mineral oil
Mineralölerzeugnis *n* mineral oil product
mineralölhaltiger Schlamm *m* mineral oil containing sludge
Mineralölprodukt *n* petroleum product, mineral oil product
miniaturisiertes Ökosystem *n* miniaturized ecosystem
Minimalausstattung *f* small-scale equipment
Minimalemission *f* minimum emission
Mischabwasser *n* mixed sewage
Mischapparat *m* mixer
mischbar miscible, mixable

Mischbarkeit *f* miscibility
Mischbecken *n* mixing basin (tank)
Mischdünger *m* compound (mixed) fertilizer
Mischdüse *f* combining nozzle
Mischer *m* mixer
Mischfarbe *f* mixed colour
Mischkultur *f* mixed culture
Mißbildung *f* malformation
Mißbrauch *m* misuse
Mist *m* manure
Mistpulver *n* poudrette
Mitochondrien *npl* mitochondria
Mitose *f* mitosis
mitotisch mitotic
  mitotische Genumwandlung *f* mitotic gene conversion
mittelflüchtige organische Schadstoffe *mpl* semi-volatile organic pollutants
Mittelwertkonzentration *f* average concentration
mixotroph mixotrophic
Mobilität *f* mobility
Modalität *f* modality
Modell *n* model
Modellanlage *f* pilot plant
Modellchemikalie *f* model chemical
modellieren model/to
Modellsand *m* moulding sand
Modellsimulation *f* model simulation
Modellsubstanz *f* model substance
Moder *m* mould
Modererde *f* humus, moult rotten earth
Modergeruch *m* rotten odour
modern moulder/to, rot/to
Modernisierungsinvestionen *fpl* investment of modernization
Modernisierungsprojekt *n* project of modernization

Modifikation *f* modification
modifizieren modify/to
Modifizierung *f* modification
Modulsystem *n* module system
Mol *n* mole
molar molar
Molekül *n* molecule
molekular molecular
Molekularbiologie *f* molecular biology
Molekularformel *f* molecular formula
Molekulargröße *f* molecular size
Molekularsieb *n* molecular sieve
Molkereibetrieb *m* dairy
Molluskizid *n* molluscicide
Molybdän *n* molybdenum
Monatsextrem *n* monthly extreme
Monatsganglinie *f* monthly output (load) curve
Monatsmittel *n* monthly average
Monitor *m* monitor
Monokultur *f* monoculture
Monomer *n* monomer
monomolekular unimolecular
Monopolwirtschaft *f* monopolism
Montageband *n* assembly line
Montagestraße *f* assembly line
Moor *n* bog, moor, peat
Moorbad *n* peat bath
Moorerde *f* peaty soil
Moorgebiet *n* swampy area
Moorwasser *n* peat water
Moos *n* moss
Morast *m* quagmire, swamp
Morbidität *f* morbidity
Morphogenese *f* morphogenesis
Morphologie *f* morphology
Mortalität *f* mortality
Mörtel *m* mortar
Motorenöl *n* lubricating oil

Motorgeräusch *n* engine noise
Mulch *m* mulch
mulchen mulch/to
Mull *m* muslin, mull
Müll *m* refuse, waste
Müll-Leichtfraktion *f* light fraction of waste
Müllabfuhr *f* garbage collection, garbage disposal, refuse collection
Müllabladeplatz *m* dumping area, refuse dump, garbage dump, garbage pit
Müllaufgabevorrichtung *f* refuse feeder, waste unloading installation
Müllbehälter *m* litter bin, refuse bin, refuse container
Müllbehälterreinigungsfahrzeug *n* cleaning vehicle for refuse bins
Müllbeseitigung *f* waste disposal
Müllcontainerdeckel *m* **mit integriertem Griff** refuse container lid with integrated handle
Mülldeponie *f* disposal site
Mülldichte *f* refuse density
Müllgemisch *n* waste mixture
Müllnotstand *m* waste emergency situation
Müllpresse *f* waste compactor
Müllsack *m* refuse sack
Müllsammelfahrzeug *n* waste collection vehicle
Müllsammelkosten *pl* waste collection costs
Mülltourismus *m* waste tourism
Müllumschlagstation *f* refuse transfer station
Müllverbrennung *f* waste incineration
Müllverbrennungsanlage *f* waste incineration plant
Müllverbrennungsschlacke *f* slag from waste incineration plants
Müllverdichterprototyp *m* prototype of waste compactor
Müllzusammensetzung *f* refuse composition
Multielementanalyse *f* multi-element analysis
multiplikative Wirkung *f* multiplicative effect
Multivitamin *n* multivitamin
Mund-zu-Mund-Beatmung *f* rescue breathing
Mundpflege *f* oral hygiene
Mundschleimhautentzündung *f* stomatitis
Munition *f* ammunition
Munitionsentsorgung *f* ordinance disposal
Muschel *f* shell
Muschelkalk *m* shell limestone
Muschelschale *f* shell
Muskel *m* muscle
Muskelfaser *f* muscle fibre
Muskelflimmern *n* muscle fibrillation
mutagen mutagenic
mutagener Effekt *m* mutagenic effect
mutagenes Risiko *n* mutagenic risk
Mutagenese *f* mutagenesis
Mutagenität *f* mutagenicity
Mutation *f* mutation
Mutationsfähigkeit *f* mutability
Mutterboden *m* native soil
Mutterkorn *n* ergot
Mutterkornvergiftung *f* ergotism
Mutterkuchen *m* placenta
Mutterlauge *f* mother liquor
Muttermilch *f* mother's milk
Muttersubstanz *f* parent substance

Mutterzelle *f* parent cell
Mutualismus *m* mutualism
Mycel *n* mycelium
Mykologie *f* mycology
Myokarditis *f* myocarditis
Myosarkom *n* myosarcoma

N

nachbehandeln retreat/to, aftertreat/to
Nachbehandlung *f* after-care, aftertreatment, retreatment
nachbrennen smoulder/to
Nachbrennkammer *f* after-burning chamber
nacheichen recalibrate/to
Nacheichung *f* recalibration
Nachexpositionszeit *f* post-exposure period
Nachfiltration *f* final filtering
nachfüllbar refillable
nachfüllen refill/to
Nachfüllpatrone *f* refillable cartridge
Nachgeruch *m* after-smell
nachgeschaltete Gasphasenreaktion *f* subsequent gas phase reaction
Nachgeschmack *m* aftertaste
nachhaltig lasting
nachkalibrieren recalibrate/to
Nachkalibrierung *f* recalibration
Nachkläranlage *f* after-clarification plant, after-purification plant
Nachklärbecken *n* after-purification basin (tank)
Nachklärung *f* after-clarification
Nachkommen *mpl* progeny
Nachkommenschaft *f* progeny

Nachkompostierung *f* after-composting
Nachlauf *m* after-run
nachprüfbar checkable
nachprüfen check/to
Nachprüfung *f* check, review
nachregeln readjust/to
Nachregistrierung *f* afterregistration
Nachreinigung *f* afterpurification, final cleaning
Nachrotte *f* afterrot
nachstellen reset/to, adjust/to
Nachstellung *f* reset
nachtblind night-blind
Nachtblindheit *f* nightblindness
nachteiliger Effekt *m* detrimental effect
Nachverbrennungszone *f* afterburning zone
nachwässern rewash/to
Nachweis *m* detection, proof, test
nachweisbar detectable, provable
nachweisen detect/to, proof/to
Nachweisführung *f* accountancy
Nachweisgrenze *f* detection limit
Nachweisreagens *n* test reagent
Nachweisreaktion *f* test reaction
Nachwirkung *f* aftereffect
Nachwuchsgeneration *f* rising generation
Nachzulassungsantrag *m* after-application admission
Nachzündung *f* retarded ignition
Nadelbiopsie *f* needle biopsy
Nadelholz *n* coniferous wood
Nadelventil *n* needle valve
Nadelwald *m* coniferous forest
Nagetier *n* rodent
Nahbestrahlung *f* short-distance irradiation

Näherung *f* approximation, approach
Näherungsrechnung *f* approximate calculation
Näherungsverfahren *n* approximate method, trial and error method
Nährboden *m* growth (nutrient) medium, matrix, medium
nähren feed/to, nourish/to
nährend alimentary, nutrient
nahrhaft alimentary, nutrient
Nährmedium *n* nutrient medium
Nährmittel *n* food, nutrient
Nährsalz *n* nutrient salt
Nährstoff *m* nutrient
nährstoffarm oligotrophic
Nährstoffeinleitung *f* **in das Meer** nutrient entry into the sea
Nährstoffeliminataion *f* nutrient elimination
Nährstoffhaushalt *m* nutrient mass balance
nährstoffreich eutrophic
Nährstoffversorgung *f* nutrient supply
Nährsubstanz *f* growth medium
Nahrung *f* diet
Nahrung aufnehmen ingest/to
Nahrungsaufnahme *f* food uptake, ingestion
Nahrungsbedarf *m* food requirement
Nahrungskette *f* food chain
Nahrungsmangel *m* innutrition
Nahrungsmittelabfall *m* food waste
Nahrungsmittelchemie *f* food chemistry
nahrungsmittelchemisch food-chemical
Nahverkehr *m* local traffic, short-distance traffic
Narkotikum *n* narcotic
narkotisch narcotic
narkotisieren narcotize/to
narkotisierend narcotizing
Narkotisierung *f* narcotization
Nasenbluten *n* nose bleed, epistaxis
Nasenhöhle *f* nasal cavity
Nasenscheidewand *f* nasal septum
Nasenspray *n* nasal spray
Naßablagerung *f* wet deposition
Naßanalyse *f* wet analysis
Naßaufbereitung *f* flotation
Naßauftragung *f* wet application
Naßdeposition *f* wet deposition
Naßelektrofilter *n* wet electrofilter
Naßentaschung *f* wet removal of ash
Naßentstaubung *f* wet dust collection
naszierend nascent
Naßmahlen *n* wet grinding
naßmechanische Bodenaufbereitung *f* wet mechanical treatment of soil
Naßoxidation *f* wet oxidation
Naßoxidation *f* **von Abwässern** wet oxidation of sewage
Naßschlammverbrennung *f* wet sludge incineration
Naßsiebung *f* wet screening
Naßverfahren *n* wet process
Naßwäscher *m* wet scrubber
nationale Prioritätenliste *f* national priority list
Nationalpark *m* national park
nativ native, natural
Natrium *n* sodium
Natur *f* nature
Natur- und Umweltschutzgruppe *f* nature and environmental protection group
Naturerscheinung *f* natural phenomenon

naturgemäß   natural
Naturhaushalt *m*   natural balance
natürlich emittierte Verbindung *f*   naturally emitted compound
natürliche Auslese *f*   natural selection
natürliche Belüftung *f*   natural ventilation
natürliche Organisation *f*   natural organization
natürlicher Kreislauf *m*   natural circuit
Naturprodukt *n*   natural product
Naturschätze *mpl*   natural resources
Naturschutz *m*   nature protection, wildlife conservation, nature conservation
Naturschutzbeauftragter *m*   representative for landscape protection
Nebel *m*   fog, mist, vapour
Nebenbestandteil *m*   minor constituent, secondary component
Nebeneffekt *m*   secondary effect, by-effect, side effect
Nebenfläche *f*   side area
Nebenwirkung *f*   side effect
Nekrose *f*   necrosis
nekrosieren   necrose/to
Nematode *m*   nematode
Nematodenbefall *m*   nematodiasis
Nematozid *n*   nematocide
Nennlast *f*   nominal load
Nennleistung *f*   nominal capacity, nominal output
Nennwert *m*   nominal value
Neodym *n*   neodymium
Neon *n*   neon
Neonlicht *n*   neon light
Neonröhre *f*   neon tube
Neoplasma *n*   neoplasm
Nephelometer *n*   nephelometer

Nephralgie *f*   nephralgia
Nephritis *f*   nephritis
Nerv *m*   nerve
Nervenbahn *f*   nervous pathway
Nervengas *n*   nerve gas
Nervengewebe *n*   nerve tissue
Nervengift *n*   neurotoxin
Nervenlähmung *f*   neuroparalysis
Nervenleiden *n*   nervous disease, neuropathy
Nesselausschlag *m*   hives
Nesselfieber *n*   nettle-rash
Nesselgift *n*   nettle poison
Nesselsucht *f*   urticaria
netzartig   reticular
Netzbarkeit *f*   wettability
netzförmig   net-shaped, reticular
Netzhaut *f*   retina
Netzwerk *n*   lattice, mesh, network, reticulum
neu pflanzen   replant/to
Neulanderschließung *f*   development of virgin land
Neubildung *f*   regeneration
neuralgisch   neuralgic
Neurochemie *f*   neurochemistry
Neurodermatose *f*   neurodermatosis
Neurohormon *n*   neurohormone
Neurologie *f*   neurology
neurologisch   neurological
Neurophysiologie *f*   neurophysiology
Neurose *f*   neurosis
neurotoxisch   neurotoxic
neutraler Abfall *m*   neutral waste
Neutralisation *f*   neutralization
Neutralisation *f* **von Industrieabwasser**   neutralization of industrial waste water
Neutralisationsverfahren *n*   neutralization method
neutralisieren   neutralize/to

Neutralisierung *f* neutralization
Neutralpunkt *m* neutral point
Neutron *n* neutron
Neutronenaktivierungsanalyse *f* neutron activation analysis
Neutronenbehandlung *f* neutron treatment
Neutronenbeschuß *m* neutron bombardment
Neutronennachweis *m* neutron detection
Neutronenquelle *f* source of neutrons
nicht nachweisbar undetectable
nicht reaktionsfähig unreactive
Nichtanwendbarkeit *f* inapplicability
nichtauslaugbar non-leachable
nichtflüchtig non-volatile
nichtflüchtige organische Schadstoffe *mpl* non-volatile organic pollutants
nichtflüchtige Stoffe *mpl* **im Abwasser** non-volatile substances in the sewage
nichtklopfend knock-free
nichtleitend non-conducting
nichtlinear non-linear
nichtlöslich indissoluble, insoluble
nichtmarkiert unlabelled
Nichtmischbarkeit *f* immiscibility
nichtrostend rustproof
nichtrußend sootless
nichtschmelzbar non-fusable
nichtsystemisch non-systemic
nichttödliche Dosis *f* sublethal dose
nichttrinkbar undrinkable
nichttrocknend non-drying
nichtwäßrig anhydrous, non-aqueous
Nichtzuständigkeit *f* incompetence

Nichtzustellung *f* non-delivery
Nickel *n* nickel
Nickelabwasser *n* nickel sewage
Nickelbad *n* nickel bath
niederbrennen burn down/to
Niederdruckkessel *m* low-pressure boiler
Niedermoor *f* fen
Niederschlag *m* precipitate, precipitation; rainfall
niederschlagen precipitate/to
Niederschlagselektrode *f* precipitation electrode
Niederschlagshöhe *f* amount of rainfall
Niederschlagsintensität *f* intensity of rainfall
Niederschlagsmenge *f* rainfall rate
Niederschlagswasser *n* rainwater
Niedertemperaturkonvertierung *f* low-temperature conversion
Niederungsgebiet *n* low plane
Niederwild *n* small game
niedrigsiedend low-boiling
Niedrigwasser *n* low tide
Niere *f* kidney
Nierenabnormalität *f* renal abnormality
Nierenbecken *n* renal pelvis
Nierenbeckenentzündung *f* pyelitis
Nierenblutung *f* renal blood flow
Nierenentzündung *f* nephritis
Nikotin *n* nicotine
Nikotingehalt *m* nicotine content
Nikotinvergiftung *f* nicotinism
Niob niobium
Nische *f* niche
Nisthilfe *f* nesting-aid
Nitrat *n* nitrate
Nitratatmung *f* nitrate breathing

Nitratauswaschung *f* nitrate leaching
Nitratelimination *f* nitrate elimination
nitratgesteuerte Denitrifikation *f* nitrate-controlled denitrification
nitratsensitive Elektrode *f* nitrate-sensitive electrode
Nitriersäure *f* nitrating acid
Nitrierung *f* nitration
Nitrifikation *f* nitrification
nitrifizierende Kläranlage *f* nitrificating clarifier
Nitrifizierung *f* nitrification
Nitrifizierungsbakterien *fpl* nitrifying bacteria
Nitrit *n* nitrite
nitritfrei nitrite-free
Nitrobakterien *fpl* nitrobacteria
Nitrocellulose *f* nitrated cellulose
Nitrosamin *n* nitrosamine
Niveaukontrolle *f* level control
Niveauunterschied *m* difference of level
Nobelium *n* nobelium
Nomenklatur *f* nomenclature
Nominalwert *m* nominal value
Normalausführung *f* standardization type
Normalbeanspruchung *f* normal stress
Normalformat *n* standardization size
Normalgröße *f* standardization size
Normalisierung *f* normalization, standardization
normen standardize/to, normalize/to
Normenaufstellung *f* standardization
Normenausschuß *m* standardization committee

Normierung *f* normalization
Normung *f* standardization
normwidrig adverse, abnormal
Notabschaltung *f* emergency shutdown
Notausgang *m* emergency door, emergency exit
Notbehelf *m* emergency device
Notbeleuchtung *f* emergency lighting
Notbrause *f* emergency shower
Notfall *m* emergency, worst-case situation
Notfallplanung *f* emergency planning
Notfunktion *f* emergency function
Notstand *f* emergency situation
Notstromaggregat *n* emergency generating set
Notstromversorgung *f* emergency power supply
novelliertes Abfallgesetz *n* - amended waste law
nuklear nuclear
nukleare Energieversorgung *f* nuclear power supply
Nuklearmedizin *f* nuclear medicine
Nuklid *n* nuclide
Nulleinstellung *f* zero adjustment
Nulleiter *m* neutral wire
Nullinie *f* zero axis
Nullpunkt *m* zero point
Nullpunktabweichung *f* zero deviation
Nullpunkteinstellung *f* zero adjustment
Nullpunktkontrolle *f* zero adjust
Nutzanwendung *f* appliance
Nutzarbeit *f* useful work
nutzbar effective
Nutzeffekt *m* efficiency

Nutzholz *n* lumber
Nutzlast *f* payload, net load
Nutzleistung *f* effective output, net efficiency
Nützlichkeit f utility
Nutzungsanspruch *m* claim of utilization
Nutzungslizenz *f* utilization licence
Nutzungspauschale *f* utilization tariff
Nutzwärme *f* effective heat

O

Oase *f* oasis
Obduktion *f* autopsy
Oberboden *m* surface soil
Oberfeuer *n* upper heat
Oberflächenabdichtung *f* surface sealing
oberflächenaktiv surface-active
oberflächenaktive Substanz *f* surfactant
Oberflächenbehandlung *f* surface treatment, surface coating
Oberflächenbehandlungsanlage *f* surface treatment plant
Oberflächenbeschaffenheit *f* nature of surface, surface finish
Oberflächenbeschichtungsanlage *f* surface coating plant
Oberflächentrockner *m* surface dryer
Oberflächenwasser *n* surface water
Oberflächenwasserausflockung *f* surface water flocculation
Obergärung *f* top fermentation
Oberkante *f* top edge
Oberschicht *f* finish, top layer
Oberwasser *n* upstream water

Oberwasserpegel *m* top water level, surface water level
objektiv objective
Objektprüfung *f* object testing
Objektschutz *m* object protection
Obstbau *m* fruit cultivation
Obstkonserve *f* canned fruit
öd desolated, waste
Ödem *n* oedema
Ödland *n* badlands, waste land
Ödlandbegrünung *f* waste land recovery
öffentliche Müllentsorgung *f* communal waste disposal
Öffentlichkeitsarbeit *f* public relations
Öffentlichkeitsbeteiligung *f* public participation
Öko-Bewußtsein *n* ecological awareness
Öko-Freak *m* ecofreak
Ökodiagnostik *f* ecodiagnostics
Ökoepidemiologie *f* ecoepidemiology
Ökofaktor *m* ecofactor
Ökokunst *f* ecological art
Ökologe *m* ecologist
Ökologie *f* ecology
ökologieorientierte Marketingstrategie *f* ecological oriented marketing strategy
ökologisch ecological
ökologisch vertretbar ecologically acceptable
ökologische Auswirkung *f* ecological effect
ökologische Chemie *f* ecological chemistry
ökologische Landwirtschaft *f* ecological agriculture

ökologische Medizin *f* ecological medicine, ecomedicine
ökologische Neurobiologie *f* ecological neurobiology
ökologische Nische *f* ecological niche
ökologische Risikoanalyse *f* ecological risk analysis
ökologische Signifikanz *f* ecological significance
ökologische Struktur *f* ecological structure
ökologische Wertigkeit *f* ecological valence
ökologische Wirksamkeit *f* ecological efficiency
ökologischer Handlungsplan *m* ecological action plan
ökologischer Schaden *m* ecological damage
ökologischer Übertragungsweg *m* ecological transfer pathway
ökologischer Zusammenbruch *m* environmental collaps
ökologischer Zustand *m* ecological state
ökologisches Gleichgewicht *n* ecological balance
ökologisches Modell *n* ecological model
ökologisches Risiko *n* ecological risk
ökomedizinisch ecomedical
ökonomisch machbar economically feasible
ökonomischer Vorteil *m* economic advantage
ökosozial ecosocial
Ökosystem *n* ecosystem
Ökosystemforschung *f* ecosystem research

Ökotoxikologie *f* ecotoxicology
ökotoxikologische Auswertung *f* ecotoxicological evaluation
ökotoxikologische Schlußfolgerung *f* ecotoxicological conclusion
ökotoxikologische Schädlichkeit *f* ecotoxicological harmfulness
Ökotyp *m* ecotype
Öl-Wasser-Trennung *f* oil-water separation
Ölabscheidung *f* oil separation
Ölabsorption *f* oil absorption
Ölauffangpfanne *f* oil catchpan
Ölbehälter *m* oil basin, oil tank
Ölbrenner *m* oil burner
Öldampf *m* oil vapour
öldicht oil-proof
Ölemulsion *f* oil emulsion
Ölfang *m* oil trap
Ölfeld *n* oil field
Ölfilm *m* oil film
ölhaltig oleiferous
ölig oily
Ölnebelabscheider *m* oil aerosol separator
Ölquellenbrand *m* oil field fire
Ölrückstand *m* oil residue
Ölschaum *m* oily scum
Öltanker *m* oil tanker
Ölteppich *m* oil slick
ölundurchlässig oil-proof
Ölvakuumpumpe *f* oil vacuum pump
ölverschmutztes Abwasser *n* oil-polluted waste water
Ölverschmutzung *f* oil pollution
ölverseuchte Erde *f* oil-contaminated soil
Ofen *m* burner, furnace, oven
Ofenabgas *n* furnace exhaust gas

Ofenbeschickung *f* charging the furnace
Ofenbetrieb *m* furnace operation
Ofenkammer *f* furnace chamber
Ofenwirkungsgrad *m* furnace efficiency
offene Flamme *f* free flame
Offsetdruck *m* offset printing
okkludieren occlude/to
Okklusion *f* occlusion
oligotroph oligotrophic
Olivenöl *n* olive oil
On-Line-Leuchtbakterientoximeter *n* on-line photobacteria toximeter
On-Site-Sanierung *f* on-site redevelopment
Opiat *n* opiate
Opium *n* opium
Optik *f* optics
optimale Heizrate *f* optimum heating rate
optimaler Arbeitsdruck *m* optimum working pressure
Optimierung *f* optimization
Optimum *n* optimum
optische Aktivität *f* optical activity
orale Verabreichung *f* oral administration
Organ *n* organ
Organiker *m* organic chemist
organisch organic
organische Chemie *f* organic chemistry
organischer Hausmüll *m* organic domestic waste
organischer Siedlungsabfall *m* organic estate waste
organischer Staub *m* organic dust
organisches Abbauprodukt *n* organic degradation product

organisches Lösemittel *n* organic solvent
Organismus *m* organism
Organkonzentration *f* **von Blei** organ concentration of lead
Organochlorverbindung *f* organochlorine compound
Organohalogenverbindung *f* organohalogen compound
Organologie *f* organology
Orientierungsdaten pl guidance data
Originalpackung *f* standardization packing, original packing
Ornithologie *f* ornithology
ortsbeweglicher Behälter *m* mobile container
ortsfest stationary
Ortung *f* orientation, location
Ortungsgerät *n* locator
Osmium *n* osmium
Osmometer *n* osmometer
Osmose *f* osmosis
Osteitis *f* osteitis
Otitis *f* otitis
Ovulation *f* ovulation
Oxid *n* oxide
Oxidation *f* oxidation
Oxidationsbeständigkeit *f* oxidation resistance, oxidation stability
Oxidationsfähigkeit *f* oxidizability
Oxidationskatalysator *m* oxidation catalyst
Oxidationsmittel *n* oxidizing agent
Oxidationsprozeß *m* oxidation process
oxidative Entfärbung *f* oxidative decolorization
Oxidbelag *m* oxide film
oxidieren oxidize/to
Oxidschicht *f* oxide layer

Ozean *m* ocean
ozeanisch oceanic
Ozeanographie *f* oceanography
ozeanographischer Parameter *m* oceanographic parameter
Ozon m(n) ozone
Ozonbelastung *f* **der Luft** ozone impact of the air
Ozongehalt *m* ozone concentration
Ozongenerator *m* ozone generator
ozonhaltig ozonic
ozonisch ozonic
Ozonisierung *f* ozonization; ozonolysis
Ozonkammer *f* ozonization chamber
Ozonkontakt *m* ozone contact
Ozonkonzentration *f* ozone concentration
Ozonloch *n* ozone hole
Ozonmesser *m* ozonometer
Ozonolyse *f* ozonolysis
Ozonometer *n* ozonometer
Ozonreaktion *f* ozone reaction
Ozonschäden *mpl* **an Pflanzen** zone damages on plants
ozonschädlich ozone-damaging
Ozonschicht *f* ozone layer
Ozonungsanlage *f* ozonization facility
Ozonwert *m* ozone value
Ozonwirkung *f* ozone effect

P

paaren pair/to, couple/to, conjugate/to
Paarungszeit *f* pairing season
Packpapier *n* packing paper, wrapping paper
Packung *f* packing, package
Packungscodierung *f* package coding
Packungsdichte *f* packing density
Packungseffekt *m* concentrative effect
Paddelrührer *m* paddle mixer, paddle stirrer
Pädiatrie *f* pediatrics
Palette *f* pallet; palette
Palisanderholz *n* palisander
Palladium *n* palladium
Pampa *f* pampas
Pankreas *n* pancreas
Pankreasenzym *n* pancreatic enzyme
Pankreasfunktion *f* pancreas function
Panne *f* failure, breakdown, defect, accident
Panzerglas *n* armoured glass
Papageienkrankheit *f* ornithosis
Papierabfall *m* paper chips, waste paper
Papierbecher *m* paper cup
Papierfabrik *f* paper mill
Papierfangzaun *m* paper intercepting fence
Papierfüller *m* paper filler
Papierproduktion *f* paper production
Papierrecyclat *n* paper recyclate
Papiersack *m* paper sack
Papiertragetasche *f* paper bag
Papierverbrauch *m* paper consumption
Papierverpackung *f* paper packing
Pappe *f* hard paper
Parabolspiegel *m* paraboloidal reflector
Paraffin *n* paraffin [wax]
Paragenese *f* paragenesis

Parallaxe *f* parallax
Paralyse *f* paralysis
paralytisch paralytic
Parameter *m* parameter
parameterabhängige Steuerung *f* parameter-dependent control
Parameterstudie *f* parameter study
Paranoia *f* paranoia
Parasit *m* parasite
Parasitenbefall *m* parasitic disease (infestation)
parasitisch parasitic
Parasitologie *f* parasitology
paratonisch paratonic
Parenchym *n* parenchyma
Paritätsveränderung *f* change of parity
Park *m* park
Parodontose *f* parodontosis
Paroxysmus *m* paroxysm
Partialdruck *m* partial pressure
Partialquerschnitt *m* partial cross section
partiell partial
Partikel *n(f)* particle
Partikelabscheidung *f* particle separation
Partikelemission *f* particle emission, particulate emission
Partikelfilter *n* particle filter
Partikelform *f* particle shape
partikelförmiger Stoff *m* particulate matter
partikelfrei particle-free
Partikelgröße *f* particle size
Partikelrückstand *m* particle residue
partikulär particular
passivieren passivate/to
Passivität *f* passivity, inactivity, inertness

Passivprobenehmer *m* diffusive sampler, passive sampler
Passivrauchen *n* passive smoking
Paste *f* paste
pasteurisieren pasteurize/to
Pasteurisierung *f* pasteurization
Pastille *f* pastille, tablet
pastös pasty
  pastöse Masse *f* paste
Pathobiologie *f* pathobiology
pathobiologisch pathobiological
pathogen pathogenic
  pathogene Pilzkultur *f* pathogenic fungus culture
Pathologe *m* pathologist
Pathologie *f* pathology
pathologisch pathological
  pathologischer Abfall *m* pathologic waste
Patientenschutz *m* patient protection
Patientenversorgung *f* patient provision
Peak *m* peak
Peakform *f* peak shape
Peakhöhe *f* peak height
Peakspitze *f* peak tip
Pech *n* pitch
pechartig pitch-like
Pegelüberwachung *f* level monitoring
Pellet *n* pellet
Pelzware *f* peltry
Penetration *f* penetration
Penicillin *n* penicillin
penicillinempfindlich sensitive to penicillin
Penicillinspiegel *m* penicillin level
penicillinunempfindlich insensitive to penicillin
Perchlorat *n* perchlorate

Periode *f* period
periodisch periodic
peripheres Nervensystem *n* peripheral nervous system
Perkolation *f* percolation
Perkussion *f* percussion
perkutan percutaneous
permanent permanent
Permanenz *f* permanence
Permeabilität *f* permeability
Permeation *f* permeation
Permeationsrohr *n* permeation tube
Permeationswiderstand *m* permeation resistance
persistent persistent
persistente Verbindung *f* persistent compound
Persistenz *f* persistence
Personendekontaminierung *f* personal decontamination
Personendosis *f* individual dose
Personenschleuse *f* personnel airlock
Personenschutz *m* personal protection
Persönlichkeitsveränderung *f* personality change
Pest *f* pestilence, plague
Pestizid *n* pesticide
Pestizidanalytik *f* pesticide analytics
Petrochemie *f* petrochemistry
Petrochemieanlage *f* petrochemical plant
Petrochemikalie *f* petrochemical
Petroleum *n* petroleum
Petroleumerzeugnis *n* petroleum product
Petrologie *f* petrology
Pfad *m* path
Pfand *n* deposit

Pfanne *f* pan, ladle
pflanzen plant/to
Pflanzenabfallkompost *m* compost of plant waste
Pflanzenabfallkompostierung *f* plant refuse composting
Pflanzenauszug *m* plant extract
Pflanzenchemie *f* phytochemistry
Pflanzengift *n* phytotoxin
Pflanzenhormon *n* phytohormone
Pflanzenkrankheit *f* plant disease
Pflanzenkrankheitserreger *m* plant germ
Pflanzennährstoff *m* plant nutrient
Pflanzenökologie *f* plant ecology
Pflanzenphysiologie *f* plant physiology
Pflanzenschädigung *m* plant pest
Pflanzenschutzgesetz *n* crop protection law, plant protection law
Pflanzenschutzmittel *n* crop protection agent, plant protecting agent
Pflanzenschutzmittelforschung *f* research of plant protection agents
Pflanzenschutzmittelrückstand *m* residue of plant protection agents
pflanzentoxisch phytotoxic
Pflanzenversuch *m* plant experiment
pflanzenverträgliche Anwendung *f* **von Insektiziden** plant-tolerable application of insecticides
Pflanzenwand *f* plant wall, green wall
pflanzlich vegetable
pflanzlicher Rohstoff *m* plant material
pflanzliches Protein *n* plant protein
Pflasterprobe *f* patch test
Pflegeleichtausrüstung *f* easy-care finishing

Pflug *m* plough
Pfropfen *m* plug, stopper, cork
pH-Elektrode *f* pH electrode
pH-Wert *m* pH value
Pharma-Außenhandel *m* pharma foreign trade
Pharmaindustrie *f* pharmaceutical industry
Pharmakochemie *f* pharmaceutical chemistry
Pharmakokinetik *f* pharmacokinetics
Pharmakologie *f* pharmacology
pharmakologisch pharmacological
Pharmaspiegelkontrolle *f* drug monitoring
pharmazeutische Chemie *f* pharmaceutical chemistry
Pharyngitis *f* pharyngitis
Phase *f* phase
Phasenänderung *f* phase change
Phasendiagramm *n* phase diagram
Phasengrenze *f* phase boundary
Phasenübergang *m* phase transition
Phenol *n* phenol
Phenolformaldehydharz *n* phenol formaldehyde resin
Phenolkonzentration *f* phenol concentration
Phon *n* phon
Phonmesser *m* phonometer
Phosphat *n* phosphate
Phosphatbelastung *f* phosphate load
Phosphatdüngemittel *n* phosphate fertilizer
Phosphateliminierung *f* phosphate elimination
Phosphateliminierungsanlage *f* phosphate elimination plant
Phosphatfällung *f* phosphate precipitation

phosphatfreies Spülmittel *n* phosphate-free washing-up liquid
Phosphor *m* phosphorus
Phosphoreszenz *f* phosphorescence
Photoabbau *m* photodegradation
photoabbaubarer Kunststoff *m* photodegradable plastic
Photoabsorption *f* photoabsorption
Photoassimilation *f* photoassimilation
photoassimilatorische Sauerstoffproduktion *f* photoassimilatory oxygen production
Photobiologie *f* photobiology
Photochemie *f* photochemistry
photochemisch photochemical
photochemischer Smog *m* photochemical smog
Photokatalyse *f* photocatalysis
Photokopierpapier *n* photocopying paper
Photomorphogenese *f* photomorphogenesis
Photorespiration *f* photorespiration
photosensitiv photo-active
physikalische Eigenschaft *f* physical property
physikalische Therapie *f* physiotherapy
physikochemische Eigenschaft *f* physicochemical property
Physiologie *f* physiology
physiologische Reaktion *f* physiological respond
Physiotherapie *f* physiotherapy
physisch physical
Phytochemie *f* phytochemistry
Phytoparasit *m* phytoparasite
Phytopathologie *f* phytopathology
Phytopharmakon *n* phytopharmaceutical

Phytopharmazie *f* phytopharmacy
Phytoplankton *n* phytoplankton
phytotoxisch phytotoxic
Phytozönose *f* phytocenosis
Piezoeffekt *m* piezoelectric effect
piezoelektrisch piezoelectric
piezoresistiver Aufnehmer *m* piezoresistant sensor (detector)
Pigment *n* pigment, coloring agent
pigmentieren pigment/to
Pigmentzelle *f* chromatophore
Pille *f* pellet, pill
Pilotanlage *f* pilot plant
Pilotanlagenmaßstab *m* pilot plant scale
Pilz *m* fungus; mushroom
Pilzgeflecht *n* mycelium
Pilzkunde *f* mycology
pilztötend fungicidal
pilztötendes Mittel *n* fungicide
Pipette *f* pipette
pipettieren pipette/to
Planetenmühle *f* planetary mill
planieren level/to, plane/to
Planiergerät *n* leveller, bulldozer
Plankton *n* plankton
Planktonalgen *fpl* plankton algae
Plasma *n* plasm[a]
Plasmaanlage *f* plasma plant
Plasmamembran *f* plasma membrane
Plasmaschweißen *n* plasma welding
Plasmazelle *f* plasma cell, plasmocyte
Plasmolyse *f* plasmolysis
Plastikabfall *m* plastic waste, plastic scrap
Plastikabfallprodukt *n* plastic waste product
Plastikabfallwiederaufbereitung *f* plastic waste recycling

Plastikfolie *f* plastic foil
Plastikrohr *n* plastic pipe
Plastikverwertung *f* disposal of plastic
Plateau *n* plateau
Platin *n* platinum
Plättchen *n* flake, platelet
Plattenwärmeaustauscher *m* plate heat exchanger
Plattieren *n* plating
Plattierung *f* plating
Plattierungsmaterial *n* cladding material
Platzbedarf *m* space requirement
Platzwechsel *m* migration, shift
Plazenta *f* placenta
plombieren seal/to
Plutonium *n* plutonium
Plutoniumbrutreaktor *m* plutonium breeder
pneumatisch pneumatic
pneumatische Förderung *f* pneumatic conveyance
Pneunomie *f* pneumonia
polare Bindung *f* polar bond
Polarimetrie *f* polarimetry
Polarisation *f* polarization
Polarisationsebene *f* plane of polarization
polarisieren polarize/to
Polarisierung *f* polarization
Polarität *f* polarity
Poliomyelitis *f* poliomyelits
Poliomyelitisimpfung *f* polio inoculation
Pollen *m* pollen
Pollenanalyse *f* pollen analysis
Pollenvorhersage *f* aeroallergen forecast
polychloriertes Biphenyl *n* polychlorinated biphenyl

polycyclisch polycyclic
polycyclischer aromatischer Kohlenwasserstoff *m* polycyclic aromatic hydrocarbon
Polyelektrolyt *m* polyelectrolyte
Polyester *m* polyester
polyhalogenierte aromatische Verbindung *f* polyhalogenated aromatic compound
Polykondensat *n* condensation polymer
Polymer *n* polymer
Polymerbeton *m* polymer concrete
Polymerisation *f* polymerization
polymerisieren polymerize/to
Population *f* population
Populationsdichte *f* population density
Populationsdynamik *f* population dynamics
Populationsgenetik *f* population genetics
Populationsökologie *f* population ecology
Populationsstatistik *f* population statistics
Populationszyklus *m* population cycle
Pore *f* pore, void
Porenflüssigkeit *f* pore fluid
Porenfüller *m* pore sealer
Porengröße *f* pore size
Porenleichtbeton *m* porous lightconcrete
Porenraum *m* pore space
Porenstruktur *f* pore structure
Porenweite *f* porosity
porig poriferous, porous
porös poriferous, porous
Porosität *f* porosity
Portion *f* portion

Porzellan *n* porcelain, china
Position *f* position
Positron *n* positron
Postulat *n* **der Irreversibilität** postulate of irreversibility
potent potent
Potential *n* potential
potentielle Gefährdung *f* potential hazard
Potentiometer *n* potentiometer
Potentiometrie *f* potentiometry
präklinische Untersuchung *f* preclinical examination
Prallmühle *f* impact mill
pränatal prenatal
präoperativ preoperativ
Präparat *n* preparation
Präparation *f* preparation
präparieren prepare/to, preserve/to
Praseodym *n* praseodym
Prävalenz *f* prevalence
präventiv preventive
Praxis *f* practice
praxisbezogen relating to practice
praxisnah practice-related
praxisorientierte Simulation *f* practice-oriented simulation
Präzipitat *n* precipitate
präzipitieren precipitate/to
Precoatschicht *f* precoat layer
Preßkuchen *m* pressed cake
Preßluft *f* compressed air
Preßlufthammer *m* pneumatic hammer
Preßmasse *f* moulding composition, moulding compound
Preßsystem *n* pressing system
Primärenergie *f* primary energy
Primärenergie *f* **aus Müll** primary energy from waste
Primärschlamm *m* primary sludge

Prinzip *n* **der geringsten Wirkung** least-energy principle
Priorität *f* priority
Prioritätenliste *f* list of priorities
private Wertstoffsammlung *f* private collection of materials of value
Proband *m* proband
Probe *f* proof, sample, specimen; test, trial, assay
Probeabzug *m* proof
Probealarm *m* practice alarm
Probebetrieb *m* trial operation
Probenahme *f* sampling
Probenahme *f* **aus der Luft** air sampling
Probenahme *f* **vom Schiff** sampling from a ship
Probenaufbewahrung *f* sample storage
Probenkörper *m* specimen
Probenstrom *m* sample flow
Probenstromförderung *f* sample flow conveyance
Probenstromzuleitung *f* sample inlet
Probezusammensetzung *f* sample composition
Problemabfall *m* waste of problematic nature
Problemanalyse *f* problem analysis
Produktdatei *f* data file of products
Produktionserlaubnis *f* production permit
Produktionsgütergewerbe *n* trade of production goods
Produktionsleistung *f* production output
Produktsicherheit *f* product safety
produktspezifischer Reststoff *m* product-specific remaining material

Produktzersetzung *f* material fouling (decomposition)
Produzentenhaftung *f* producer liability
Profilanalyse *f* profile analysis
Prognoseergebnis *n* prognosis result, forecast result
Prognosesystem *n* forecast system, prognosis system
Prognoseverfahren *n* prognosis method
Proliferation *f* proliferation
prolifieren proliferate/to
Promethium *n* promethium
prophylaktisch prophylactic
Prophylaxe *f* prophylaxis, prevention
Protactinium *n* protactinium
Proteinabbau *m* protein degradation
Proteinanalyse *f* protein analysis
Proteolyse *f* proteolysis
Proton *n* proton
Protonenbeschleuniger *m* proton accelerator
Protonenbeschuß *m* proton bombardment
Protoplasma *n* protoplasm
Prototyp *m* prototype
Prozeßabwasser *n* process water, sewage water
Prozeßanalyse *f* process analysis
Prozeßautomatisierung *f* process automation
Prozeßautomatisierungsaufgabe *f* process automation task
Prozeßbeobachtung *f* process observation
Prozeßbild *n* process image
Prozeßbrennstoff *m* process fuel
Prozeßdaten pl process data

Prozeßdatenabbild *n* image of process data
Prozeßdatenbank *f* process data bank
Prozeßdatenerfassung *f* process data acquisition
Prozeßdatenübermittlung *f* process data communication
Prozeßdatenverarbeitung *f* data processing, processing of [process] data
Prozeßelement *n* process element
Prozeßfließbild *n* process flow sheet
Prozeßführung *f* process management, process control
prozeßgeführte Ablaufsteuerung *f* processed sequencing control
prozeßgeführte Verriegelung *f* process-controlled interlocking
Prozeßkontrolle *f* process control
Prozeßleitsystem *n* process control system
Prozeßoptimierung *f* process optimization
Prozessor *m* processor
Prozeßrechner *m* process computer
Prozeßstabilität *f* process stability
Prozeßstation *f* process station
Prozeßstörung *f* process malfunction, process disturbance
Prozeßtechnologie *f* process technology
Prozeßüberblick *m* process overview
Prozeßwasser *n* process water
Prüfaerosol *n* test aerosol
Prüfanstalt *f* testing institute
Prüfapparat *m* testing apparatus
prüfbar testable
Prüfbericht *m* test report
Prüfbescheinigung *f* für Tankfahrzeuge test certificate for tank trucks
Prüfbett *n* test bed
Prüfdaten pl test data
prüfen test/to, check/to, examine/to, gauge/to, inspect/to
Prüfer *m* tester, examiner
Prüfergebnis *n* test result
Prüffeld *n* test field
Prüfkammer *f* testing chamber
Prüfling *m* sample
Prüfmaterial *n* testing material
Prüfmethode *f* test method
Prüfmodell *n* test model
Prüfröhrchen *n* test tube
Prüfroutine *f* test routine
Prüfschein *m* test certificate
Prüfschema *n* testing scheme
Prüfsonde *f* test probe
Prüfstreifen *m* test strip
Prüfung *f* test, check, examination, inspection, verification
  Prüfung *f* der biologischen Abbaubarkeit biodegradation testing
Prüfvorrichtung *f* testing arrangement
Prüfvorschrift *f* direction for testing
Prüfwesen *n* testing
Prüfzeugnis *n* test certificate
Pseudokrupp *m* pseudocroup
Psychologie *f* psychology
psychomotorischer Test *m* psychomotor test
Psychopharmakon *n* psycho-drug, psychochemical
Psychosedativum *n* tranquilizer
psychosomatisch psychosomatic
Psychotherapie *f* psychotherapy
Puder *m* powder
Puffer *m* buffer
Pufferlösung *f* buffer solution
Pufferung *f* buffering

Pufferwirkung *f* buffering action
Pulver *n* powder
Pulverbeschichtung *f* powder coating
pulverisierte Aktivkohle *f* pulverized charcoal
Pulverkohle *f* powdered coal
Pulverlöscher *m* powdered fire extinguisher
Pulvermischer *m* powder blender
Pumpeffekt *m* pumping effect
Pumpenschacht *m* pump shaft
Pumpensumpf *m* pump sump
pumpfähig pumpable
punktförmig pointshaped
punktschweißen spot-weld/to
Punktschweißen *n* spot (point) welding
Punktschweißmaschine *f* spot welder
Pustel *f* pustule
Pyrolyse *f* pyrolysis
Pyrolyse *f* **von Polymerabfall** pyrolysis of polymer waste material
Pyrolyse *f* **von Siedlungsabfällen** pyrolysis of estate wastes
Pyrolysegas *n* pyrolysis gas
Pyrolysereaktor *m* pyrolysis reactor
Pyrolyseverbrennungsanlage *f* pyrolysis incineration plant
Pyrolyseverfahren *n* pyrolysis method
pyrolysieren pyrolyze/to
pyrolytisch pyrolytic
Pyrometer *n* pyrometer
pyrophor pyrophoric

Q

Quadrant *m* quadrant
qualitative Analyse *f* qualitative Analysis
Qualitätsanforderung *f* requirement on quality standard
Qualitätskontrolle *f* quality control
Qualitätssicherung *f* quality assurance
Qualitätssicherungsmaßnahme *f* quality security measure
Qualm *m* smoke
Quant *n* quantum
Quantenbiologie *f* quantum biology
Quantenchemie *f* quantum chemistry
Quantenzahl *f* quantum number
quantitative Ausbeute *f* quantitative yield
Quantum *n* quantity, quantum
Quarantäne *f* quarantine
Quarz *m* quartz
Quarzglas *n* quartz glass
quarzhaltig quartziferous
Quarzkörner *npl* quartz grains
Quarzlampe *f* quartz lamp
Quarzmehl *n* quartz powder
quasielastisch quasielastic
quasistabil quasistable
quasistatisch quasistatic
Quecksilber *n* mercury
Quecksilberbatterie *f* mercury battery
Quecksilberrecycling *n* mercury recycling
Quecksilbertoxizität *f* mercury toxicity
Quelle *f* spring, well; source, origin, root
quellen steep/to, swell/to

Quellverhalten *n* swelling behaviour
Quellvermögen *n* swelling capacity, swelling power
Quellwasser *n* spring water, well water
Querempfindlichkeit *f* cross sensitivity
Querschnitt *m* cross section, profile

R

racemisch racemic
Rachenkatarrh *m* pharyngitis
Rachitis *f* rachitis, rickets
Radar *m* radar
Radarfalle *f* radar trap
Radarkontrolle *f* radar control
Radarschirm *m* radar screen
Radartechnik *f* radar technology
Radar-Wetterbeobachtung *f* weather observation by radar
Radialgebläse *n* radial blast
radikal radical
Radikal *n* radical
Radikalfänger *m* scavenger
radioaktiv radioactive
  radioaktive Chemikalie *f* radioactive chemical
  radioaktive Strahlung *f* radioactive radiation
  radioaktive Verseuchung *f* radioactive contamination
  radioaktives Abwasser *n* radioactive effluent, radioactive waste water
Radioaktivität *f* radioactivity
  Radioaktivität *f* **der Luft** air radioactivity
Radiobiologie *f* radiobiology
Radiochemie *f* radiochemistry

Radioisotop *n* radioactive isotope, unstable isotope
Radiologie *f* radiology
radiologisch radiological
radiomarkiert isotope-labelled, radiolabelled
  radiomarkierte Verbindung *f* radiolabelled compound
Radionuklid *n* radionuclide
Radioökologie *f* radioecology
Radiosonde *f* radio probe
Radiotoxicology *f* radiotoxikology
Radium *n* radium
Radiumtherapie *f* radium therapy
radiumverseucht radium-contaminated
Radlader *m* wheel loader
Radon *n* radon
Radonpegel *m* radon level
Raffination *f* refining
Raffinationsabfall *m* refinery waste
Raffinationsanlage *f* refinery
Raffinerie *f* refinery
Raffinerieabfall *m* refinery waste
Raffinerieabwasser *n* refinery sewage
Radonquelle *f* radon spring
raffinieren refine/to
raffiniert refined
Raffinierung *f* refining
Rahmenbestimmung *f* **zur Entsorgung** regulation for the disposal
Rahmenfilter *n* frame filter
Rahmenfilterpresse *f* frame filter press
Raketenabschußbasis *f* rocket launching site
Raketenkopf *m* rocket head
Raketentreibstoff *m* rocket propellant

Randbedingung *f* boundary condition
Randbevölkerung *f* fringe population
Ranzigkeit *f* rancidity, rancidness
Rapsöl *n* rape oil
Rasen *m* lawn
Rasterelektronenmikroskop *n* scanning electron microscope
Rastern *n* screening
Rasterschirm *m* scanning screen
Rastersystem *n* screen system
rationelle Energieverwendung *f* economical energy utilization
Rattengift *n* rat poison
Rattentest *m* rat test
rattenverseucht ratty
Raubbau *m* overexploitation
Raubfischbestand *m* predatory fish population
Raubgier *f* rapacity
Raubvogel *m* bird of prey
Rauch *m* fume, smoke, flue
Rauchabzugsöffnung *f* chimney hole
Rauchbekämpfung *f* smoke abatement
Rauchbelästigung *f* smoke nuisance
Rauchbildung *f* smoke formation, smoke generation
Rauchdichte *f* smoke density
rauchdurchsetzter Nebel *m* smog
Raucherhusten *m* smoker's cough
Räucherkammer *f* smoking chamber
rauchfreie Entsorgung *f* smokeless disposal
Rauchgasfiltertechnik *f* flue gas filter technology
Rauchgasnaßentschwefelung *f* wet desulphurization of flue gas

Rauchgasreinigung *f* flue gas purification
Rauchgasreinigungsanlage *f* flue gas purification (cleaning) plant
Rauchgasreinigungsreststoff *m* residue of flue gas purification
Rauchgasströmung *f* flue gas flow
Rauchgasteilstrom *m* partial flow of flue gas
Rauchgasweg *m* pathway of flue gas
Raumanzug *m* space suit
Raumausdehnung *f* spatial distribution
Raumbedarf *m* spatial requirement
Raumersparnis *f* economy in space, saving in space
Raumfahrtmedizin *f* space medicine
Raumformel *f* space formula
räumliche Verteilung *f* spatial distribution
Räummaschine *f* broaching machine
Raumordnungsgesetz *n* law of environmental planning
raumsparend room-saving, compact
Rauschgift *n* dope, drug, intoxicant, narcotic
Rauschgiftsucht *f* drug addiction
Rauschgiftsüchtiger *m* drug addict
Rauschwirkung *f* intoxicant effect
Reagens *n*, **Reagenz** *n* reagent
Reagenzglas *n* test tube
reagieren react/to
Reaktion *f* reaction, response
Reaktionsablauf *m* reaction sequence
Reaktionsbeschleunigung *f* catalysis
Reaktionschemikalie *f* reaction chemical
Reaktionsgrad *m* degree of reaction

Reaktionsluft *f* reaction air
Reaktionsmechanismus *m* reaction mechanism
Reaktionsprodukt *n* reaction product
reaktionsträge inert
reaktive Massengüter *npl* reactive mass goods
reaktivieren reactivate/to
Reaktivierung *n* reactivation
Reaktivität *f* reactivity
Reaktor *m* reactor, converter
Reaktorabfall *m* reactor waste
Reaktorabschirmung *f* reactor shielding
Reaktorbrennstoff *m* reactor fuel
Reaktorkern *m* reactor core
Reaktorleistung *f* reactor power
Reaktormantel *m* reactor shell
Reaktorvolumen *n* reactor volume
Realisierung *f* realization
Reanimationstechnik *f* technique of resuscitation
Reanimationszentrum *n* centre of resuscitation
Rechen *m* rake
Rechenanlage *f* screening plant
Rechengitter *n* rake grid
Rechengut *n* rake screenings, rake good
Rechengutpresse *f* rake material press, press for rake good
Rechenkorb *m* rake cage
Rechenmaterial *n* rake material
Rechenrückstand *m* screening residue
Rechensystem *n* rake system
rechnergestützt computer-assisted, computer-aided
rechnerunterstützt computer-assisted, computer-aided

recyclierbar recyclable
recyclieren recycle/to
Recycling *n* recycling
Recyclinganlage *f* recycling plant
Recyclinganlage *f* **für Kunststoffabfall** recycling plant for plastic waste
recyclingfähig recyclable
recyclingfähiger Abfall *m* recyclable waste
Recyclingmaterial *n* recycling material
Recyclingmodell *n* recycling model
Recyclingrate *f* recycling rate
Reduktion *f* reduction
Reduktionskraft *f* reducing power
Reduktionsmittel *n* reducing agent
Reduktionsverfahren *n* reduction method
Reduktionsvorgang *m* reducing process
Reduktionszone *f* reduction zone
reduzieren reduce/to
Reduzierstück *n* adapter, fitting reducing
Reduzierung *f* reduction
Referenzchemikalie *f* reference chemical
Referenzmaterial *n* reference material
Referenzmethode *f* reference method
Reflektor *m* reflector
Reflexbewegung *f* reflex movement
Reflexion *f* reflection, reflexion
reformieren reform/to
Reformierung *f* reformation, regeneration
Reformkost *f* health food
Refraktion *f* refraction
Refraktometer *n* refractometer

Refugium *n* refuge
Regelbarkeit *f* controllability
Regelkreis *m* closed control loop
regeln regulate/to, control/to
Regelung *f* regulation, control
Regelungstechnik *f* control engineering
Regelventil *n* control valve
Regelvorrichtung *f* control device
Regen *m* rain, rainfall
Regenbecken *n* rain[fall] basin
regendicht raintight
Regenerat *n* regenerate, regenerated material
Regeneratgummi *n* reclaimed rubber
Regeneration *f* regeneration, reclaiming, recovery
Regenerationsfähigkeit *f* regeneration power
Regenerationsvermögen *n* recovery power
Regenerationsvorgang *m* regenerative process
regenerative Energie *f* regenerative energy
regenerative Energiequelle *f* regenerative energy source
regenerative Wärmerückgewinnung *f* regenerative heat recovery
regeneratives System *n* regenerative system
Regeneratmedium *n* regenerated medium
Regenerieranlage *f* regeneration plant
regenerieren regenerate/to, reclaim/to, renew/to, recover/to
Regenerierung *f* regeneration
Regenerierungsanlage *f* regeneration plant

Regenerierungszeit *f* regeneration time (period)
Regenhöhe *f* rainfall rate (level)
Regenmesser *m* rain gauge
Regenmessung *f* rainfall measurement
Regenrückhaltebecken *n* rain storage reservoir
Regenschauer *m* shower
Regentropfen *m* raindrop
Regenüberlaufbecken *n* rain spillway basin (reservoir), rainwater overflow basin
Regenwald *m* rain forest
Regenwasser *n* rainwater
Regenwasserkanal *m* rainwater canal
Regenwasserverschmutzung *f* rainwater pollution
regionaler Umweltausschuß *m* regional environmental committee
Registriergerät *n* recording instrument
Registriergeschwindigkeit *f* chart speed
Registrierpapier *n* recording paper
Registrierung *f* registration, recording
Regler *m* regulator, controller
Regression *f* regression
Regulation *f* regulation
Regulationszyklus *m* regulation cycle
regulierbar regulable
Regulierbarkeit *f* controllability, adjustability
regulieren regulate/to, control/to, adjust/to
regulierende Maßnahme *f* regulating measure
Regulierung *f* regulation

Rehabilitation f rehabilitation
Rehabilitationszentrum n rehabilitation centre
rehabilitieren rehabilitate/to
Rehabilitierung f rehabilitation
Reibung f friction, abrasion
Reibungsabnutzung f frictional wear
Reibungsfläche f friction surface
Reibungskoeffizient m coefficient of friction
Reibungsschicht f friction layer
Reibungswärme f friction heat
Reichgas n rich gas
Reifengummichips mpl scrap tyre rubber
Reifenrunderneuerung f tyre recapping
Reihenversuch m successive experiment
Reimplantation f reimplantation
Reinheit f purity
Reinheitsgrad m degree of purity
Reinheitsprüfung f purity test
reinigen clean/to, cleanse/to, purge/to, purify/to
Reinigen n cleaning, cleansing, purifying
Reiniger m cleaner, purifier
Reinigung f cleaning, purification
Reinigungsanlage f purification plant
Reinigungsanlage f für Altlasten purification plant for old-sites
Reinigungsapparat m purifier
Reinigungsaufbereitung f von Boden extraction clean-up of soil
Reinigungsausrüstung f cleaning kit
Reinigungsautomatisierung f cleaning automation
Reinigungsbecken n cleaning basin

Reinigungsbeseitigung f von Ölverschmutzungen oil spill clean-up
Reinigungsgraben m cleaning trench
Reinigungsmittel n cleaner; detergent
Reinigungsprozeß m purification process
Reinigungsrückstand m cleaning residue
Reinigungsstufe f purification stage
Reinigungszeit f cleaning time
Reinkultur f pure culture
Reinluftgebiet n area of pure air
Reinsubstanz f pure substance
Reinwasser n pure water
reißfest tear-proof
Reißfestigkeit f rupture strength, tear resistance
Reißwolf m shredder, shredding machine
reizbar irritable, touchy
Reizbarkeit f irritability, touchiness
Reizgas n sneeze gas, tear gas
Reizklima n bracing climate
Reizmittel n stimulant
Reizstoff m irritant, stimulating substance
Reizsubstanz f stimulating substance
Reizung f irritation
Reizwirkung f stimulating effect
Reklamation f objection, complaint, reclamation
Reklamationsantwort f reply to complaints
reklamieren complain about/to, reclaim/to
Rekombination f recombination
Rekombinationsreaktion f recombination reaction

rekombinieren recombine/to
rekonstruieren reconstruct/to
Rekonstruktion f reconstruction, restoration
Rektifikation f rectification
Rektifizierapparat m rectifier
rektifizieren fractionate/to
Rektifizierung f rectification
Rekultivierung f recultivation
Rekultivierungsschicht f recultivation layer
Relation f relation
relative Luftfeuchtigkeit f relative humidity
relativer Fehler m relative error
Relativität f relativity
Relevanz f relevance
Relikt n relic
Renaturierung f renaturation
renovieren renew/to, renovate/to
Renovierung f renovation
Renovierungsphase f renovation phase
Reproduktionsrate f reproduction rate
Reproduzierbarkeit f reproducibility
Reservat n reservate, reservation
Reserve f reserve
Reservetank m reserve tank
Reservierung f reservation
Reservoir n reservoir
resistent resistant, immune
Resistenz f resistance
Resonanz f resonance; mesomerism
resorbierbar reabsorbable
resorbieren reabsorb/to, resorb/to
Resorption f reabsorption, resorption
Respirator m respirator
Ressource f resource

Ressourcenausnutzung f resource exploitation
Ressourcenschonung f protection of resources
restaurieren restore/to, renovate/to
Restaurierung f restoration, renovation
Restbestand m remaining stock
Restfracht f residual load
Restgas n residual gas
Restgasentsorgung f residual gas disposal
Restmonomer n residual monomer
Restmüll m residual waste
Restmüllbehandlungsanlage f processing plant for residual waste
Restschwefeldioxid n residual sulphur dioxide
Reststoff m residual (remaining) material, residual (remaining) substance, remainder, residual matter
Reststoffminimierung f minimization of residual substances
Reststofftransport m transportation of remaining materials
Retention f retention
Retentionsfaktor m retention factor
Retentionsindex m retention index
Retentionszeit f period of retention, retention time
Retorte f retort
Retortengraphit m retort graphite
retten rescue/to, save/to
Rettung f rescue, salvage
reversibel reversible
Reversibilität f reversibility
reversible Reaktion f reversible reaction
Revision f revision
Revitalisierung f revitalization
Rezeptor m receptor

reziproker Wert *m*   reciprocal
Rezyklierung *f*   recycling
Rhenium *n*   rhenium
Rheologie *f*   rheology
Rhesus-negativ   rhesus-negative
Rhesus-positiv   rhesus-positive
Rhesusaffe *m*   rhesus monkey
Rhesusfaktor *m*   rhesus factor
rheumatisch   rheumatic
Rheumatismus *m*   rheumatism
Rhodium *n*   rhodium
Richtigbefund *m*   verification
Richtungsfokussierung *f*   directional focus[s]ing
Richtungsgröße *f*   directional quantity
Richtwert *m*   index value
Rieselanlage *f*   irrigation plant
Rieselfeld *n*   sewage field, sewage farm
Rieselfilter *n*   trickling filter
rieseln   trickle/to, irrigate/to
Rieselwasser *n*   irrigation water, trickling water
Riff *n*   reef
Rinde *f*   bark
Rindenaufbereitung *f*   bark processing
Rindenkompost *m*   bark compost
Rinderdung *m*   cattle manure
Rinderpest *f*   cattle plague
Ringelektrode *f*   ring electrode
Ringofen *m*   annular furnace
Ringöffnung *f*   ring fission
Ringschaltung *f*   loop system
Ringschluß *m*   ring closure
Rinnsteinbesen *m*   gutter broom
Risikoabschätzung *f*   risk assessment
Risikoanalyse *f*   risk analysis
Risikobetrachtung *f*   risk assessment
Risikobeurteilung *f*   risk valuation
Risikobewertung *f*   risk valuation
Risikoerfassung *f*   risk registration
Risikoermittlung *f*   risk ascertainment
Risikofeststellung *f*   risk assessment
Risikopotential *n*   risk potential
Rißbildung *f*   crack formation
Robbensterben *n*   seal mortality, death of seals
Robbensterblichkeit *f*   seal mortality
Rodentizid *n*   rodenticide
roden   clear/to, root out/to
Rodung *f*   clearing
Rohabwasser *n*   crude waste water, raw sewage, untreated sewage
Rohbenzin *n*   virgin gasoline, crude gasoline
Rohdichte *f*   raw density
Roheisen-Schrott-Verfahren *n*   pig-and-scrap process
Rohfaser *f*   crude fibre
Rohgas *n*   crude gas, raw gas
Rohgummi *n*   crude rubber
Rohrbruch *m*   pipe burst
Röhrenofen *m*   tubular furnace, tube furnace
Rohrerneuerung *f*   pipe renewal
Rohrförderer *m*   tubular conveyor
Rohrgewinde *n*   pipe thread
Rohrleger *m*   pipe fitter
Rohrleitung *f*   piping, tubing
Rohrleitungsbau *m*   pipeline construction
Rohrleitungsnennweite *f*   nominal width of pipeline
Rohrleitungsnetz *n*   piping network
Rohrleitungssystem *n*   pipeline system, pipework system
Rohrverlegung *f*   pipe installation
Rohsoda *f*   crude soda

Rohstoff *m* raw material, crude material, stuff
Rohstoffaufbereitung *f* raw material processing, crude material processing
Rohstoffgewinnung *f* raw material recovery
Rohstoffmangel *f* raw material shortage
Rohstoffquelle *f* source of raw material
Rohstoffrückgewinnung *f* raw material recovery
Rohwasserressource *f* raw water resource
röntgen X-ray/to
Röntgenabsorptionsspektrum *n* X-ray absorption spectrum
Röntgenanalyse *f* X-ray analysis
Röntgenapparatur *f* X-ray apparatus
Röntgenaufnahme *f* roentgenogram, X-ray pattern, X-ray photograph
Röntgenbefund *m* radiological result (finding)
röntgenologisch radiological
Röstanlage *f* roasting plant
Rostansatz *m* deposit of rust
rostbeständig non-corrosive, rust-resistant, stainless
rostbeständiger Stahl *m* stainless (rust-free) steel
Rösten *n* roasting
Rostentferner *m* rust remover
Rostfestigkeit *f* corrosion resistance
rostfrei rust-free, rust-proof, antirust, stainless
rostfreier Edelstahl *m* stainless steel, rust-proof steel
Röstgas *n* roasting gas

rostgeschützt rust-proof
Röstgut *n* roasted material
rostig rusty, corroded
Rostmittel *n* rust inhibitor, anticorrosive agent
Rostschlacke *f* rust slag
Rostschutz *m* rust protection
Rostschutzfarbe *f* rust preventive paint
Röststaub *m* dust of roasted ore
Röstung *f* roasting
Rotationsdiffusionskoeffizient *m* rotational diffusion coefficient
Rotationspumpe *f* rotary pump
Rotationstrockner *m* rotational dryer
Rotationsverdampfer *m* rotary evaporator
Rotationszerstäuber *m* rotary atomizer
Rote Liste *f* red data book
Rotglut *f* redness
Rotte *f* rot
Rotteausgangsstoff *m* rot starting material
Rottedauer *f* rot period
Rotteerwärmung *f* rot warming
Rottegrad *m* degree of rot process
Rottekompost *m* rot compost
Rotteplatte *f* rot plate
Rottesystem *n* rot system
Rotteturm *m* rot tower, silo digester
Rottevorgang *m* rot procedure
Rottezelle *f* rot cell (chamber)
Routineablauf *m* routine course
Routineanalyse *f* routine analysis
Routineüberwachung *f* routine monitoring
Rubidium *n* rubidium
Rüböl *n* rape oil

Rückdiffusionstest *m* back diffusion test
Rückenmarkentzündung *f* myelitis
Rückentwicklung *f* regresssion
Rückfall *m* relapse
Rückfluß *m* back flow, reflux
Rückflußkühler *m* reflux condenser, return condenser
Rückführbarkeit *f* recyclability
rückführen recycle/to
Rückführschlamm *m* return sludge
Rückführung *f* feedback
Rückgewinnung *f* regeneration, reclaiming, recovery, reclamation, recycling, regaining
Rückgewinnungsanlage *f* regeneration plant, regenerative plant
Rückhaltefaktor *m* retention factor
Rückhaltezeit *f* retention time
Rücklauf *m* reflux
Rücklaufschlamm *m* return sludge
Rücklaufverhältnis *n* backflow ratio
Rückleitung *f* recirculation
Rücknahme *f* repurchase, taking back
Rücknahmepflicht *f* obligation to take back, repurchase obligation
Rücknahmeverpflichtung *f* repurchase obligation
Rückreaktion *f* back reaction, reverse reaction
Rückschlagventil *n* flow back valve, safety valve
rückspülbares Filter *n* backflushable filter
Rückstand *m* remainder, residue, dregs
Rückstandsanalyse *f* residue analysis
Rückstandsbestimmung *f* residue determination
Rückstandsrecycling *n* residue recycling
rückstandsfrei residue-free
Rückwasser *n* backwash, backwater
Rückwirkung *f* feedback, reaction
Ruhebelastung *f* static load
Ruhelage *f* position of rest
ruhende Luftschicht *f* static air layer
ruhendes Feststoffbett *n* static solid bed
Ruhr *f* dysentery
Ruhrbazillus *m* dysentery bacillus
rühren agitate/to, stir/to, touch/to
Rührer *m* stirrer, mixer
Rührkessel *m* stirring vessel
Rührwerk *n* stirrer
Ruine *f* ruin
runderneuern recap/to, retreat/to
Runderneuerung *f* recapping
Ruß *m* carbon black, soot
Rußbestimmung *f* soot determination
Rußbildung *f* formation of soot
rußen soot/to
Rußkohle *f* sooty coal
Rußzahl *f* soot index
Rüstungsindustrie *f* war material industry
Ruthenium *n* ruthenium
Rüttelfilter *n* vibration filter
Rüttelsieb *n* vibrating screen

S

Saatbeizmittel *n* seed dressing
Saatgut *n* seed
Saatgutaufbereitungsanlage *f* seeds processing plant
Saatzucht *f* seed growing

Sackfilter *n* bag filter
Saline *f* salt works, saline
Salinität *f* salinity
Salmonelle *f* salmonella
Salmonellenerkrankung *f* salmonellosis
Salpeter *m* saltpetre
Salz *n* salt
Salzablagerung *f* salt deposit
Salzader *f* salt vein
Salzaerosol *n* salt aerosol
Salzbedarf *m* salt requirement
Salzbelastung *f* salt freight (load)
Salzbergwerk *n* [rock] salt mine
Salzbildner *m* halogen, salt former
salzen salt/to
Salzfracht *f* salt freight (load)
salzhaltig saliferous
salzig saline
Salzsäure *f* hydrogen chloride
Salzsee *m* salt lake
Salzstock *m* salt dome
Samarium *n* samarium
Samen *m* seed; sperm
Samenbank *f* sperm bank
samentötend spermicidal
Samenzelle *f* sperm cell
Sammelbecken *n* collecting basin
Sammelbehälter *m* collecting vessel, collection vessel
Sammelcontainer *m* collection container
Sammelleitung *f* collecting pipe
Sammelnetzwerk *n* **für Altglasrecycling** collection network for waste glass recycling
Sammelphase *f* collecting medium, collecting phase
Sammelraum *m* collecting chamber
Sammelrohr *n* collecting pipe
Sammelstelle *f* collection place

Sammelware *f* collection goods
Sammler *m* collector
Sammlung *f* collecting, collection
Sammlungsbehälter *m* collection vessel
Sand-Öl-Schlamm *m* sand-oil sludge
Sandbad *n* sand bath
Sandboden *m* sandy soil
Sandfang *m* sand trap
Sandfangausrüstung *f* sand trapping equipment
Sandfanggebläse *n* sand trapping blower
Sandfangsystem *n* sand trapping system
Sandfangzaun *m* sand fence, sand trapping fence
Sandfilter *n* sand filter
Sandfiltersystem *n* sand filtering system
Sandfiltration *f* sand filtering
sandführend sand-bearing
sandig sandy
Sandklassierer *m* sand classifier
Sandmergel *m* sandy merl
Sandsack *m* sand bag
Sandschicht *f* layer of sand
Sandschiefer *m* sandy shale
Sandstein *m* sandstone
sanieren redevelop/to, clean-up/to, rehabilitate/to
Sanierung *f* rehabilitation, redevelopment, remediation, remedial, revival
Sanierungsaktion *f* remedial action
Sanierungsgebiet *n* redevelopment area
Sanierungskonzept *n* concept of redevelopment

Sanierungsmaßnahme *f* measure of redevelopment
Sanierungstechnik *f* redevelopment technology
Sanierungstechnologie *f* technology of redevelopment
Sanierungsverfahren *n* remediation method
Sanierungsziel *n* aim of redevelopment
sanitär sanitary
Sanitärtechnik *f* sanitary engineering
Saprobiensystem *n* saprobiotic system
Sarkom *n* sarkoma
Satellitenbild *n* satellite picture
sättigen saturate/to, concentrate/to
Sättigung *f* saturation
Sättigungsfeuchte *f* saturation humidity
Sättigungsgrenze *f* saturation limit
Sättigungskonzentration *f* saturation concentration
Sättigungstemperatur *f* saturation temperature
Sättigungswert *m* saturation value
Satztrockner *m* batch dryer
sauber clean, pure
säubern clean/to, cleanse/to, purify/to
sauer acidic, sour
säuern acidify/to, sour/to
Sauerstoff *m* oxygen
Sauerstoff-Belebtschlammverfahren *n* oxygen-activated sludge process
Sauerstoffabgabe *f* release of oxygen
Sauerstoffabscheider *m* oxygen trap
sauerstoffangereicherte Belüftung *f* oxygen-enriched aeration
sauerstoffarm oxygen-poor
Sauerstoffatmung *f* oxygen breathing
Sauerstoffbedarf *m* oxygen demand
Sauerstoffbelüftung *f* oxygen ventilation
Sauerstoffbildung *f* formation of oxygen
Sauerstoffbleiche *f* oxygen bleaching
Sauerstoffeintragsaggregat *n* unit (aggregate) for oxygen entry
Sauerstoffeintragsleistung *f* oxygen entry capacity, oxygen input capacity
Sauerstoffeintragsregelung *f* oxygen entry control
Sauerstoffentzug *m* deoxygenation
Sauerstoffgehalt *m* oxygen content
Sauerstoffgewinnung *f* oxygen production
Sauerstoffkonzentration *f* oxygen concentration
Sauerstoffkreislauf *m* oxygen cycle
Sauerstoffmangel *m* oxygen deficiency
Sauerstoffregelkreis *m* oxygen control loop [system]
sauerstoffreich oxygen-rich
Sauerstoffsättigungswert *m* oxygen saturation value
Sauerstoffverbrauch *m* oxygen consumption
Sauerstoffversorger *m* oxygen supplier
Sauganlage *f* suction plant
Saugeinrichtung *f* suction device
Säugetier *n* mammal
Saugfilter *n* suction filter

Saugfiltration *f* suction filtration
Saugflasche *f* suction bottle
Saugkraft *f* suction force
Saugleistung *f* suction power
Saugleitung *f* suction pipe
Säuglingsnahrung *f* baby food
Säuglingssterblichkeit *f* infant mortality
Saugluft *f* suction air
Saugrohr *n* suction pipe
Saugröhrchen *n* suction capillary
Saugwirkung *f* suction effect
Säule *f* column, pile, tower
Säulenbeladung *f* column loading
Säulenbelegung *f* column coating
Säulenbeschichtung *f* column coating
Säulencharakteristik *f* column characteristic
Säulenchromatographie column chromatography
Säure *f* acid
Säure-Base-Stoffwechsel *m* acid-base metabolism
Säureabfall *m* acid sludge
Säureabscheider *m* acid separator
säurebeständig acid-proof, acid-resistant
Säurebeständigkeit *f* resistance to acids
Säureempfindlichkeit *f* sensitivity to acid
säurefest acid-proof, acid-resistant
säurefrei acid-free
säurehaltig acidic, acidiferous
saurer Abfall *m* acidic waste
säureresistent acid-resistant
säurewidrig antiacid
Schabeeinrichtung *f* scraping device
Schaber *m* scraper
Schacht *m* shaft; pit

Schachtanlage *f* shaft plant
Schachtofen *m* shaft furnace
Schachtrieseltrockner *m* shaft trickling dryer
Schachtstein *m* shaft brick
Schachttrockner *m* tunnel dryer
Schadeinheit *f* pollution unit
Schaden *m* **durch Mikropartikel** microparticulate damage
Schadenersatz leisten pay damages/to
Schadenersatzanspruch abwehren reject claim for damages/to
Schadenserfassung *f* assessment of damage
Schadgas *n* noxious gas
Schadgasabscheidung *f* separating of noxious gas
schadhaft defective
schadhafte Stelle *f* defect
schädliche Substanz *f* harmful substance
schädliche Wirkung *f* detrimental (harmful) effect
schädlicher Effekt *m* detrimental (harmful) effect
Schädling *m* pest
schädlingsbekämpfend pesticidal
Schädlingsbekämpfung *f* pest control
Schädlingsbekämpfungsmittel *n* pesticide, biocide
Schadstoff *m* harmful substance, noxious substance, pollutant
Schadstoffanreicherung *f* accumulation of harmful substances
Schadstoffausbreitung *f* propagation of pollutants
Schadstoffaustritt *m* emission (release) of a pollutant

Schadstoffbeseitigung *f* pollutant disposal
Schadstoffemission *f* pollutant emission, emission of pollutant[s]
Schadstoffentsorgung *f* pollutant disposal
Schadstofffließanalyse *f* pollutant flow analysis
Schadstofffracht *f* pollutant freight (load)
schadstoffhaltiger Abfall *m* **aus Haushalten** waste from households containing harmful substances
Schadstoffkonzentration *f* pollutant concentration
Schadstoffminimierung *f* **bei Verbrennungsprozessen** minimization of pollution in incinerating
Schadstoffwanderung *f* pollutant migration
Schall *m* sound
Schallabsorption *f* sound absorption
Schallbeugung *f* sound diffraction
Schalldämpfung *f* silencing
Schallisolierung *f* sound insulation
Schallmeßgerät *n* phonometer
Schallpegel *m* noise level, sound level
Schaltuhr *f* switch clock, time switch
Schamotte *f* fireclay
Schaufelrad *n* impeller, paddle wheel
Schaufelrührer *m* paddle stirrer
Schaufeltrockner *m* blade dryer, paddle dryer
Schauloch *n* observation hole
Schaum *m* foam, scum
Schaumbeton *m* porous concrete
Schäumen *n* foaming
Schaumfalle *f* foam trap

Schaummittel *n* expanding agent
Schaumschutz *m* foam protection
Schaumteppich *m* foam carpet
Schaumverhütung *f* foam prevention
Scheibenrührer *m* disk mixer
scheiden separate/to, clear/to, screen/to, part/to
Scheidesieb *n* separating sieve
Scheidetrichter *m* separating funnel
Scheidewand *f* septum
Scherbeanspruchung *f* shearing force
Scherkraft *f* shearing force
Scherung *f* shearing
Scheuerfestigkeit *f* abrasion resistance
schichtartig lamellar
Schichtdicke *f* film thickness
Schichtdickenmessung *f* film thickness measurement
Schichtenbildung *f* lamination, stratification
Schichtengitter *n* layer lattice
Schieber *m* slide valve, slider, sluice valve
Schieberstation *f* [slide] valve station
Schieberstellung *f* slide valve position, gate valve position
Schiefer *m* shale, slate
Schieferöl *n* shale oil
Schieferplatte *f* slate slab
Schiffbeschlagnahme *f* **durch den Staat** embargo
Schiffbruch *m* shipwreck
Schilddrüse *f* thyroid gland
Schilddrüsenfunktion *f* thyroid activity, thyroid function
Schilddrüsenhormon *n* thyroxine
Schilddrüsenkrebs *m* thyroid cancer

Schilddrüsenunterfunktion *f* hypothyrosis, thyropenia
Schimmel *m* mould
Schimmelpilz *m* mould fungus
Schimmelpilzbildung *f* mould formation
Schirm *m* screen, shield
Schirmwirkung *f* screening effect
Schizothymie *f* schizothymia
Schlachthof *m* slaughterhouse
Schlacke *f* cinder, slag
Schlackenaufbereitung *f* slag processing
Schlackenbeton *m* slag concrete
Schlackenbrecher *m* slag crusher
Schlackenhalde *f* slag dump
Schlackenmühle *f* slag mill
Schlackeschmelzverhalten *n* slag smelting behaviour
Schlackeverwertung *f* slag utilization
Schlafkrankheit *f* sleeping sickness
Schlagelastizität *f* impact elasticity
Schlagfestigkeit *f* impact resistance, impact strength
Schlagwetter *n* fire damp
Schlamm *m* mire, mud, scum, silt, slime, sludge, slurry, slush
Schlammabbauleistung *f* sludge decomposition efficiency
Schlammablagerung *f* mud deposit, sludge deposit[ion]
Schlammabscheider *m* mud (sludge) trap
Schlammaktivierung *f* sludge activation
Schlammanalyse *f* sludge analysis
Schlammalter *n* sludge age
Schlammbecken *f* sludge basin
Schlammbehandlung *f* sludge treatment (processing)

Schlammbeseitigung *f* sludge removal
Schlammbildung *f* sludge formation
Schlammbindemittel *n* sludge binder
Schlammblähung *f* bulking of sludge
Schlammdesinfizierung *f* sludge disinfection
Schlammeindickung *f* sludge thickening
Schlammeindickungsmethode *f* sludge thickening method
Schlammentkeimung *f* sludge disinfection
Schlammentwässerung *f* sludge dewatering
Schlammfang *m* sludge trap
Schlammgehaltsmessung *f* measurement of sludge content
schlammig miry, muddy, slimy, sludgy
Schlammkuchen *m* sludge cake
Schlammkuchenpelletierung *f* sludge cake pelletizing
Schlammnachbehandlung *f* sludge aftertreatment
Schlammpyrolyse *f* sludge pyrolysis
Schlammsiebung *f* sludge screening
Schlammtransportleitung *f* sludge transportation pipe
Schlammtrocknung *f* sludge drying
Schlämmung *f* elutriation, decantation, sedimentation
Schlammverbrennung *f* sludge incineration
Schlammverbrennungskapazität *f* capacity of sludge incineration
Schlammverklappung *f* sludge barging

Schlämmverfahren *n* elutriation process, sedimentation
Schlammvoreindickung *f* pre-thickening of sludge
Schlammwasserableitung *f* sludge water discharge
Schlangenkühler *m* coil condenser
Schlauchwiderstandsthermometer *n* tubing resistance thermometer
Schlaufenrührer *m* loop agitator
Schlechtwetterperiode *f* period of poor weather
schleichendes Gift *n* slow poison
Schleifmittel *n* abrasive, polish
Schleifmittelindustrie *f* abrasive industry
Schleifschlamm *m* grinding sludge
Schleifstaub *m* abrasive dust
Schleim *m* slime, mucus
schleimabsondernd mucous
Schleimabsonderung *f* mucous secretion
Schleimhaut *f* mucosa, mucous membrane
schleimig mucous, slimy, viscid
Schleuse *f* sluice
Schleusenkanal *m* sluice
Schlick *m* mud, silt
Schlitzauslaß *m* slit-shaped outlet
Schlitzdüse *f* slit-nozzle
Schluckimpfstoff *m* oral vaccine
Schlußabnahme *f* final acceptance
schmaler Peak *m* narrow peak
schmarotzend parasitic
Schmarotzer *m* parasite
schmarotzerhaft parasitic
Schmelzanlage *f* smelting plant
schmelzbar meltable
Schmelze *f* fusion, melt, melting
schmelzen fuse/to, melt/to, smelt/to

Schmelzen *n* fusion, melt, melting, smelt, smelting
Schmelzfluß *m* melt flux, smelting flux
Schmelzflußelektrolyse *f* electrolysis of fused salts, igneous electrolysis, smelting flux electrolysis
Schmelzindex *m* melting index hot-melt adhesive
Schmelzpunkt *m* melting point, point of fusion
Schmelzwärme *f* heat of fusion
Schmelzwasser *n* meltwater
Schmierapparat *m* lubricator
Schmiere *f* lubricant
Schmiereffekt *m* lubricating effect
Schmieren *n* lubricating
schmieren grease/to, lubricate/to
Schmierfähigkeit *f* lubricity
Schmierfett *n* lubricating grease
Schmierflüssigkeit *f* lubricating liquid
schmierig oily
Schmiermittel *n* lubricant
Schmieröl *n* lubricating oil
Schmierspritze *f* lubricating oil syringe
Schmierstoff *m* lubricant, grease
Schmierstoffadditiv *n* lubricant additive
Schmirgelstaub *m* emery dust
Schmutz *m* dirt, mud
Schmutzfänger *m* dirt trap, tail flap
schmutzig impure, muddy, turbid
schmutzlösend dirt dissolving
Schmutzstoff *m* **in der Luft** air pollutant
Schmutzstoffbelastung *f* pollution load, pollution impact
Schmutzwasser *n* dirty (turbid, waste) water

Schneckenförderer *m* worm conveyor
Schneckenrad *n* screw gear
Schneckentrockner *m* screw dryer
Schneefall *m* snowfall
Schneewasser *n* snow water
Schneidöl *n* cutting oil
Schnellanalyse *f* proximate analysis, rapid analysis
Schnellaufladung *f* rapid charge
schnellbindender Zement *m* quick-setting cement
Schnellbinder *m* quick-setting cement
Schnellbrutreaktor *m* fast-breeding reactor
Schnellfilteranlage *f* rapid filtration plant
Schnelltest *m* rapid test
Schnelltestverfahren *n* rapid testing method
Schnittstelle *f* **zur Prozeßleittechnik** interface to the process control
Schockabkühlung *f* shock cooling
Schonung *f* **natürlicher Ressourcen** preservation of natural resources
Schornstein *m* chimney, stack
Schotter *m* rubble
Schotterbeton *m* ballast concrete
Schrägsortiermaschine *f* inclined sorting machine
Schredder *m* shredder
Schredder-Sortieranlage *f* shredder classification plant
Schredderabfall *m* shredder waste
Schrott *m* scrap [metal]; scrap iron
Schrottaufbereitungsanlage *f* scrap preparation plant
Schrotthändler *m* scrap dealer
Schrotthaufen *m* scrap heap
Schrottplatz *m* scrap yard
Schrottverhüttung *f* scrap melting
Schrottverwertung *f* scrap recovery
Schrumpffolie *f* shrink film
Schubkraft *f* shearing force, thrusting force
Schutt *m* rubbish, refuse, debris, rubble
Schüttdichte *f* bulk density
Schüttelmaschine *f* mechanical shaker
Schüttelsieb *m* shaker screen
Schüttgewicht *n* bulk weight
Schüttgut *n* bulk goods
Schüttgutoberfäche *f* surface of bulky goods
Schutthaufen *m* refuse dump
Schütthöhe *f* height of bed
Schüttler *m* shaker
Schüttmaterial *n* bulky material
Schüttschicht *f* fixed bed
Schutzanforderung *f* protective requirement
Schutzanstrich *m* protective coating, seal
Schutzanzug *m* protective suit
Schutzausrüstung *f* protective equipment
Schutzbeschichtung *f* protection coating
Schutzbestimmung *f* protection regulation
Schutzbrille *f* goggles
Schutzgas *n* inert gas
schutzgeimpft vaccinated
Schutzkleidung *f* protective clothing
Schutzschicht *f* protective layer
schwach alkalisch subalkaline
schwach basisch subalkaline
schwach bestrahlt lightly irradiated
schwach sauer subacid

Schwachlastbelebung *f* low loaded aeration
Schwachlastbetrieb *m* light-load operation
Schwachlaststufe *f* light-load step
Schwächung *f* attenuation
schwammig spongy; porous
schwanger pregnant
Schwangerschaft *f* gestation, pregnancy
Schwangerschaftstest *m* pregnancy test
Schwangerschaftsunterbrechung *f* interruption of pregnancy
Schwankungsbereich *m* range of variations
Schwebefähigkeit *f* suspensibility
schweben float/to, suspend/to
schwebend suspended
Schwebestaubmasse *f* suspended particulate matter
Schwebstoff *m* suspended matter
schwebstoffrei free from suspended matter
Schwefel *m* sulphur
Schwefeldioxid *n* sulphur dioxide
schwefelfrei non-sulphur
schwefeln . sulphur/to
Schwefelwasserstoffkonzentration *f* hydrogen sulphide concentration
Schweinefett *n* lard
Schweinefleisch *n* pork
Schweinerotlauf *m* erysipeloid
Schweißarbeit *f* welding
Schweißelektrode *f* welding electrode
schweißen weld/to
Schweißen *n* welding
Schweißgasabsauganlage *f* welding gas extraction (exhausting) plant

Schweißmaschine *f* welding machine
Schweißnaht *f* welding seam, welded joint
Schweißtechnik *f* technique of welding
Schweißverfahren *n* welding process
Schwelbrand *m* smouldering fire
Schwelen *n* smouldering
Schwelfeuer *n* smouldering fire
Schwelgas *n* smouldering gas
Schwellendetektor *m* threshold detector
Schwellenwert *m* threshold value
Schwellenwertmessung *f* threshold value measurement
Schwelzone *f* smouldering fire zone
Schwemmverfahren *n* flotation
Schwemmwasser *n* wash water
schwer abbaubare Verbindung *f* heavily degradable compound
schwerer Wasserstoff *m* deuterium
schweres Wasser *n* heavy water
schwerflüchtig low-volatile
schwerflüchtiger Brennstoff *m* non-volatile fuel
Schwerindustrie *f* heavy industry
Schwerkraft *f* force of gravity, gravity force
Schwerkraftablagerung *f* gravitational deposition
Schwerkraftölabscheider *m* gravity oil separator
Schwermetall *n* heavy metal
schwermetallarm poor in heavy metals
Schwermetallauslaugungsrate *f* leaching rate for heavy metals
Schwermetalleintrag *m* impact of heavy metals

Schwermetallgehalt *m* heavy metal content
Schwermetallkonzentration *f* heavy metal concentration
Schwermetallmobilität *f* mobility of heavy metals
Schwerwasserreaktor *m* heavy-water pile
Schwimmkran *m* floating crane
Schwimmsand *m* quicksand
Schwimmstoff *m* floating material
Schwingkabel *n* vibrational cable, oscillating cable
Schwingsonde *f* vibrational probe, vibratory probe, oscillating probe
Schwingungsdämpfung *f* vibration damping
Schwund *m* atrophy, leakage, shrinkage, wastage, fading
Sediment *n* sediment
sedimentär sedimentary
Sedimentation *f* sedimentation
Sedimentationsanalyse *f* sedimentation analysis
Sedimentationsbecken *n* sedimentation basin
Sedimentgestein *n* sedimentary rock
sedimentieren sediment/to
Sedimentierung *f* sedimentation
See *m* lake
Seebeben *n* seaquake
Seegras *n* aquatic weed, sea grass
Seehafen *m* port
Seeigelpopulation *f* sea urchin population
Seekatastrophe *f* sea (marine) disaster
Seeklima *n* marine climate
Seeluft *f* sea air
Seesanierung *f* lake remediation
Seeseite *f* sea-shore

Seetang *m* seaweed, kelp
Seeverbrennung *f* ocean combustion
Seeverunreinigung *f* marine pollution
Seewasserentsalzung *f* sea water desalination
Segmentblende *f* segmental orifice
Seife *f* soap
Seifenblasenzähler *m* soap bubble flow meter
Seifenlauge *f* soap suds, soap solution
Seilsonde *f* cable probe
Seismogramm *n* seismogram
Seismograph *m* seismograph
Seitenkette *f* side chain
Sekretion *f* secretion
Sekundärluft *f* secondary air
Sekundärlufteinrichtung *f* equipment for secondary air
Sekundärrohstoff *m* secondary raw material
Selbstentladung *f* self-discharge, unassisted discharge
Selbstentstehung *f* autogenesis
Selbstentzündung *f* self-ignition, spontaneous ignition, spontaneous inflammation
Selbstentzündungstemperatur *f* self-ignition temperature
selbsterhaltend self-sustaining
selbstgemacht home-made
selbsthaftendes Säulenetikett *n* self-adhesive column label
Selbstregulation *f* self-regulation, auto-regulation
Selbstreinigung *f* self-cleaning
Selbstreinigung *f* **der Atmosphäre** self-cleaning of the atmosphere
Selbstreinigungskraft *f* **der Gewässer** self-cleaning power of waters

Selbstreinigungsvermögen *n* self-cleaning capability
selektive Reduktion *f* selective reduction
Selektivfilter *n* selective filter
Selektivität *f* selectivity
Selen *n* selenium
Selenzelle *f* selenium cell
Seltenerdmetall *n* rare-earth metal
semipermeabel semipermeable
Senfgas *n* yperite, mustard gas
Sensibilisierung *f* sensitization
Sensibilität *f* sensitivity
sensitiv sensitive
Sensitivitätsanalyse *f* sensitivity analysis
Sensor *m* sensor
Sensoranwendung *f* sensor application
Sensorauswahl *f* sensor selection
sensorisch sensory
Sensorkabel *n* sensor cable
Sensorüberwachung *f* sensor monitoring
separate Tonne *f* **für Kunststoffabfall** separate bin for plastic waste
Separation *f* separation
Septum *n* septum
sequentielle Prüfung *f* sequential testing
Serie *f* series
seriell serial
 serielle Abläufe *mpl* serialized procedures
Serienfertigung *f* series production
serienmäßig standard
Seriennummer *f* serial number
Serologie *f* serology
seronegativ sero-negative
seropositiv sero-positive
Serum *n* serum
Serumkonzentration *f* serum concentration
Servoantriebssystem *n* servo drive system
Seuche *f* pestilence, plague, epidemic
Sicherheit *f* **der Abfallentsorgung** safety of the waste disposal
Sicherheit *f* **geologischer Entsorgung** safety of geological disposal
Sicherheitsabschaltung *f* rapid shut-down, safety shut-down
Sicherheitsabstand *m* safety zone
Sicherheitsanforderung *f* safety requirement
Sicherheitsausrüstung *f* safety equipment
Sicherheitsbarriere *f* safety barrier
Sicherheitsberater *m* security adviser
Sicherheitsberatung *f* security consulting
sicherheitsbezogene Ausrüstung *f* safety-related equipment
Sicherheitsdeponie *f* safety disposal (landfill) site, safety refuse dump
Sicherheitseinrichtung *f* safety device, security device
Sicherheitsfaktor *m* safety factor
Sicherheitskluft *f* safety harness
Sicherheitskontrolle *f* security (safety) check
Sicherheitskonzept *n* **für Deponien** safety concept for disposal sites
Sicherheitsstandard *m* safety standard
Sicherheitstechnik *f* safety engineering
sicherheitstechnische Anforderung *f* technical safety demand

Sicherheitsüberprüfung *f* security test[ing]
Sicherheitsuniform *f* safety harness
Sicherheitsuntersuchung *f* safety investigation
Sicherheitsverriegelung *f* failsafe interlock
Sicherstellung *f* **von Daten** backup of data, data backup
Sicherung *f* safeguarding; fuse *{electrical}*
Sicherungsmaßnahme *f* safety measure
Sicherungsroutine *f* safety routine, security routine
sichtbares Alarmzeichen *n* visible sign of alarm
Sichtbarkeit *f* visibility
Sichtbild *n* display
Sichtbildgerät *n* display unit
Sichter *m* classifier
Sichtungsanlage *f* separating plant
Sickerbecken *n* leaching basin, percolating basin
Sickergrube *f* soakaway, sewage pit
Sickerkonzentrat *n* leaching concentrate
sickern trickle/to, seep/to
Sickeröl *n* percolating oil
Sickerung *f* percolating
Sickerungsstrecke *f* percolating range
Sickerwasser *n* leakage (leaching, percolating) water, soakage
Sickerwasserbehandlung *f* treatment of leaching water
Sickerwasserdränage *f* leaching water drainage
Sickerwasserentsorgung *f* leaching water disposal, disposal of leaching water
Sickerwasserrate *f* rate of percolating water
Sickerwasserreinigungsanlage *f* leaching water purification plant
Sickerwassersammlung *f* leaching water collection
Sickerwasserverringerung *f* decrease of leaching water
Sieb *n* screen, sieve
Siebanalyse *f* sieve analysis
Siebbandpresse *f* screening belt press
Siebbereich *m* screening area
Siebboden *m* sieve plate, sieve tray
Siebeffekt *m* screening effect
Siebfilter *n* sieving filter, filtering screen
Siebgut *n* screen rejects
Siebmaschine *f* sieving machine
Siebrechen *m* sieve rake
Siebtrommel *f* sieve drum
Siedepunkt *m* boiling point, point of boiling
Siedetemperatur *f* boiling temperature
Siedeverzug *m* delay in boiling
Siedewasserreaktor *m* boiling water reactor
Siedlungsabfall *m* estate waste
Siedlungsabfallentsorgung *f* disposal of estate wastes
Siedlungsfläche *f* settlement area
Siedlungsstruktur *f* housing society
Signaleinrichtung *f* signal alarm
Sielhaut *f* sewer slime
Signallampe *f* signal lamp
Signalüberlagerung *f* signal superimposition
Signifikanzzahl *f* significance level
Silage *f* silage
Silber *n* silver

siliciumhaltiger Staub *m* silicon-containing dust
Silikat *n* silicate
Silikose *f* silicosis
Silizium *n* silicon
Silo *m* silo
Silofutter *n* silage fodder
Simulation *f* simulation
Simulationsmodell *n* simulation model
Simulator *m* simulator
simulieren simulate/to
simultan simultaneous
Simultanbetrieb *m* simultaneous mode
Simultanfällung *f* simultaneous precipitation
Sinkscheideverfahren *n* sink and float process
Sinkstoff *m* sediment, deposit, precipitate
Sinnesorgan *n* organ of sense, sense organ
sintern sinter/to
Sinterung *f* sintering, baking, fusing
Sinusitis *f* sinusitis
Siphon *m* siphon
Siphonüberlauf *m* siphon spillway
Sirene *f* siren, hooter
Skandium *n* scandium
Sklerose *f* sclerosis
Smog *m* smog
smogbildend smog-forming
Smogkatastrophe *f* smog disaster
Soda *f* soda
Soforthilfe *f* immediate aid
Sofortmaßnahme *f* immediate measure
Sofortmaßnahme *f* **zur Phosphateliminierung aus Abwasser** immediate measure for the elimination of phosphate
Software *f* software
softwarekontrolliert software-controlled
Softwarepaket *n* software package
Sohle *f* bottom
Sohlenabdichtung *f* bottom sealing
Sol *n* sol
solar solar
Solarenergie *f* solar energy
Solarkraftwerk *n* solar power plant
Solarstrahlung *f* solar radiation
Solarzelle *f* solar cell
Soleentsorgung *f* **durch Entsalzung** brine disposal by desalination
Sollwert *m* set point
Sollwertvorgabe *f* set point selection
solvatisieren solvate/to
Solvens *n* solvent
somnolent somnolent
Somnolenz *f* somnolence
Sonde *f* probe
Sondenkopf *m* probe head
Sonderabfall *m* hazardous waste, waste to be especially supervised (observed)
Sonderabfallanlieferung *f* special waste delivery
Sonderabfallartenkatalog *m* catalogue of special waste types
Sonderabfalldeponie *f* special waste dump, hazardous site
Sonderabfallentsorgung *f* disposal of special waste
Sonderabfallerzeuger *m* producer of hazardous waste
Sonderabfallgemisch *n* hazardous waste mixture

Sonderabfallgesellschaft *f* company for special waste, company for waste especially to be supervised
Sonderabfalltransport *m* special waste transport
Sonderabfallverbrennung *f* special waste incineration
Sonderabfallverbrennungsanlage *f* incineration plant for special waste
Sonderabfallwirtschaft *f* hazardous waste management
Sonderabfallzusammensetzung *f* composition of hazardous (special) waste
Sonderanfertigung *f* special construction
Sonderausführung *f* special design
Sondergröße *f* special size
Sonderkonstruktion *f* special construction
Sondermüll *m* hazardous waste, special waste
Sondermüllverbrennung *f* incineration of special waste
Sondervorschrift *f* special rule
Sonnenbestrahlung *f* solar radiation
Sonnenenergie *f* solar power
Sonnenfleck *m* sun spot
Sonnensimulationsanlage *f* sun simulation plant
Sonnenstrahlung *f* solar radiation
Sonnenwärme *f* solar heat
Sorption *f* sorption
Sorptionseigenschaft *f* sorption property
Sorptionsgleichgewicht *n* sorption equilibrium
Sorptionsrückstand *m* sorption residue
sortenrein type-specific
sortenreine Erfassung *f* von Bioabfällen collection of biowaste by type
sortenreine Trennung *f* bei der Biomüllkompostierung type-specific separation in biological waste composting
Sortieranlage *f* classification plant, sorting equipment, sorting (separating) plant, grading plant
Sortieren *n* sorting, grading, classifying, sizing
sortieren sort/to, grade/to, classify/to, size/to
Sortierfähigkeit *f* sorting capability
Sortiermaschine *f* sorting machine
Sortiermerkmal *n* sorting criterion
Sortiersystem *n* sorting system
Sortiertechnik *f* classifying technique
Sortiment *n* assortment
Sozialarbeit *f* social work
Sozialarbeiter *m* health inspector
sozialmedizinisches Problem *n* social-medical problem
Sozialpolitik *f* social policy
spaltbar fissile, cleavable, separable
Spaltmaterial *n* fission material
Spaltprodukt *n* fission product
Spaltstoffabfall *m* reactor waste
Spanholz *n* chip wood
Speicherbecken *n* storage reservoir
Speicherbehälter *m* hold tank
Speicherkapazität *f* storage capacity
speichern accumulate/to, store/to
Speicherofen *m* storage heater
Speichertechnik *f* storage technique
Speiseleitung *f* feed line
speisen feed/to, load/to
Speisesalz *n* table salt

Speisewasser n feed water
Speisewasservorwärmung f feedwater preheating
spektral spectral
Spektrometrie f spectrometry
Spektrum n spectrum
Spender m donor
Sperma n sperm
Sperre f aus Mörtel grout barrier
Sperrflüssigkeit f sealing liquid
Sperrmüll m bulky refuse
Sperrmüllbeseitigung f disposal of bulky refuse
Sperrschicht f blocking layer
Spezialcontainer m special container
Spezialfahrzeug n special-purpose vehicle
Spezialkompost m special compost
Spezifikation f specification
spezifische Ausstrahlung f emittance
spezifische Wärme f specific heat
Sphäre f sphere
sphärisch spherical
Spirometer n spirometer
Spitzenbelastung f peak load
Spitzenleistung f top performance
Spitzenstrom m peak current
splitlose Einspritzung f splitless injection
splitterfrei non-splintering, shatterproof
Spore f spore
Sporenbildung f spore formation
Spray n spray
Spraydose f spray can
Spraydosenschredder m spray can shredder
sprengen blow up/to, blast/to; water/to

Sprengmittel n explosive
sprießen sprout/to
Sprinkler m sprinkler
Sprinkleranlage f sprinkler system
Spritzapparat m sprinkler
Spritze f syringe
Spritzkabine f spray booth
Spritzkopf m die head
Spritzmaschine f extruder
Sproß m sprout, shoot
Sprühabsorption f spray absorption
Sprühabsorptionsverfahren n spray absorption technique (process)
Sprüharm m spray arm
Sprühdose f spray, aerosol can
Sprühelektrode f corona discharge electrode
sprühen spray/to
Sprühnebel m wet fog
Sprühtrockner m spraying dryer
Sprungtemperatur f transition temperature
Spülanlage f washing plant, cleansing plant
spülen purge/to, rinse/to, flush/to
Spülen n lavage
Spülfilter n scavenging filter
Spülflüssigkeit f rinsing liquid
Spülgas n cleansing gas
Spüllauge f flushing liquor
Spülmittel n detergent, washing-up liquid
Spülprogramm n washing programme, purging programme
Spülsystem n für Abwasserkanäle purging system for sewers
Spülung f rinsing, flush[ing], lavage
Spülventil n scour valve, flushing gate
Spülvorrichtung f purging device
Spülwirkung f purging effect

Spurenchemie *f* tracer chemistry
Spurenelement *n* trace element
Spurengas *n* trace gas
Spurenmetall *n* trace metal
Spurennachweis *m* trace detection
Spurenstoff *m* trace substance
Spurenstoffinertisierung *f* trace substance inertization
stabiles Gleichgewicht *f* stable equilibrium
Stabilisator *m* stabilizer
stabilisieren stabilize/to
Stabilisierung *f* stabilization
Stabilisierungsgitter *n* stabilization grid
Stabilisierungsgrenze *f* stability limit
Stabilisierungszeit *f* stabilization time
Stabilität *f* stability
Stabilität *f* **der Ökosphäre** stability of the ecosphere
Stabilität herabsetzen destabilize/to
Stabilitätsbeziehung *f* stability relationship
Stabilitätsgrad *m* degree of stability
Stabsonde *f* bar probe
Stadtbevölkerung *f* urban population
Stadtgas *n* town gas
städtische Population *f* urban population
städtisches Abwasser *n* municipal effluent
städtisches Abwasserkanalsystem *n* municipal sewage pipe system
Stadtplanung *f* urban planning
Stahlbeton *m* reinforced concrete
Stahlerzeugung *f* steel production

Stahlflasche *f* steel bottle
Stahlprobe *f* steel sample
Stahlschrott *m* steel scrap
Stahlspundwand *f* steel sheet piling
Stammlauge *f* mother liquor
Stammlösung *f* stock solution
Stammsubstanz *f* parent substance
Stammverbindung *f* parent substance
Stammzelle *f* parent cell
Stampfer *m* tamper
Standanzeiger *m* level indicator
Standardabweichung *f* standard deviation, relative standard deviation
Standardeigenschaft *f* standard property
Standardfehler *m* standard error
standardisieren standardize/to
Standardisierung *f* standardization
Standardmaterial *n* certified reference material
Standardmethode *f* standard method
Standardsäule *f* standard column
Standardschacht *m* standard shaft
Ständer *m* support
Standort *m* location, site
Standortalternative *f* location alternative
Standorteinschränkung *f* location restriction
standortspezifisch location-specific
Standortwahl *f* selection of location, choice of location
Standsäule *f* stationary column
Standzeit *f* shelf life, can stability
Stapelförderer *m* stack conveyor
stark radioaktiv highly radioactive
Stärkemehl *n* farina, starch
Stärkepolymer *n* starch polymer
Starrkrampf *m* tetanus
Startreaktion *f* initial reaction

Stationsleitgerät *n* station control unit
stationäre Phase *f* stationary phase
stationärer Zustand *m* steady state
statisches Gleichgewicht *n* static balance
statistisch verteilt randomly distributed
statistische Güteziffer *f* statistical quality index
statistische Kennziffer *f* statistical index number
statistische Sicherheit *f* statistical certainty
Staub *m* dust, powder
Staubabscheidung *f* dust separation
Staubaufnahme *f* dust uptake
Staubbelästigung *f* dust nuisance
staubdicht dust-proof
Staubentfernung *f* dust removal
Staubentwicklung *f* formation of dust
Staubfilter *n* dust filter
Staubförderer *m* dust conveyor
Staubfracht *f* dust load
staubfreie Abfüllanlage *f* dust-free filling station
staubig dusty
Staubrückgewinnung *f* dust recovery
Staubsammeltrichter *m* dust collection funnel
Staudamm *m* barrage, dam
Stauklappe *f* baffle plate
Staumauer *f* dam, retaining wall
Stauwasser *n* backwater
Stechtorf *m* cut peat
Steigrohr *n* delivery pipe, feedpipe, ascending tube, rinsing pipe
Steingut *n* pottery, stoneware
Steinkohle *f* hard coal, pit coal, mineral coal
Steinkohlenbenzin *n* coal gasoline
Steinkohlenteer *m* bituminous coal tar
Steinmörtel *m* cement
Steinschutt *m* rubble
Steinwolle *f* rock wool
Steppe *f* steppe
sterbender Wald *m* dying forest
Sterblichkeit *f* mortality
Sterblichkeit *f* **nach der Geburt** postnatal mortality
Sterblichkeitsrate *f* mortality rate
Sterblichkeitsziffer *f* mortality rate
steril sterile, aseptic
Sterilfiltration *f* sterile filtration
Sterilisation *f* sterilization
sterilisieren sterilize/to, pasteurize/to
sterische Hinderung *f* steric hindrance
Stethoskop *n* stethoscope
Stichprobe *f* random sample, spot sample
Stickstoff *m* nitrogen
Stickstoffausscheidung *f* nitrogen excretion
Stickstoffdioxid *n* nitrogen dioxide
Stickstofffreisetzung *f* nitrogen release
stickstoffhaltig nitrogenous
Stickstoffkreislauf *m* nitrogen cycle
Stickstoffoxid *n* nitrogen oxide
Stickstoffproblematik *f* nitrogen problem
Stickstoffquelle *f* source of nitrogen
Stickstoffumsetzung *f* nitrogen conversion
Stillegung *f* closure
Stillstand *m* standstill

Stillzeit *f* lactation
Stimulans *n* stimulant
stimulieren stimulate/to
stimulierend stimulating
Stöchiometrie *f* stoichiometry
stöchiometrisch stoichiometric
stöchiometrische Menge *f* stoichiometric amount
Stoffaufnahme *f* uptake of a substance
Stoffaustausch *m* mass transfer
Stoffbilanz *f* material balance
Stoffdatei *f* data file of substances
Stoffgruppe *f* substance group
Stoffumsetzung *f* conversion of substances
Stoffwechselendprodukt *n* final product of the metabolism
Stoffwechselleistung *f* **einer Zelle** metabolic performance of a cell
Störfall *m* case of accident, case of disturbance, case of trouble
Störfallverordnung *f* regulation in case of disturbance (accident)
Störmeldung *f* fault signal, fault indication
Störstelle *f* point of disturbance
Störung *f* disturbance, failure, interference, perturbation; fault
störungsfrei trouble-free
Störüberwachung *f* disturbance monitoring
stoßdämpfend shock-absorbent
stoßempfindlich sensitive to shock
stoßfest shock-proof
Stoßquerschnitt *m* collision cross section
strahlen radiate/to
Strahlenbiologie *f* radiation biology
Strahlenchemie *f* actinochemistry

strahlend radiant, radiating, radiative
Strahlendosis *f* dose of radiation, radiation dosage
strahlendurchlässig radiolucent
Strahlenemission *f* radiation
strahlenempfindlich radiosensitive
strahlenfrei free of radiation
Strahlengefahr *f* radiation hazard
Strahlengefährdung *f* danger of radiation
Strahlengenetik *f* radiation genetics
strahlengeschädigt damaged by radiation
Strahlenkunde *f* radiology
Strahlenschaden *m* injury by radiation
Strahlenschutz *m* radiation protection
Strahlenschutz *m* **für Beschäftigte** occupational radiation protection
Strahlenschutzüberwachung *f* radiation protection service
strahlenundurchlässig radiopaque
Strahlgaswäscher *m* jet gas scrubber
Strahlpumpe *f* jet pump
Strahlung *f* radiation
Strahlungsabsorption *f* radiation absorption
Strahlungsdichte *f* radiant flux density
Strahlungsdosis *f* radiation dose, radiation dosage
Strahlungsenergie *f* radiation energy
Strahlungsgürtel *m* radiation belt
Strahlungsverhalten *n* **der Atmosphäre** radiation behaviour of the atmosphere
Strahlwäscher *m* jet washer

Straßenbau *m* road construction
Straßenbelag *m* road surface
Straßenmarkierung *f* road marking
Straßenverkehrsgesetz *n* traffic law
Straßenverkehrssicherheit *f* road safety
Straßenverkehrsunfall *m* road accident
Stratosphäre *f* stratosphere
Streß *m* stress
Streuungswinkel *m* angle of scattering
Strippung *f* stripping
Strohlehm *m* straw clay
Strohleichtlehm *m* straw light-clay
Strohverbrennung *f* straw combustion
Strohverwertung *f* straw utilization
Strombelastung *f* power load
Stromerzeugung *f* generation of current
Stromkreis *m* electric circuit
Stromleitung *f* power line
Stromquelle *f* power source
Stromschwankung *f* current fluctuation
Strömung *f* flow, current
Strömungslehre *f* hydrodynamics
Strömungsrichtung *f* direction of flow
Stromverbraucher *m* power consumer
Strontium *n* strontium
Struktur-Funktions-Beziehung *f* structure-funktion relationship
Strukturanalyse *f* structural analysis
Strukturchemie *f* structural chemistry
Struktureinheit *f* structural unit
Strukturelement *n* structural element

strukturelle Integrität *f* structural integrity
strukturempfindlich structure-sensitive
strukturlos structureless
Strukturschema *n* configuration
Strukturveränderung *f* structural change
Stufenkaskade *f* square cascade
Stufenreaktion *f* step reaction
Stützbalken *m* brace
Styrol *n* styrene
subakute Toxizität *f* subacute toxicity
subchronische Studie *f* **an Tieren** subchronical study on animals
subchronische Wirkung *f* subchronic effect
subkutan subcutaneous
subletale chronische Konzentration *f* sublethal chronic concentration
subletale Toxizität *f* sublethal toxicity
Sublimation *f* sublimation
Substanzaufnahme *f* uptake of a substance
substanzbezogene Wirkung *f* substance-related effect
substanzbezogener Effekt *m* substance-related effect
Substanzklasse *f* substance group
Substanzmenge *f* amount of substance
substanzspezifische Reaktion *f* substance-specific reaction
Substanzverlust *m* loss of material
Substitution *f* substitution, replacement
Substitutionsreaktion *f* replacement reaction, substitution reaction

Substrat *n* substrate
Substrathemmung *f* substrate inhibition
subtherapeutische Dosis *f* subtherapeutical dose
subtropisch subtropical
subzellulare Lokalisierung *f* subcellular localization
Suchstrategie *f* search strategy
Sucht *f* addiction, mania, passion, rage
Sulfid *n* sulphide
Sulfit *n* sulphite
Summenformel *f* total molecular formula
Summenparameter *m* sum[mation] parameter
Sumpf *m* marsh, swamp, bog, mire
Sumpfgas *n* marsh gas
sumpfig marshy, boggy, miry
sumpfiger Flußarm *m* bayou
Sumpfland *n* swamp land
Sumpfpflanze *f* marsh plant
supraflüssig superfluid
supraflüssiger Zustand *m* superfluidity
Supraflüssigkeit *f* superfluidity
supraleitend superconducting
supraleitfähig superconducting
suspendieren suspend/to
Suspension *f* suspension
Suszeptibilität *f* susceptibility
Süßwasserökosystem *n* freshwater ecosystem
Symbiose *f* symbiosis
Symmetriezentrum *n* centre of symmetry
symmetrischer Peak *m* symmetrical peak
Symptom *n* symptom
Symptomatik *f* symptomatics
symptomatisch symptomatic
Symptomkombination *f* symptom combination
Synergismus *m* synergism
Synergist *m* synergist
synoptisches Modell *n* synoptic model
Synthese *f* synthesis
synthetisches Harz *n* synthetic resin
synthetisches Rohöl *n* syncrude
synthetisieren synthesize/to
Synökologie *f* synecology
Synökosystem *n* synecosystem
Systemanalyse *f* system analysis
systematisch systematic
Systementwicklung *f* system development
systemisch systemic
Szintillationszähler *m* scintillation counter

T

Tablettenidentifizierungsprogramm *n* tablet identification programme
Tablettiermaschine *f* tabletting machine
Tablettierung *f* tabletting
Tagebau *m* opencast mining, open workings
Tagebaubetrieb *m* quarrying
Tagebauförderung *f* quarrying
Tagebaugewinnung *f* opencast working
Tagesganglinie *f* daily output curve, daily load curve
tägliche orale Dosis *f* daily oral dose
tägliche Schwankung *f* daily fluctuation

täglicher Austausch *m* daily exchange
Talg *m* tallow
talgartig tallow-like
talgig sebaceous
Talk *m* talcum, talc
Talsperre *f* barrage, riverdam
Talsperrenbelüftung *f* barrage aeration, dam aeration
Talsperrenwasser *n* barrage water
Tang *m* seaweed; tang
Tank *m* tank, container, storage vessel
Tankanstrich *m* tank coating
Tankbiologie *f* tank biology
Tankcontainer *m* tank
Tanker *m* tanker
Tankfahrzeug *n* tank truck
Tankschiff *n* tanker
Tankstelle *f* filling station, petrol station, service station
Tankwagen *m* tank waggon
Tantal *n* tantalum
Taschenfilter *n* bag filter
Tau *n* dew; rope
Tauchbelüfter *m* immersion aerator
Tauchdekontamination *f* immersion decontamination
Tauchelektrode *f* immersion electrode
Tauchpumpe *f* immersion pump, submerged pump
Tauchrohr *n* immersion stem
Taupunkt *m* dew point, saturation temperature
Technetium *n* technetium
Technikakzeptanz *f* acceptance of technology
technische Aufbereitung *f* technological processing

technische Maßnahme *f* technical measure
technische Umweltchemikalie *f* technical environmental chemical
technischer Umweltschutz *m* technical environmental protection
technischer Umweltschutz *m* technological environmental protection
Technologietransfer *m* technology transfer
Teer *m* tar
Teerabscheider *m* tar extractor
Teeranstrich *m* coat of tar
teerartig tarry
Teerasphalt *m* tar asphalt
Teerband *n* tarred tape
Teerbildung *f* tar formation
Teeren *n* tarring
Teergehalt *n* tar content
Teergrube *f* tar pit
Teerpappe *f* tarred board
Teerpech *n* tar pitch
Teich *m* pond
Teichbelüftung *f* pond aeration
Teilchen *n* particle
Teilchenbeschleuniger *m* particle accelerator
Teilchenbeschuß *m* particle bombardment
Teilchendurchmesser *m* particle diameter
Teilchenexplosion *f* particle explosion
Teilchenform *f* particle shape
Teillast *f* partial load
Teilverdampfung *f* partial evaporation
teilverträglich partially compatible
Telekommunikation *f* telecommunication
Telemeter *m* telemeter

Tellur *f* tellurium
Temperaturabfall *m* temperature drop
Temperaturabhängigkeit *f* temperature dependence
Temperaturgradient *m* temperature gradient
temperaturprogrammierte Analyse *f* temperature-programmed analysis
tempern temper/to, anneal/to
Temperverfahren *n* annealing process
temporäre Härte *f* temporary hardness
Tensid *n* tenside
teratogene Wirkung *f* teratogenic effect
teratogener Effekt *m* teratogenic effect
teratogenes Risiko *n* teratogenic risk
Terbium *n* terbium
Termite *f* termite
Terrain *n* terrain
Terrasse *f* terrace
terrestisches Feldsystem *n* terrestrial field system
terrestrische Vegetation *f* terrestric vegetation
terrestrisches Modellökosystem *n* terrestric model ecosystem
terrestrisches System *n* terrestric system
Territorialgewässer *npl* territorial waters
Test *m* test, assay
Testaerosol *n* test aerosol
Testdaten pl test data
testen test/to
Tester *m* tester

Testkörper *m* test body, test piece
Testmethode *f* test method
Testschadstoff *m* test pollutant
Teststreifen *m* test strip
Testsystem *n* test system
Tetanus *m* tetanus
Tetanusbazillus *m* tetanus bacillus
Tetanusschutzimpfung *f* tetanus inoculation
Textilabfall *m* textile waste
Textilfaser *f* textile fibre
textilindustrielles Abwasser *n* textile industrial waste water
Thallium *n* thallium
Thallophyten *mpl* thallophytes
theoretische Chemie *f* theoretical chemistry
Thermalquelle *f* thermal spring
Thermalschwefelbad *n* thermal sulphur bath
Therme *f* thermal spring
thermionisch thermionic
thermisch thermal
thermische Abluftreinigung *f* thermal purification of exhaust air
thermische Behandlung *f* thermal treatment
thermische Entsorgung *f* thermal disposal
thermische Konvertierung *f* thermal conversion
Thermoanalyse *f* thermal analysis
thermobiotisch thermobiotic
Thermochemie *f* thermochemistry
thermochemisch thermochemical
Thermodiffusion *f* thermal diffusion, thermodiffusion
Thermodynamik *f* thermodynamics
thermoelastisch thermoelastic
thermokinetisch thermokinetic
Thermometer *n* thermometer

thermophile Faulung *f* thermophilic fouling (digestion)
thermophysikalisch thermophysical
thermophysikalische Eigenschaft *f* thermophysical property
Thermoplast *m* thermoplastic
thermoplastisch thermoplastic
Thermosflasche *f* vacuum bottle, thermos bottle
Thermospannung *f* thermoelectric power
thermostabil thermoresistant
thixotroper Schlamm *m* thixotropic sludge
Thymusdrüse *f* thymus gland
Thyroxin *n* thyroxine
Tide *f* tide
Tiefbrunnen *m* deep well
Tiefenbelüftungssystem *n* deep-ventilation system
Tiefenfiltration *f* deep-bed filtration
Tiefenmesser *m* sea gauge
Tiefenwasser *n* hypolimnic water
Tiefenwirkung *f* penetration effect
Tiefkühleinrichtung *f* deep-freeze installation
Tiefkühlung *f* deep cooling
Tiefsee *f* deep sea
Tiefseedeponierung *f* deep sea dumping
Tiegel *m* crucible, pan
Tierexperiment *n* animal experiment
tierexperimenteller Befund *m* animal experimentation finding, finding of animal experiments
Tierfett *n* animal fat
Tierfutter *n* animal feed
Tierklinik *f* veterinary hospital
Tierkörper *m* animal carcass
Tiermagen *m* maw
Tiermedizin *f* veterinary medicine

Tierparasit *m* zooparasite
Tierschutz *m* animal protection
Tinktur *f* tincture
Titan *n* titanium
Titer *m* titre
Titersubstanz *f* titrant
Titerwert *m* titre value
Titrans *n* titrant
Titration *f* titration
Titrationsanalyse *f* volumetric method
titrieren titrate/to
Titrierung *f* titration
Titrimetrie *f* titrimetric analysis, titrimetry, volumetry
Tochtergeschwulst *n* metastasis
tödliche Dosis *f* lethal dose
tödliche Menge *f* lethal dose
Tokologie *f* tocology
tolerabel tolerable
Toleranz *f* tolerance
Toleranzbereich *m* allowable limits
Toleranzdosis *f* threshold dose, tolerance dose
Toleranzgrenze *f* control limit
Toleranzwert *m* tolerance value
tolerieren tolerate/to
Tollkirschenextrakt *m* extract of belladonna
Tollwut *f* rabies
Tomographie *f* tomography
Ton *m* clay
tonartig argillaceous
Tonboden *m* clay soil
Tonentsorgung *f* clay disposal
Tonerde *f* clay
Tonfilter *n* clay filter
Tonfrequenz *f* sound frequency
tonhaltig argillaceous
tonimmobilisiertes Pestizid *n* clay-immobilized pesticide

Tonmaterial *n*   argillaceous material
Tonmehl *n*   clay powder
Tonschicht *f*   clay layer
Topfzeit *f*   can life, pot time
Topographie *f*   topography
Torf *m*   peat
Torfbrikett *n*   peat briquet
Torfgas *n*   peat gas
Torfplatte *f*   peat plate
Torf-Rinden-Kompost *m*   muck-bork compost
Torsionsbeanspruchung *f*   torsion stress
Totgeburt *f*   stillbirth
Totgeburtenrate *f*   stillbirth rate
Totvolumen *n*   stagnant volume
Totwasser *n*   dead water
toxikodynamisch   toxicodynamic
Toxikokinetik *f*   toxicokinetics
Toxikologie *f*   toxicology
toxikologisch   toxicological
toxikologisch relevanter Stoff *m*   toxicological relevant substance
toxikologische Chemie *f*   toxicological chemistry
Toxikum *n*   toxicant
Toxin *n*   toxin
toxisch   toxic, poisonous
toxisch belastetes Abwasser *n*   toxic impacted waste water
toxisch wirkender Stoff *m*   toxicant
toxische Reaktion *f*   toxic reaction
toxische Verbindung *f*   toxicant
toxische Wirkung *f*   toxic effect
toxischer Abfall *m*   toxic waste
toxisches Abbauprodukt *n*   toxic degradation product
toxisches Intervall *n*   toxic interval
toxisches Potential *n*   toxic potential

toxisches Äquivalent *n*   toxic equivalent dose
Toxizität *f*   toxicity
Toxizitätsgrenze *f*   toxicity treshold
Toxizitätsmesssung *f*   toxicity measurement
Toxizitätsprüfung *f*   toxicity testing
Trabantenstadt *f*   satellite town
Trächtigkeit *f*   gestation
tragbares pH-Meter *n*   portable pH-meter
Trägergas *n*   carrier gas
Trägergasleitung *f*   carrier gas line
Trägerkatalysator *m*   supported catalyst
Trägersubstanz *f*   carrier substance
Trägersubstanz *f*   vehicle
Tragfähigkeit *f*   bearing strength
Trägheit *f*   inactivity
Trägheitskraftabscheider *m*   inertial force separator
trägheitslos   inertialless
Trakt *m*   tract
Tränendrüse *f*   lachrymal gland, lacrimal gland
tränken   soak/to, saturate/to, steep/to; water/to
Tranquilizer *m*   tranquilizer
Transfer *m*   transfer
Transformator *m*   transformer
Transformatorenöl *n*   transformer oil
transformieren   transform/to
Transmission *f*   transmission
Transplantation *f*   transplantation
transplantieren   transplant/to
Transport *m*   shipment, transp[ation]
Transport *m* **gefährlicher Stoffe**   transport of dangerous substances
transportabel   transportable

Transportband *n* belt conveyor
Transporteinheit *f* transportation unit
transportfähig transportable
transportieren transport/to, move/to
Transportkontrolle *f* transport control
Transportzulassung *f* transport permission
Treiben *n* drift
Treibgas *n* aerosol propelling gas, propellant
Treibhaus *n* greenhouse
Treibhauseffekt *m* greenhouse effect
Treibhausgas *n* greenhouse gas
Treibmittel *n* blowing agent, expanding agent, propellant
Treibsand *m* quicksand, shifting sand, silt
Treibstoff *m* fuel, petrol
Trendkontrolle *f* trend control
Trenndüse *f* jet separator
Trennen *n* dividing, partition, separation, separating
trennen divide/to, part/to, separate/to
Trennleistung *f* separation efficiency
Trennlinie *f* parting line, separating line
Trennmittel *n* parting agent
Trennschicht *f* interlayer
Trennschritt *m* separating stage
Trennschärfe *f* selectivity, separation efficiency
Trennsystem *n* separating system
Trennsäule *f* separating column, separation column
Trennung *f* dissolution, parting, separation

Trennung *f* **durch Schwerkraft** gravitational separation
Trennungsfläche *f* interface
Trennungsschicht *f* separating layer
Trennverfahren *n* separation process
Trichter *m* funnel
Triebkraft *f* driving force, impuls
Triebwerk *n* drive, gear
trinkbar drinkable, potable
Trinkwasser *n* drinking water, potable water
Trinkwasseraufbereitung *f* drinking water preparation (processing)
Trinkwasserbiologie *f* drinking water biology
Trinkwasserfluoridierung *f* drinking water fluoridizing
Trinkwassergewinnungsanlage *f* drinking water plant
Trinkwasserproduktion *f* drinking water production
Trinkwasserqualität *f* drinking-water quality
Trinkwasserschutzgebiet *n* drinking (potable) water protection area
Trinkwasserreservoir *n* drinking-water reservoir
Trinkwassertransportsystem *n* drinking water transport system
Trinkwasserverordnung *f* drinking water decree
Trinkwasserverseuchung *f* drinking water contamination
Trinkwasserversorgung *f* potable water supply
Trinkwasserwerk *n* drinking waterworks
Tripelpunkt *m* triple-point
Trittschall *m* running noise

Trittschalldämmung *f* running noise insulation
Trittschallpegel *m* running noise level
Trockenablagerung *f* dry deposition
Trockenbatterie *f* dry battery
Trockenbeet *n* drying bed
Trockenbestandteil *m* solid constituent
Trockenblech *n* tray
Trockendeposition *f* dry deposition
Trockendestillation *f* dry distillation
Trockenentgasung *f* dry distillation
Trockengebiet *n* dry area
Trockenhefe *f* yeast powder
Trockenheit *f* dryness
Trockenkammer *f* stove, drying chamber
trockenlegen impolder/to
Trockenlegung *f* drainage
Trockenmauer *f* dry wall
Trockenmittel *n* siccative
Trockenreinigungsanlage *f* dry purification plant
Trockenschale *f* tray
Trockenschlammdeponie *f* dry sludge disposal site
Trockenschleifen *n* dry grinding
Trockenschleuder *f* drying centrifuge
Trockenschrank *m* drying cupboard
Trocknung *f* drying
Trocknungsanlage *f* drying equipment
Trocknungsenergie *f* drying energy
Trocknungskreislauf *m* drying circuit
Trocknungsmittel *n* drying agent
Trocknungsverfahren *n* drying process

Trommeldrehfilter *n* rotary drum filter
Tropen *f* the tropics
Tropenfestigkeit *f* tropical stability
Tropenklima *n* tropical climate
Tropenmedizin *f* tropical medicine
Tropenwald *m* tropical forest
Tropfenabscheider *m* drop separator, mist eliminator
Tropfflasche *f* dropping bottle
Tropfkörper *m* trickling filter, biological film
tropisch tropical
tropischer Regenwald *m* tropical rain forest
Troposphäre *f* troposphere
Trübung *f* turbidity, cloudiness
Trübungsmessung *f* turbidity measurement
Trübungsmeßeinrichtung *f* turbidimeter, nephelometer
Trübungsmeßgerät *n* nephelometer
Trübungspunkt *m* cloud point
Tuberkulose *f* tuberculosis
Tumor *m* tumour
Tumorbildung *f* tumourigenesis
Tumordosis *f* tumour dose
Tumorlokalisierung *f* tumour localization
Tumorwachstum *n* growth of the tumour
Tumorzelle *f* tumour cell
Tümpel *m* pond
Turbogebläse *n* turbo blower
turbulente Strömung *f* turbulent flow
Turbulenz *f* turbulence
Turmbiologie *f* tower biology
Turmsäure *f* tower acid
Typhus *m* typhus
Typhusbakterien *fpl* typhoid bacilli

U

übelriechend malodorous, evil-smelling, ill-smelling
Überalterung *f* supcrannuation
überbeanspruchen overstrain/to
Überbeanspruchung *f* overstress
Überbehandlung *f* overtreatment
Überbelastung *f* overloading
Überbevölkerung *f* overpopulation
überdosieren overdose/to
Überdosierung *f* overdosage
Überdruck *m* excess pressure, overpressure
Überdruckventil *n* pressure relief valve, blow off valve
Überdüngung *f* eutrophication
überempfindlich allergic, oversensitive
Überempfindlichkeit *f* allergy, hypersensitivity, ultrasensitivity
Überernährung *f* overfeeding, superalimentation
überfließen flow over/to, overflow/to
Überfluten *n* flooding
Überfunktion *f* hyperfunction, overactivity
Überfüllsicherung *f* overfilling protection
Überfütterung *f* overfeeding
Übergangsbestimmung *f* interim regulation
Übergangslösung *f* provisional solution
Übergangsmetall *n* transition metal
Übergangsregelung *f* temporary arrangement
Übergangstemperatur *f* transition temperature
Übergewicht *n* overweight

überhitzen overheat/to, superheat/to
überkritisch supercritical
überkritische Flüssigkeit *f* supercritical fluid
überkritischer Druck *m* supercritical pressure
überladen overload/to, overburden/to, supercharge/to
Überladung *f* overloading
Überlagerung *f* overlap, superimposition
Überlandleitung *f* landline
Überlappung *f* overlap
überlasten stress/to
Überlauf *m* spillway
überlaufen overflow/to
Überlaufkanal *m* overflow drain
Überlaufkontrolle *f* spill control
Überlaufwehr *n* overflow dam
Überleben *n* survival
Überlebenschance *f* survival chance
Überlebenswahrscheinlichkeit *f* survival probability
Überlebenszeit *f* survival time
übermäßig nährstoffreich polytrophic
Übernahmeprüfung *f* acceptance test
Übersättigung *f* oversaturation
Übersäuerung *f* hyperacidity, overacidification, superacidity
Überschußdampf *m* excess vapour
Überschußschlamm *m* excess sludge
Überschwemmung *f* flood
Übersichtsuntersuchung *f* gross examination
Überstandswasser *n* excess water
überströmen flow over/to, overflow/to

Übertage-Sonderabfalldeponie *f* aboveground special waste dump
übertragbare Krankheit *f* communicable disease
überwachen monitor/to, observe/to, supervise/to, watch/to
Überwacher *m* monitor
Überwachung *f* monitoring, observation, supervision, surveillance
Überwachung *f* **aus niedriger Höhe** low-altitude survey
Überwachungsbehörde *f* surveillance office
Überwachungsmaßnahme *f* control measure
Überwachungspflicht *f* surveillance obligation
Überweidung *f* overgrazing
Überzug *m* film, overlay
ubiquitär ubiquitous
ubiquitäre Konzentration *f* ubiquitous concentration
ubiquitärer Stoff *m* ubiquitous substance
ubiquitäres Auftreten *n* ubiquitous occurence
Ubiquität *f* ubiquity
Ufer *n* bank; shore
Uferausbildung *f* bank formation
Uferbefestigung *f* bank stabilization
Uferböschung *f* embankment
Ufererosion *f* shore erosion
Uferfiltration *f* bank filtration
ufernah near-shore
Ultrafiltration *f* ultrafiltration
Ultrafiltrationsrohr *n* ultrafiltration tube
ultrafiltrierter Probenstrom *m* ultrafiltered sample flow
ultrarein ultrapure
Ultraschall *m* ultrasonics

Ultraschallbehandlung *f* ultrasonic wave treatment
Ultraschallgenerator *f* ultrasonic generator
Ultraschallsensor *m* ultrasonic sensor
ultraviolett ultraviolet
Ultraviolettspektrum *n* ultraviolet spectrum
Ultraviolettstrahlung *f* ultraviolet radiation
Ultrazentrifuge *f* ultracentrifuge
Umbau *m* reconstruction
umbauen rebuild/to, reconstruct/to
umbilden transform/to
Umgangsgenehmigung *f* handling licence
Umgebungsdruck *m* ambient pressure
Umgebungseinfluß *m* influence of the surroundings
Umgebungsempfindlichkeit *f* environmental sensibility
Umgebungsluft *f* ambient air, ambient atmosphere
Umgebungstemperatur *f* ambient temperature
Umgehung *f* bypassing
umhüllen wrap/to, jacket/to
Umhüllung *f* wrapping, jacket
umkehrbar reversible
Umkehrbarkeit *f* reversibility
Umkehrosmose *f* reverse (reversed) osmosis
Umkehrung *f* reversal
Umladeanlage *f* reloading plant
umlagern rearrange/to
Umlagerung *f* rearrangement
Umlauf *m* cycle
Umlaufkühlwasser *n* circulating cooling water

Umlaufpuffer *m* cyclic buffer, circulating buffer
Umlaufpumpe *f* circulating pump
Umlaufreaktor *m* circulating reactor
Umlaufschmierung *f* circulating system lubrication
Umlauftrockner *m* rotary dryer
Umlaufverdampfer *m* circulating air evaporator
Umlaufwärmetauscher *m* circulation heat exchanger
Umlenkabscheider *m* diverting separator
Umluft *f* circulating air
Umluftförderanlage *f* circulating air conveyor
Umluftverfahren *n* air circulation system
Umordnung *f* rearrangement
Umpflanzung *f* transplantation, retransplantation
umrühren agitate/to, stir/to
umschalten switch/to
Umschlagszeit *f* shelf time
umschmelzen recast/to, remelt/to
Umsetzungsprozeß *m* conversion process
Umspanner *m* transformer
Umwälzleistung *f* circulating performance
Umwälzpumpe *f* recirculation pump
umwandeln convert/to, rearrange/to, transmute/to
Umwandlung *f* conversion, transformation, change, metamorphosis, mutation
Umwandlungsprodukt *n* reaction product
Umwandlungsreaktion *f* transformation reaction
Umwandlungstemperatur *f* transition temperature
Umwelt *f* environment
Umweltabkommen *n* environmental agreement
Umweltanalytik *f* environmental analytics
Umweltauswirkung *f* environmental effect
umweltbedingtes Konzentrationsniveau *n* environmental concentration level
Umweltbeeinflussung *f* environmental impact
Umweltbelastung *f* environmental burden, environmental impact
Umweltbewußtsein *n* ecological awareness, environmental sensibility
Umweltchemikalie *f* environmental chemical
Umweltengel *m* environmental angel (label)
umweltfeindlich environmentally harmful
Umweltforschung *f* environmental research
Umweltfragen *fpl* environmental matters
umweltfreundlich environmetally beneficial
umweltgefährdend environmentally endangering
umweltgefährdende Abfallbeseitigung *f* environmental endangering waste disposal
umweltgefährdende Altlast *f* environmental endangering old-site
umweltgefährlich dangerous to the environment
umweltgerecht environmentally

acceptable, environmentally compatible (neutral)
umweltgerechte Verbrennung *f* environmentally acceptable incineration
umweltgerechtes Herstellungsverfahren *n* environmentally compatible manufacturing process
Umweltgesetzgebung *f* environmental legislation
Umwelthysterie *f* environmental hysteria
Umweltkompartiment *n* environmental compartment
Umweltkriminalität *f* environmental crime
Umweltkrimineller *m* environmental criminal
Umweltmedium *n* environmental medium
umweltökonomisch environmental economical
umweltorientiert environmental oriented
umweltorientierter Schwerpunkt *m* environmentally oriented priority
Umweltpolitik *f* environmental policy
umweltpolitische Kompetenz *f* environmental political competence
umweltpolitische Maßnahme *f* environmental political measure
umweltpolitische Regelung *f* environmental political provision
Umweltpolizei *f* environmental policy
Umweltprobe *f* environmental sample
Umweltprogramm *n* environmental programme

Umweltrelevanz *f* environmental relevance
Umweltrisiko *n* environmental risk
Umweltschaden *m* environmental damage
umweltschonende Abfallentsorgung *f* environmentally acceptable waste disposal
umweltschonende Technologie *f* environmental preserving technology
Umweltschutz *m* environmental protection
Umweltschutzaspekt *m* environmental aspect
Umweltschutzaufwendungen *fpl* expenditure for environmental protection
Umweltschutzerfordernis *n* environmental need
Umweltschutzinvestitionen *fpl* investments for environmental protection
Umweltschutzorganisation *f* organization for environmental protection
Umweltschutzpapier *n* recycling paper
Umweltschutzplanung *f* planning of environmental protection
Umweltschutzpreis *m* environmental protection award
umweltschädigender Stoff *m* environmental damaging substance
Umweltsimulationstechnik *f* environmental simulation technology
Umweltsimulationsverfahren *n* environmental simulation method
Umweltsituation *f* environmental situation
Umweltsteuer *f* environmental tax
Umwelttechnik *f* environmental engineering (technology)

Umwelttechnikmarkt *m* environmental technology market
Umwelttechnologie *f* environmental technology
Umweltverbrechen *n* environmental crime
Umweltverbrecher *m* environmental criminal
Umweltverhalten *n* environmental behaviour
Umweltverschmutzer *m* polluter
umweltverträglich environmetally acceptable
umweltverträgliche Alternative *f* environmentally acceptable alternative
umweltverträgliche Entsorgung *f* environmentally compatible disposal
umweltverträgliche Technologie *f* environmetally acceptable technology
umweltverträgliches Produkt *n* environmentally compatible product
Umweltüberwachungsprogramm *n* environmental monitoring programme
Umweltverträglichkeit *f* environmental compatibility
Umweltverträglichkeitsprüfung *f* environmental compatibility test
Umweltzeichen *n* environmental sign
umweltzerstörend ecocidal
unabhängige Analysenmethode *f* independent analysis method
unaufgelöst undissolved, unresolved
unauflösbar unsolvable
unausrottbar ineradicable
unbeabsichtigte Exposition *f* accidental exposure

unbebaubar uncultivable
unbebaut untilled, waste
Unbedenklichkeit *f* absolute safety
Unbedenklichkeitsbescheinigung *f* security clearance
Unbedenklichkeitsschwelle *f* limit of absolute safety
unbegrenzt haltbar keep for an indefinite period
unbekannte Zusammensetzung *f* unknown composition
unbelebt inanimate, lifeless
unbelebte Natur *f* inanimate nature
Unbelebtheit *f* inanimation
unbelüftete Zone *f* unaerated zone
unbemannte Kläranlage *f* unmanned sewage treatment plant
unbeständig inconstant, labile, transient
Unbeständigkeit *f* inconstancy
unbrennbar non-flammable
undefinierbar indefinable
undicht leaking, leaky, untight
Undichtheit *f* leakiness
Undichtigkeit *f* leak, leakiness
undurchdringlich impenetrable, impermeable
undurchlässig impermeable, proof, tight
unempfindlich insensitive
unerprobt untried
unerschlossen undeveloped
unersetzlich irretrievable
unerwünschter Effekt *m* undesired effect
Unfall *m* accident
Unfalldiagnose *f* accident diagnosis
Unfallgefahr *f* accident risk
Unfallmerkblatt *n* accident instruction sheet

Unfallstation *f* accident (emergency) ward, casualty (ambulance) station
Unfallverhütung *f* prevention of accidents
Unfallverhütungsprogramm *n* accident prevention programme
unfiltrierbar non-filterable
unfruchtbar infertile, infructuous, sterile, unfruitful
Unfruchtbarkeit *f* infertility, unfruitfulness
Unfruchtbarmachung *f* sterilization
ungefiltert unfiltered
ungefährlich harmless, safe
ungeklärt raw
ungekocht raw, uncooked
ungelüftet unventilated
ungelöst undissolved
ungenießbar unpalatable
ungeordnete Müllkippe *f* disordered refuse dump
ungeordneter Zustand *m* disordered state
ungeschützt uncovered, unguarded, unprotected
Ungeschütztheit *f* vulnerability
ungesund unhealthy, unsound
ungesättigt unsaturated
ungetestet untested
Ungeziefer *n* vermin, pest
Ungleichgewicht *n* disequilibrium
unheilbar incurable, irreparable
Unheilbarkeit *f* incurability
unhygienisch unhygienic, insanitary
unkontrollierter Deponiegasaustritt *m* uncontrolled gas leakage of disposal sites
Unkraut *n* weed
Unkrautbekämpfung *f* weed control

Unkrautvernichtungsmittel *n* herbicide
unkultivierbar uncultivable
Unland *n* fallow [land]
unmischbar immiscible
unpasteurisiert unpasteurized
unrein unclean, impure
Unreinheit *f* contamination
unsauber unclean, impure
unschmackhaft unpalatable
unschädlich harmless, innoxious
Unschädlichkeit *f* harmlessness
Unschärferelation *f* uncertainty relation
Unstetigkeitsstelle *f* discontinuity position
unsystemisch non-systemic
unterbevölkert underpopulated
Unterbrechungsbad *n* stop bath
Unterdruckkammer *f* vacuum chamber
Untereinheit *f* subunit
Unterentwicklung *f* dysplasia, underdevelopment
Unterernährung *f* hypoalimentation, malnutrition, subnutrition, undernourishment, undernutrition
Unterfamilie *f* subfamily
Untergewicht *n* underweight
Untergrund *m* underground
Untergrundverseuchung *f* underground contamination
Untergruppe *f* section, subgroup
unterirdisch underground
unterkritisch subcritical
unterkühlen supercool/to
Unterrichtungspflicht f information duty
Untersuchung *f* **nach dem Tode** post-mortem examination

Untersuchung *f* **wird durchgeführt** study is performed
Untersuchungsrahmen *m* frame of a study
Untertagebau *m* underground mining (working)
Untertagedeponie *f* underground waste disposal
Untertage-Sonderabfalldeponie *f* underground special waste dump
Untertagespeicherung *f* underground storage
Untertagevergasung *f* underground gasification
untertägige Ablagerung *f* underground deposition
Unterwassereinkapselung *f* **von Abfall** underwater encapsulation of waste
Unterwasserlager *n* submerged bearing
untrennbar inseparable
unumkehrbar irreversible, non-reversible
unverbleit unleaded
unverbraucht unreacted
unverbrennbar incombustible, non-combustible
Unverbrennbarkeit *f* incombustibility
unverdünnt undiluted
unvereinbar incompatible
Unvereinbarkeit *f* incompatibility
unverfälscht genuine, undiluted
unvermeidbarer Abfall *m* unavoidable waste
unvermeidbares Risiko *n* unavoidable risk
unvermischbar immiscible
Unvermischbarkeit *f* immiscibility
unvermischt absolute, pure, unalloyed, virgin
unverrottbar unrottable
unverrottbar rot-proof
unverseucht uncontaminated
unverträglich incompatible
Unverträglichkeit *f* incompatibility
unvollständige Verbrennung *f* incomplete (partial) combustion, restricted combustion
unwirksam inactive, ineffective, inefficient
Unwirksamkeit *f* inactivity, inefficiency
unwirtschaftlich uneconomic
unwissenschftlich unscientific
Urämie *f* uraemia
Uran *n* uranium
urbar machen cultivate/to
Urbarmachung *f* reclamation, cultivation
Urin *m* urine
Urwald *m* jungle, primeval forest
UV-Bestrahlungseinheit *f* UV radiation unit
UV-Entkeimungsanlage *f* UV disinfection plant
UV-Strahlung *f* UV radiation

V

Vakuole *f* vacuole
Vakuumbandfilter *n* vacuum belt filter
Vakuumdestillation *f* vacuum distillation
vakuumdicht vacuum-tight
Vakuumfett *n* vacuum grease
Vakuumfilter *n* vacuum filter

Vakuumfiltration *f* vacuum filtration
Vakuumflotation *f* vacuum flotation
Vakuumkehrmaschinenbesen *m* vacuum broom sweeper
Vakuumofen *m* vacuum furnace
Vakuumpumpe *f* vacuum pump
vakzinieren vaccinate/to
Valorisation *f* valorization
Vanadium *n* vanadium
Vandalismus *m* vandalism
Variationskoeffizient *m* coefficient of variation
Vegatationsgebiet *n* vegetation area
Vegetarier *m* vegetarian
Vegetation *f* vegetation
Vegetationsgeographie *f* vegetation geography
vegetationsloses Land *n* badland[s]
Vegetationsperiode *f* vegetation period
Vegetationszeit *f* vegetative period
vegetativ vegetative
vegetieren vegetate/to
Ventilation *f* ventilation
Ventilator *m* fan
Ventilatorantrieb *m* fan drive
ventilieren ventilate/to
Venturiwäscher *m* venturi scrubber
Veranlagung *f* habitude
Verarbeitbarkeit *f* processability
Verarbeitungsbedingung *f* operational condition
Verarbeitungstemperatur *f* process temperature
veraschen incinerate/to
Veraschungsschälchen *n* incineration dish
Verbandskasten *m* first-aid box
Verbleib *m* **von Chemikalien** fate of chemicals
verbleien lead/to
Verbrauch *m* consumption
Verbraucher *m* consumer
Verbrauchererorganisation *f* consumer organization
Verbraucherverband *m* consumer cooperative
Verbrauchsprognose *f* consumption forecast
verbrennen burn/to, incinerate/to
Verbrennung *f* burning, combustion, incineration
Verbrennung *f* **auf See** ocean combustion, ocean incineration
Verbrennungsaggregat *n* incineration aggregate
Verbrennungsanlage *f* incineration plant
Verbrennungsanlage *f* **für Krankenhausmüll** incineration plant for clinical waste
Verbrennungsgeschwindigkeit *f* combustion rate
Verbrennungskammer *f* combustion chamber
Verbrennungsluftvorwärmung *f* incineration air preheating
Verbrennungsrückstand *m* combustion residue, incineration residue
Verbrennungssystem *n* incinerator system
Verbundglas *n* safety glass
Verdachtsflächenbewertung *f* valuation of areas of suspicion
verdampfbar evaporable, vaporizable
Verdampfbarkeit *f* vaporizability
verdampfen evaporate/to, vaporize/to
Verdampfer *m* vaporizer

Verdampfung 314

Verdampfung *f* evaporation, vaporization
Verdampfungsenthalpie *f* evaporation enthalpy
Verdampfungsgleichgewicht *n* equilibrium of evaporation
Verdampfungskurve *f* evaporation curve
Verdampfungsreaktor *m* boiling-water reactor
Verdaulichkeit *f* digestibility
Verdauungskanal *m* alimentary canal
Verdauungsorgan *n* digestive organ
Verdauungstrakt *m* alimentary tract, digestive tract, intestinal tract
Verderb *m* deterioration, decay, ruin, spoilage
verderben deteriorate/to, ruin/to, spoil/to
verdichten compact/to, compress/to, concentrate/to
Verdichtungsleistung *f* compacting capacity
verdicken thicken/to
Verdickungsmittel *n* thickening agent
verdorren wither/to, dry up/to
Verdrängungsreaktion *f* replacement reaction
Verdunstbarkeit *f* vaporizability
verdunsten evaporate/to, vaporize/to
Verdunstung *f* evaporation, vaporization
Verdunstungsintensität *f* intensity of evaporation
Verdunstungskühlturm *n* evaporative cooling tower
verdünnen dilute/to
Verdünner *m* thinner

Verdünnung *f* dilution, diluting
Verdünnungseffekt *m* dilution effect
Verdünnungsgesetz *n* diluting law
Verdünnungsmittel *n* diluent, diluting agent, thinner
Verdünnungswärme *f* diluting heat
veredeln process/to, refine/to
Veredelung *f* processing
Vererbung *f* heredity
Verfahrenskombination *f* method combination
Verfahrenstechnik *f* process engineering
verfahrenstechnische Auslegung *f* process dimensioning (design, rating)
verfahrenstechnisches Grundelement *n* basic process element
verfahrentechnische Realisierung *f* process realization
Verfall *m* decay, decline, deterioration
verfallen decay/to
verfaulen rot [away]/to, decay/to, putrefy/to
verfestigen solidify/to
Verfestigung *f* solidification
Verflüchtigung *f* evaporation, sublimation, volatilization
Verflüssigung *f* liquefaction
Verflüssigungsapparat *m* liquefier
Verfügbarkeit *f* availability
vergasbar vaporizable
vergasen gasify/to, vaporize/to
Vergasung *f* gasification
Vergeudung *f* waste
vergiften poison/to, intoxicate/to, toxify/to
vergiftet poisoned
Vergiftung *f* poisoning, intoxication

Vergiftungsdiagnose *f* diagnosis of poisoning
Vergiftungsursache *f* cause of poisoning
Vergiftungsverdacht *m* suspicion of poisoning
Vergiftungzeichen *n* symptom of poisoning
verglasen vitrify/to
Verglasung *f* vitrification, glazing
Vergleichsprobe *f* control sample
Vergrößerungsfaktor *m* magnification factor
Verqußmasse *f* sealing compound
Vergällung *f* denaturation
Vergärung *f* fermentation
Verholzung *f* lignification
verhungern starve/to, die of hunger/to
verhüten prevent/to
verhütten *n* smelt/to
Verhüttung *f* smelting
Verhütung *f* **von Arbeitsunfällen** prevention of occupational accidents
verkalken calcify/to
verkapseln encapsulate/to
Verkehr *m* traffic
Verkehrschaos *n* traffic chaos
Verkehrslärm *m* traffic noise
Verkehrsordnung *f* traffic regulations
Verkehrssicherheit *f* traffic safety
Verkehrsstauung *f* traffic jam
verkieseln silicify/to
Verkitten *n* sealing
Verklappung *f* **auf See** ocean dumping
Verklappung *f* **von Säure** dumping of acid
Verklappungsplatz *m* area for ocean dumping

Verkleidung *f* cladding
Verkleidungsmaterial *n* cladding material
Verknöcherung *f* ossification
Verkohlung *f* carbonization
Verkokung *f* carbonization
verkümmert degenerated
Verkümmerung *f* atrophy, degeneration
Verlandung *f* silting up, aggradation
Verletzung *f* **durch Bestrahlung** radiation injury
Vermarktung *f* marketing
Vermeidung *f* **der Umweltbelastungen** avoidance of the environmental pollution
Vermeidung *f* **von Emissionen** avoidance of emissions
Vermeidungskonzept *n* concept of avoidance
vermindertes Wachstum *n* reduced growth
vermodern mould[er]/to
Vermoderung *f* mouldering
vernetztes Makromolekül *n* cross-linked macromolecule
Vernetzung *f* cross-linkage, cross-bonding
Vernetzungsstelle *f* cross-linking site
Verpackung *f* package, pack[ag]ing, wrapping
Verpackungsabfall *m* waste from packings
Verpackungsaufwand *m* packing expenditure
Verpackungseinheit *f* packing unit
Verpackungsmaschine *f* packaging (wrapping) machine
Verpackungsmaterial *n* packaging (packing) material

verpestet polluted, foul
verpflanzen transplant/to, replant/to
Verpflanzung *f* transplantation
verrosten rust/to
verrottbarer Stoff *m* putrescible substance
verrotten rot [away]/to
Verrottung *f* rotting
Versalzung *f* **des Bodens** salination of the soil
Versandgefäß *n* mailing case
Versandschachtel *f* packing case
Verschiedenheit *f* **der Lebensgewohnheiten** diversity of habits of life
Verschlechterung *f* deterioration
Verschleißfestigkeit *f* wear resistance
Verschleiß *m* wear [and tear]
Verschleißprüfung *f* wear test
Verschlußkappe *f* cap
Verschlußzange *f* crimper
Verschmelzen *n* fusion, melt
Verschmelzung *f* coalescence, melting
verschmutzen pollute/to, contaminate/to
verschmutzt polluted
verschmutzte Zone *f* polluted zone
Verschmutzung *f* pollution
Verschmutzungsgrad *m* degree of pollution, pollution level
Verschmutzungsherd *m* source of pollution
Verschwelung *f* carbonization process, smoldering
Verschwendung *f* waste
verschärfte Gesetzgebung *f* **im Umweltschutz** more stringent legislation in the environmental protection
Versenken *n* immersion

verseuchen infect/to, contaminate/to
Verseuchung *f* contamination, pollution
Verseuchungsbekämpfung *f* contamination control
versickern lassen trickle/to
Versickerung *f* percolating
Versickerungsbereich *m* percolating area
Versickerungsmenge *f* percolating amount
Versiegeln *n* sealing
versiegeln seal/to
versorgen provide/to, supply/to
Versorgungsnetz *n* supply network
Versorgungsnetzautomatisierung *f* supply network automation
verstäuben spray/to
Versteppung *f* steppization
Versuch *m* experiment, test, trial, try
Versuch *m* **mit Ratten** rat test
Versuchsanlage *f* pilot plant, testing plant
Versuchsanordnung *f* experimental arrangement
Versuchsbedingung *f* test condition
Versuchsbericht *m* test report
Versuchsdaten pl test data
Versuchsdauer *f* experimental time, test time
Versuchsdurchführung *f* test practice
Versuchsergebnis *n* experimental result
Versuchsfehler *m* experimental error
Versuchsinstallation *f* pilot installation
Versuchsmodell *n* pilot model
Versuchsperson *f* proband
Versuchspflanzung *f* trial plantation

Versuchsprotokoll *n* test protocol
Versuchssubstanz *f* substance under investigation
Versuchssystem *n* test system
Versuchstier *n* test animal
Versuchsvorschrift *f* experimental direction
versuchsweise tentative
Verteiler *m* dispenser
Verteilerroutine *f* distribution routine
Verteilung *f* distribution
Verteilung *f* **in der Umgebungsluft** distribution in the ambient air
Verteilungschromatographie *f* distribution chromatography
Verteilungsmodell *n* distribution model
Verteilungskoeffizient *m* partition coefficient
Verteilungsverhalten *n* distribution behaviour
verträglich compatible
verunreinigen contaminate/to, pollute/to
verunreinigende Substanz *f* contaminant, pollutant
verunreinigt feculent, polluted
Verunreinigung *f* contamination, impurity, pollution
Verunreinigung *f* **der Ausrüstung** contamination of the equipment
Verunreinigungen *fpl* **in Reagenzien** impurities in reagents
Verunreinigungsgrad *m* impurity level, degree of pollution
verursachendes Agens (Mittel) *n* causative agent
Verweilzeit *f* retention time, hold-up time, residence (residential) time
verwelken wilt/to; fade/to

Verwendbarkeit *f* usability, usefulness
Verwendbarkeitsdauer *f* working life
Verwendungsbereich *m* range of use
Verwendungsfähigkeit *f* usability, usefulness
Verwendungsgebiet *n* field of application
verwerfen reject/to
Verwerfung *f* rejection
verwertbar utilizable
verwerten utilize/to, exploit/to
Verwertung *f* utilization
Verwertungskonzept *n* concept of utilization
Verwertungsmöglichkeit *f* possibility of utilization
Verwertungsverfahren *n* utilization method
Verwertungvorrang *m* recycling priority
verwesen putrefy/to, decompose/to
Verwesung *f* putrefaction, decomposition
verwildert rank
verwittern weather/to
Verwitterung *f* weathering
Verwundbarkeit *f* vulnerability
verwüstet desolated
Verwüstung *f* devastation, ravage
Verzehrung *f* consumption
Verzinken *n* zinc plating
verzinnen tin[-plate]/to
Verzweigungsstelle *f* branch point
Verzögerungsfaktor *m* retention factor
Verzögerungsmittel *n* retardant
Veränderliche *f* variable
Veränderlichkeit *f* mutability
verödet waste

Veterinär m veterinary
Veterinärmedizin f veterinary medicine
Veterinärmedizinrückstände mpl residues of veterinary drugs
Vibration f vibration
vibrieren vibrate/to
Vieh n livestock, cattle
Viehbestand m livestock, stock
Viehfutter n forage
Viehzucht f stock breeding
Viehzüchter m stock breeder
Vielfalt f diversity, multitude
vielfarbig multicoloured, polychrome
vielzellig multicellular
Virologe m virologist
Virologie m virology
virotoxisch virotoxic
Virozid n virocide
Virus n virus
Virusimpfstoff m virus vaccine
Virusinfektion f viral infection
Viruskrankheit f virosis, virus disease
Viruskultur f virus culture
Viruskunde f virology
viskos viscid, viscous
Viskosität f viscosity
visuell visual
visuelle Reaktionszeit f visual reaction time
vital vital
Vitalisierung f vitalization
Vitamin n vitamin
Vitaminmangel m poverty in vitamines, vitamin deficiency
Vogelkunde f ornithology
Vogelschutz m bird protection
Vogelzug m bird migration
Volksgesundheit f national health

Volkswirtschaft f national economy, political economy
Vollanalyse f complete analysis
Vollanzeige f full indication
Volldünger m complete fertilizer
Vollmilch f rich milk
vollsaugen soak/to
vollständige Reaktion f complete reaction
vollständiger Test m im Innen- und Außenbereich complete indoor and outdoor test
vollständiges Meßverfahren n complete measuring method
Vollwertnahrung f whole food
Volumenprozent n percent by volume
Volumenreduzierung f volume reduction
Volumetrie f titrimetric analysis, volumetric analysis, volumetry
volumetrisch titrimetric, volumetric
voluminös voluminous
Vor-Ort-Untersuchung f on-site investigation
Vorbehandeln n pretreatment, preliminary treatment, preconditioning
vorbehandeln pretreat/to, precondition/to
Vorbehandlung f von Abfällen pretreatment of waste
vorbeugen prevent/to
vorbeugend prophylactic, preventive
Vorbeugung f prevention, prophylaxis
Vorbeugungsstrategie f preventive action strategy
Vorbrecheranlage precrushing plant

Vorentwurf *m* preliminary design
vorfabrizieren prefabricate/to
vorfertigen prefabricate/to
Vorfilter *n* prefilter
Vorfilterung *f* preliminary filtering
Vorfluter *m* preclarifier, outlet ditch
vorgepackt prepacked
vorgereinigtes Abwasser *n* prepurified sewage, prepurified waste water
vorgeschaltete Denitrifikation *f* series-connected denitrification, incoming denitrification
Vorhersage *f* forecast, prediction, prognosis
Vorhersagedienst *n* forecast service
vorhersagen forecast/to, predict/to, prognosticate/to
Vorkehrung *f* precaution
vorkeimen pregerminate/to
vorklinische Vergiftungserscheinung *f* preclinical sign of poisoning
Vorklärbecken *n* preclarifier basin
vorklären preclarify/to
Vorklärung *f* preclarification
vorkühlen precool/to
Vorlagebecken *n* receiver basin
Vorläufer *m* precursor
vorläufige Diagnose *f* tentative diagnosis
vormischen premix/to
Vorozonisierung *f* preozonation
Vorprobe *f* quick test
vorprogrammiert preprogrammed
Vorrat *m* reserve, reservoir, stock
Vorratsflasche *f* stock bottle
vorreinigen preclean/to, prepurify/to
Vorreiniger *m* precleaner
Vorreinigung *f* precleaning, preliminary cleaning

Vorrichtung *f* appliance, device, facility
Vorrotte *f* prerot
Vorrottefläche *f* prerot area
Vorrottetrommel *f* prerot drum
vorschmelzen prefuse/to
Vorschub *m* feed
Vorsicht *f* caution, precaution
Vorsichtsmaßnahme *f* precaution
Vorsichtsmaßregel *f* precaution rule
Vorsorge *f* precaution
Vorsorgeprinzip *n* precaution principle
Vorsorgeuntersuchung *f* medical check-up
Vorsortieren *n* preclassifying
Vorstoß *m* adapter
Vortrocknung *f* predrying
Voruntersuchung *f* preliminary investigation
Vorvakuum *n* initial vacuum
Vorverbrennung *f* precombustion
Vorverrottung *f* prerotting
Vorversuch *m* preliminary experiment
Vorwärmung *f* preheating
Vorzerkleinerung *f* precrushing
Vulkanisation *f* vulcanization
Vulkanisierung *f* vulcanization
Völkerrecht *n* law of nations, international law

W

Waage *f* balance
Waagschale *f* scale
Wabenkatalysator *m* honeycomb catalyst
Wachs *n* wax
Wachstum *n* growth, proliferation

Wachstumsbedingung *f* growth condition
Wachstumsfaktor *m* growth factor
wachstumsfördernd growth-promoting
wachstumshemmend growth-inhibiting, paratonic
Wachstumshemmer *m* growth inhibitor
wachstumshindernd growth-inhibiting
Wachstumshormon *n* growth hormone
Wachstumskontrolle *f* growth control
Wachstumskurve *f* growth curve
Wachstumsperiode *f* vegetative period
Wachstumsrate *f* growth rate
Wachstumsregulation *f* growth regulation
Wachstumsregulator *m* growth regulator
Wägegenauigkeit *f* weighing accuracy
Wägevorrichtung *f* weighing appliance
Wägung *f* weighing
wahre Konzentration *f* true concentration
wahrer Wert *m* true value
wahrnehmbar observable
wahrscheinliche Abweichung *f* probable deviation
Wahrscheinlichkeit *f* probability
Wahrscheinlichkeitsgesetz *n* probability law
Wahrscheinlichkeitskurve *f* probability curve
Wahrscheinlichkeitsrechnung *f* probability calculus, theory of chances
Wahrscheinlichkeitsverteilung *f* probability distribution
Wal *m* whale
Wald *m* forest, wood
Waldbestand *m* forest stand
Waldboden *m* forest soil
Waldbrand *m* forest fire
Waldfläche *f* wooded area
Waldgebiet *n* forest area
Waldgrenze *f* forest border
Waldland *n* woodland
Wald-Ökosystem *n* forest ecosystem
Waldrand *m* edge of the wood
Waldregeneration *f* forest regeneration
Waldschaden *m* forest damage
Waldschutzgebiet *n* forest protection area
Waldvegetation *f* forest vegetation
Waldverwüstung *f* forest devastation (destruction)
Walzentrockner *m* rolling dryer
Wanderung *f* migration
Wandfarbe *f* wall paint
Wandler *m* converter
Wandverkleidung *f* wall covering
Warmblüter *m* warm-blooded animal
Warmblütertoxikologie *f* warm-blooded animal toxicology
Wärme *f* heat, warmth
Wärmeanstieg *m* heat rise
Wärmeaufnahmefähigkeit *f* caloric receptivity
Wärmeaustausch *m* interchange of heat
Wärmebehandlung *f* heat treatment
Wärmebetrag *m* amount of heat
Wärmebilanz *f* heat balance

Wärmedämmplatte *f* **aus Abfallpapier** thermal insulating panel manufactured from waste paper
Wärmedämmwirkung *f* good heat insulating effect
Wärmeentwicklung *f* development of heat
Wärmeisolierungsystem *n* heat insulation system
Wärmekapazität *f* thermal (heat) capacity
Wärmerückgewinnungssystem *n* heat recovery system
Wärmetauscher *m* heat exchanger
Wärmetauscherwirkungsgrad *m* degree of heat exchange efficiency
Wärmeverbrauch *m* heat consumption
wärmeverbrauchend heat-consuming
Wärmezufuhr *f* heat supply
Warndosimeter *n* alarm dosimeter
Warneinstellung *f* alarm setting
Warnfarbe *f* warning colour
Warngerät *n* warning device
Warnlicht *n* warning light
Warnmeldesytem *n* warning and report system
Warnsignal *n* caution signal, danger signal
Wartung *f* service, maintenance
Wartungsprogramm *n* service routine
Wartungsservice *m* maintenance
Wartungssystem *n* service system
Waschanlage *f* washing equipment, washing plant
Waschanlage *f* **für wiederzuverwendende Folien** washing plant for reusable foils
waschbar washable

Waschbarkeit *f* washability
Waschbeständigkeit *f* fastness to washing
waschecht washable
Wascheffekt *m* cleaning effect
waschen rinse, wash
Wäscher *m* washer
Wäscherkreislauf *m* washer cycle
Waschfestigkeit *f* wet abrasion resistance
Waschflasche *f* wash bottle
Waschflüssigkeit *f* washing liquid
Waschkraft *f* washing efficiency
Waschmittel *n* detergent, washing agent, washing powder
Waschsuspension *f* washing suspension
Waschturm *m* washing tower
Waschverfahren *n* washing method
Waschvorrichtung *f* washer
Waschwasser *n* wash water
Waschwasserbehälter *m* water container, water vessel
Waschwirkung *f* washing efficiency
Wasser *n* water
Wasserabfluß *m* discharge of water
Wasserabscheidung *f* water separation
Wasserabspaltung *f* elimination of water
wasserabstoßend hydrophobic, water-repellant
Wasserader *f* water vein
Wasseranalyse *f* water analysis
Wasseranalytik *f* water analytics
Wasseraufbereitung *f* water preparation, water processing
Wasserbedarf *m* water requirement
Wasserbeladung *f* water treatment
wasserbeständig water resistant

Wasserbewirtschaftung *f* water management
Wasserbilanz *f* water balance
wasserbindend hydrophilic
Wasserchemie *f* water chemistry
Wasserdampf *m* water vapour
wasserdampfflüchtig steam-volatile
wasserdicht waterproof, watertight, impermeable
wasserdurchlässig water permeable
wasserdurchlässiger Damm *m* **für Regenwasser** *n* water permeable dam for rainwater
Wassereinsparung *f* water saving
Wasserenthärtungsanlage *f* water softening plant
wasserfester Anstrich *m* water-resistant painting
wasserfreies Salz *n* anhydrous salt
wassergefährdender Stoff *m* water-endangering material (substance)
wasserglasvergütet waterglass-improved
Wassergüte *f* water quality
Wasserhaushalt *m* water balance
wasserhärtbar water-curable
wasserhärtbares Harz *n* **zur Abfallentsorgung** *f* water-curable resin for waste disposal
Wasserhärte *f* water hardness
Wasserkreislauf *m* water cycle, hydrological cycle
Wasserleitung *f* mains
Wassermangel *m* water famine
Wassermeldekabel *n* water alarm cable
wässern water/to, hydrate/to, soak/to, steep/to; dilute/to,
Wasserniveau *n* water level
Wasserprobe *f* water sample

Wasserqualitätsproblem *n* problem in water quality
Wasserreinigung *f* water purification
Wasserressource *f* water resource
Wasserschutzgebiet *n* water protection area
Wasserspeicher *m* water reservoir
Wasserspeicherfähigkeit *f* water storage capability
Wasserspiegel *m* water level
Wasserstoff *m* hydrogen
Wasserstoffperoxid *n* hydrogen peroxide
Wassertoxikologie *f* aquatic toxicology
wasserundurchlässig water-impermeable, watertight
wasserundurchlässige Tonschicht *f* water impermeable clay layer
Wasserverbandsrecht *n* water authority law
Wasserverbrauch *m* water consumption
Wasserverdunstung *f* water evaporation
Wasserversorgung *f* water supply
Wasserversorgung *f* **in der chemischen Industrie** water suply in the chemical industry
Wasserversorgungsanlage *f* water supply plant
Wasserversorgungsnetz *n* water supply network
Wasserverteilungsanlage *f* water distribution plant
Wasserverteilungssystem *n* water distribution system
Wasservorratstank *m* water supply tank

Wasserwiederverwendung *f* water reuse, water reutilization
Wasserwirtschaft *f* water-supply and distribution management
wäßrig aqueous
wäßrige Lösung *f* aqueous solution
wäßriger Auszug *m* water leacheate
wäßrig-organische Lösung *f* aqueous-organic solution
Wattenmeer *n* mud flats
Wechselbeanspruchung *f* alternative stress
Wechselbeziehung *f* correlation, interrelation, relationship
Wechselwirkungen *f* zwischen Umweltchemikalien interaction between environmental chemicals
Weg *m* um Dateneingabe zu überprüfen way to check the data input
Wegwerf- throwaway, disposable
Wegwerfpackung *f* disposable package
Wegwerfverfahren *n* throw-away method
weibliche Ratte *f* female rat
weich soft
 weich machen soften/to, plasticize/to
Weichholz *n* softwood
Weichmachen *n* softening
 Weichmachen *n* von Wasser softening of water
Weichmacher *m* plasticizer, softener, softening agent
weichmacherfrei unplasticized
Weichtiere *npl* molluscs
Weichwerden *n* softening
Weide *f* pasture, meadow
Weiher *m* pond
Weinbau *m* wine growing

Weinberg *m* vineyard
Weise *f* manner, mode, style, type
Weißblechdose *f* tinplate can
Weißblechrecycling *n* tinplate recycling
weißes Blutkörperchen *n* leucocyte
Weißglas *n* white glass
Weißglut *f* white heat
weißmachen whiten
Weite *f* width
Weiterbehandlung *f* further treatment
Weiterbenutzung *f* continued use
weiterbrennen burn away/to
weiterleiten convey/to
Weiterverwendung *f* further utilization, subsequent utilization
 Weiterverwendung *f* des dekontaminierten Erdreichs further utilization of the contaminated soil
weitgehend substantial
Weithalskolben *m* wide-mouth flask
weitmaschig coarse-meshed
weitverbreitet widespread
welken wilt/to, wither/to
Welle *f* wave
Wellenbrecher *m* wave breaker
Wellenenergieanlage *f* wave energy plant
Wels *m* wels
Welt *f* world
Weltall *n* cosmos
Weltenergieverbrauch *m* world energy consumption
Weltgesundheitsorganisation *f* world health organization (WHO)
Weltklima *n* world climate
Weltmarktsituation *f* in der Abfallwirtschaft situation of the world market in the waste management
Weltraum *m* space

Weltraumforschung *f* space research
Weltraumzeitalter *n* space age
weltumfassend ecumenical, global, world-wide
weltumfassende Konzentrationen *fpl* global concentrations
Weltvorrat *m* world's supply
weltweit world-wide
Wendepunkt *m* point of inflection
Wendigkeit *m* mobility, versatility
Werden *n* development, genesis
Werk *n* piece, work
Werkstoff *m* material
Werkzeug *n* tool, appliance, instrument
Wert *m* value
Wert *m* **ist fest eingestellt** value is adjusted
Wert *m* **liegt nahe bei** value lies proximate to
Wert *m* **liegt um** value lies around
Wertangabe *f* declaration of value
Wertbestimmung *f* valuation
Wertminderung *f* reduction in value
Wertordnung *f* system of value
Wertsteigerung *f* increase in value
Wertstoff *m* substance of value, material of value, valuable material, recyclate
Wertstoffrecycling *n* recycling of recyclates
Wertstoffsortierung *f* sorting of reyclates, sorting of materials of value
Wertstoffverfahren *n* recyclate method
Werturteil *n* judgement of value
wertvoll precious, valuable
Wesen *n* nature
wesentlich essential, important, intrinsic, significant, prime

Wettbewerb *m* competition
Wettbewerb *m* **im Binnenmarkt** *m* competition on the inner market
Wetter *n* weather
Wetteramt *n* met office
wetterbeständig weather-resistant, weatherproof
Wetterbeständigkeit *f* weather resistance
Wettereinfluß *m* weather influence
wetterfest weatherproof
wetterfester Anstrich *m* weatherproof painting
Wetterfestigkeit *f* exterior durability
Wetterkunde *f* meteorology
Wetterlage *f* weather conditions
Wetterschlüssel *m* climatic data
wichtig important, major, significant
Wichtigkeit *f* importance
wider die Natur *f* contrary to nature
widernatürlich unnatural
widersinnig absurd
Widerspruch *m* antithesis, discrepancy, protest
Widerstand *m* resistance
Widerstandsdraht *f* resistance wire
widerstandsfähig resistant
Widerstandsfähigkeit *f* **gegen Schadstoffe** resistance to pollutants
Widerstandsthermometer *n* resistance thermometer
widerstehen resist
Wiederanlauf *m* restart
Wiederanpassung *f* readjustment
wiederaufarbeiten reprocess/to, recycle/to
Wiederaufarbeitung *f* recycling, reprocessing, reclamation

Wiederaufarbeitungsanlage *f* reprocessing plant
Wiederaufbau *m* reconstruction
Wiederaufbereitung *f* **von Polyurethanabfall** *m* recycling of polyurethane waste
Wiederaufbereitung *f* **von Reifenchips** *n* reclamation of tire scrap
wiederaufforsten reafforest/to
Wiederaufforstung *f* reafforestation, reforestation
Wiederaufgreifen *n* revival
Wiederaufleben *n* revival
wiederbeladen recessive
Wiederbelebung *f* reanimation, revitalization, revival
Wiederbesiedlung *f* repopulation
Wiedereinfuhr *f* reimportation
Wiedereinpflanzung *f* replantation
Wiedereinschmelzen *n* recasting
wiedereinschmelzen remelt/to
Wiedereinsetzung *f* rehabilitation
Wiederfindung *f* recovery
Wiedergesundung von Boden *f* recovery of soil
wiedergewinnen recover/to, regain/to, regenerate/to, reclaim/to
Wiedergewinnung *f* reclaiming, recovery, salvage, reparation
Wiedergewinnungsmethode *f* reclamation method
wiederherstellbar recoverable, restorable
wiederherstellen rebuild/to, rehabilitate/to, renovate/to, restore/to
Wiederherstellung *f* reconditioning, reconstruction, regeneration, reparation, restoration
Wiederholbarkeit *f* repeatability
wiederholen repeat
wiederholt repeated, replicate

wiederholte Exposition *f* repeated exposure
Wiederholung *f* duplication, repetition, replicate
Wiederholungsexperiment *n* replicate test
Wiederinbetriebnahme *f* restart
Wiederkehr *f* return
Wiederkäuer *m* ruminant
Wiedervereinigung *f* recombination
wiederverwenden reuse/to, recycle/to
Wiederverwendung *f* reutilization, repeated use, reuse
Wiederverwendungssystem *n* reutilization system, reuse system
Wiederverwendungsverfahren *n* recycling method
wiederverwertbar recyclable, reusable
Wiederverwertbarkeit *f* recyclability, reusability
wiederverwerten reuse/to, recycle/to
Wiederverwertung *f* reutilization
Wiederverwertung *f* **von Bauschutt** *m* recycling of debris
Wiese *f* lea, meadow
wilde Bauschuttkippe *f* unauthorized rubble dump
Wildleben *n* wildlife
Wind *m* wind
Windeinfluß *m* **auf den Transport** influence of wind on the transportation
Winderosion *f* erosion by wind
Windgeschwindigkeitsmesser *m* anemometer
Windkraft *f* windpower
Windmesser *m* wind gauge
Windschutzscheibe *f* windscreen

Windstoß m blast
Winter m winter
Winterschlaf m hibernation
Wirbel m vertebra
Wirbelbewegung f turbulence
wirbellos invertebrate
Wirbelschichtasche f fluidized-bed ash
Wirbelschichtofen m fluidized-bed furnace
Wirbelschichtreaktor m fluidized-bed reactor
Wirbelschicht-Verbrennungsanlage f fluidized-bed incineration plant
Wirbeltier n vertebrate
Wirkstoffentwicklung f development of ingredients
wirkungsbezogen effect-related
Wirkungsdauer f duration of effect, persistence
Wirkungsfähigkeit f effectiveness
Wirkungsgrad m effectiveness, efficiency
Wirkungsgradverschlechterung f impairment of the efficiency degree
wirkungslos inactive, inert, non-effective
wirkungsvoll effective
wirtschaftliche Wiederverwertung f economic reutilization
Wirtschaftlichkeit f economy
Wismut n bismuth
witterungsbeständig resistant to weathering, weather-resistant
Witterungsbeständigkeit f outdoor durability
Wochenextrema npl weekly maxima
Wochenganglinie f weekly load curve, weekly output curve
Wohlgeruch m fragrance
wohlriechend fragrant, odorous

Wolke f cloud
Wolkenbruch m cloudburst
Wrack n wreck
wuchern proliferate/to
Wucherung f proliferation; tumour
Wuchsstoff m growth promoter
Wundstarrkrampf m tetanus
Wurm m worm
wurmvertilgend worm-destroying
Wurzel f root
Wurzelaufnahme f root uptake
Wurzelsystem n root system
Wüste f desert
Wüstenboden m desert soil
Wüstenpflanze f desert plant
Wüstentier n desert animal

X

Xenobiose f xenobiosis
xenobiotisch xenobiotic
xenobiotische Chemikalie f xenobiotic chemical
Xenon n xenon
Xerographie f xerography
Xerosole mpl xerosols

Y

Ytterbium n ytterbium
Yttrium n yttrium
Yermosole mpl yermosols

Z

zähflüssig semifluid, viscous
Zähigkeit *f* tenacity, toughness
Zähigkeitstest *m* toughness test
Zählerstand *m* counter content
Zählwerk *n* counter, register, counting device, meter
Zahnfüllung *f* tooth filling
Zahnfäule *f* dental caries
Zapfsäule *f* petrol pump
Zaun *m* fen
Zeche *f* mine, coal pit, pit
Zechenstillegung *f* pit closure
Zeigerablesung *f* indicator reading
Zeit *f* **nach der Exposition** post-exposure period
zeitabhängig time-dependent
Zeitfaktor *m* time factor
zeitliche Folge *f* time sequence
zeitlicher Abschnitt *m* period
Zellaufbau *m* cell structure
Zellbestandteil *m* cellular constituent
zellbiologisches Testmodell *n* cell-biological testing model
Zelle *f* cell; compartment; segment
Zellenabsterben *n* necrobiosis
Zellenaktivität *f* cell activity
zellenbildend cytogenic
zellenförmig cellular
Zellmasse *f* cellular substance
Zellmembran *f* cellular membrane
zellphysiologisch cytophysiologic
Zellstoff *m* cellulose
Zellstoffbleichen *n* cellulose bleaching
Zellsubstrat *n* cell substrate
Zement *m* cement
Zementeinspritzung *f* cementation
Zementfabrik *f* cement factory
zementgebundenes Füllmaterial *n* cement-bound filling material
Zementstaub *m* cement dust
Zentralcomputer *m* central computer
zentrale Aufbereitungsanlage *f* central processing plant
zentrale Vorbehandlungsanlage *f* central pretreatment plant
Zentralheizung *f* central heating
Zentrallabor *n* **für Abfalluntersuchung** central laboratory for waste investigation
Zentralnervensystem *f* central nervous system
Zentralsteuerung *f* centralized control
Zentrifugalgebläse *n* centrifugal blast
Zentrifugallüfter *m* centrifugal aerator
Zentrifuge *f* centrifuge
zentrifugieren centrifuge/to
Zentripetalkraft *f* centripetal force
Zeolith *m* zeolithe
zeolithhaltig zeolitic
Zerfall *m* decay, ruin, disintegration, decomposition
zerfallen break down/to, decompose/to, disintegrate/to, dissociate/to
Zerfallsfolge *f* decay sequence
Zerfallsgeschwindigkeit *f* rate of decay
Zerfallskonstante *f* radioactive constant, transformation constant
Zerfallsprodukt *n* decomposition product
Zerfaserer *m* shredder
zerfasern shred/to, fray/to
Zerkleinerer *m* shredder, disintegrator

zerkleinern grind/to, shred/to
Zerkleinerungsanlage *f* crushing plant
Zerkleinerungsmaschine *f* crusher
Zerkleinerungstechnik *f* crushing technology
zerlegen decompose/to, split up/to, degrade/to, resolve/to, take apart/to
Zerlegung *f* decomposition, disintegration, separation
zermahlen mill/to, pulverize/to
zerreiben grind/to, pulverize/to
zerreißfest tear-proof
Zerreißprobe *f* tensile test
zerschnitzeln shred/to
zersetzen decompose/to, degrade/to, disintegrate/to
Zersetzer *m* decomposing agent
Zersetzung *f* decomposition, degradation, decay, dissociation, disintegration
Zersetzungsgeschwindigkeit *f* rate of decomposition
Zersetzungsprodukt *n* decomposition product
Zersetzungsreaktion *f* decomposition reaction
Zersetzungsrückstand *m* decomposition residue
Zersiedlung *f* spoil by development
zerspalten split/to, cleave/to
zersplittern crack/to, shatter/to, splinter/to
zerstreuen disperse/to, scatter/to, dissipate/to, diffuse/to
Zerstreuung *f* scattering, dispersion, dissipation
zerstäuben spray/to, atomize/to, disperse/to
Zerstäuber *m* sprayer, atomizer, diffuser

Zerstäubung *f* atomization, dispersion
Zerstäubungsmechanismus *m* dispersion mechanism
Zerstäubungstrocknung *f* spray drying
zerstören destruct/to, ruin/to, destroy/to, demolish/to
Zerstörung *f* ruin, demolition, destruction, deterioration
zerstörungsfreie Messung *f* non-destructive measurement
zerstörungsfreier Test *m* non-destructive test
Zerstörungswut *f* vandalism
Zeugungsfähigkeit *f* potency
Ziegel *m* brick, tile
Zielatom *n* target atom
Zielorgan *n* target organ
Zink *n* zinc
Zinkblech *n* sheet zinc
Zinn *n* tin
Zirkonium *n* zirconium
Zirkulation *f* circulation
Zirkulationsströmung *f* circulation flow
Zirkulationssystem *n* circulation system
Zirkulationsverhältnis *n* circulation ratio
zirkulieren circulate/to
Zirrhose *f* cirrhosis
Zisterne *f* tank, well
Zivilschutz *m* civil defence
zivilrechtliche Sonderabfallgesellschaft *f* civil law company for waste especially to be supervised
Zonenschmelzen *n* zone melting
Zoochemie *f* zoochemistry
Zoologie *f* zoology
zoologisch zoological

Zooplankton *n* zooplankton
Zooökologie *f* zooecology
Zucht *f* cultivation
Züchten *n* breeding; incubation
Zucker *m* sugar
Zuckergehalt *m* sugar content
Zuckerkrankheit *f* diabetes
Zuckerrohr *n* sugar cane
Zudosierung *f* dosage
Zufall *m* accident, random cause, chance
Zufallsfehler *m* accidental error, random error
Zufallsfolge *f* random series
Zufallsgröße *f* random variable
Zufluß *m* inflow, influx
Zufuhr *f* feed, supply, inlet, addition
zuführen add/to, feed/to
Zuführung *f* feeding
zufällige Schwankung *f* random fluctuation
zufälliger Fehler *m* accidental error, random error
Zugabe *f* allowance
Zugang *m* access
Zugänglichkeit *m* accessibility
Zugkraft *f* tension
Zugvogel *m* migrant bird
Zukunftserwartung *f* future expectation
zulässige Abweichung *f* permissible tolerance
zulässige Belastung *f* permitted load level
zulässiger Expositionsgrenzwert *m* tolerable level of exposure
zulässiger Wert *m* permissible value
Zulassung *f* admission, admittance, licensing, permit
Zulassungsentscheidung *f* decision of admission
Zulassungssystem *n* system of admission
Zulauf *m* inflow
Zuleitung *f* inlet, supply
Zuluft *f* inlet air
Zumischung *f* addition
Zündung *f* ignition
Zündvorgang *m* ignition
zurückbehalten retain/to
zurückbilden reconvert/to
zurückfließen flow back/to
Zurückfließen *n* reflux
Zurückfluten *n* reflux
zurückführen recycle/to
zurückgewinnen recover/to, reclaim/to
Zurückgewinnung *f* recovery, recycling, reclamation
zurückhalten reserve/to, restrain/to
Zurückhaltung *f* retention
Zurücknahme *f* withdrawal
zurückrufen recall/to
zurückstrahlen reflect/to
zurückströmen flow back/to
zurückverwandeln reconvert/to
zurückweisen reject/to, withdraw/to
Zurückweisung *f* rejection
Zurückziehung *f* withdrawal
zusammenballen agglomerate/to
Zusammenballung *f* agglomeration
Zusammenbruch *m* breakdown, crash, debacle
zusammendrücken compact/to, compress/to, press/to
zusammenschmelzen fuse together/to
Zusammensetzung *f* composition
Zusammenwirken *n* synergism
zusammenziehen contract/to

Zusatz m addition, additive, supplement
Zusatzfeuerung *f* additional firing
Zusatzmittel *n* additive
Zusatzstoff *m* added substance, additive
Zuschlag *m* additional charge
zuschmelzen seal/to
zusetzen add/to
Zustandsanzeige *f* status display, state indication
Zustandsänderung *f* change of state
Zuverlässigkeitskriterium *n* reliability criterion
Zuwachs *m* increase, increment
Zweig *m* **der Abfallwirtschaft** branch of the waste management
zweikernig binuclear
Zweikomponenten-Injektionsharz *n* two-component injection resin
Zweikomponentenkleber *m* mixed adhesive
Zweipunktregelung *f* on-off control
Zwergwuchs *m* nanosomia, stunted growth, dwarfism
zwergwüchsig nanous
Zwilling *m* twin

Zwischenfall *m* accident
Zwischenklärung *f* intermediate clarification
Zwischenlagerung *f* intermediate storage
Zwischenreaktion *f* intermediate reaction
Zwischenschicht *f* interlayer, intermediate layer
zyklisch cyclic
zyklischer Kohlenwasserstoff *m* cyclic hydrocarbon
Zyklonabscheider *m* cyclone separator
Zyklonentstauber *m* cyclone dust separator
Zyklonfilter *n* cyclone filter
Zyklus *m* period
Zykluszeit *f* cycle time
zymogen zymogenic
Zyste *f* cyst
zytogen cytogenic
zytogenetischer Effekt *m* cytogenetic effect
Zytologie *f* cytology
Zytoplasma *n* cytoplasma
Zytostatikum *n* cytostatic substance

Gerhard Wenske

parat Dictionary of Chemistry
English/German

parat Wörterbuch Chemie
Englisch/Deutsch

1991. Ca. 1550 Seiten. Gebunden.
DM 345,-. ISBN 3-527-26428-0

Eine unglaubliche Menge an Information:

Dieses Werk enthält mehr als 100.000 Einträge der Chemie, der chemischen Verfahrenstechnik und angrenzender Gebiete in englisch-deutscher Sprachrichtung mit zahlreichen Hinweisen auf Synonyme und Anwendungsgebiete. Viele Summenformeln sind angegeben, auf IUPAC-Bezeichnungen wird besonders Bezug genommen, veraltete oder seltene Bezeichnungen sind besonders gekennzeichnet.

Der Autor, selbst ein ausgezeichneter Terminologie-Experte, hat detaillierte Erläuterungen zu den Einträgen zusammengestellt. Es ist nicht übertrieben zu behaupten, daß das Werk seinesgleichen auf der Welt sucht. Es ist unerläßlich für jeden, der mit der Übersetzung oder der Lektüre anglo-amerikanischer Fachtexte der Chemie, der chemischen Verfahrenstechnik oder verwandter Gebiete, wie Ökologie, Biochemie und physikalische Chemie zu arbeiten hat.

VCH • Postfach 101161 • D-6940 Weinheim